Promoting Climate Change Mitigation and Resilience Strategies Through Human Resource Management

Qamar Ali
https://orcid.org/0000-0002-8178-7591
Virtual University of Pakistan, Pakistan

Muhammad Ashfaq
The University of Faisalabad, Pakistan

Muhammad Tariq Iqbal Khan
Government College University, Pakistan

IGI Global
Scientific Publishing
Publishing Tomorrow's Research Today

Vice President of Editorial	Melissa Wagner
Director of Acquisitions	Mikaela Felty
Director of Book Development	Jocelynn Hessler
Production Manager	Mike Brehm
Cover Design	Jose Rosado

Published in the United States of America by
IGI Global Scientific Publishing
701 East Chocolate Avenue
Hershey, PA, 17033, USA
Tel: 717-533-8845 | Fax: 717-533-7115
Website: https://www.igi-global.com E-mail: cust@igi-global.com

Library of Congress Cataloging-in-Publication Data

Names: Ali, Qamar, 1988- editor http://id.loc.gov/rwo/agents/n2025003896 | Ashfaq, Muhammad, 1960- editor http://id.loc.gov/rwo/agents/n2025022892 | Iqbal Khan, Muhammad Tariq, 1988- editor http://id.loc.gov/rwo/agents/n2025003898

Title: Promoting climate change mitigation and resilience strategies through human resource management / [edited by] Qamar Ali, Muhammad Ashfaq, Muhammad Tariq Iqbal Khan.

Description: Hershey, PA : IGI Global Scientific Publishing, [2026] | Includes bibliographical references and index. | Summary: "This book is related to the problem of climate change, and its influence on the major areas of the world's economy and people's lives, including agriculture, water, food, health, and infrastructure"-- Provided by publisher.

Identifiers: LCCN 2025011210 (print) | LCCN 2025011211 (ebook) | ISBN 9798337324791 hardcover | ISBN 9798337324807 paperback | ISBN 9798337324814 ebook

Subjects: LCSH: Sustainable development http://id.loc.gov/authorities/subjects/sh92005743 | Climatic changes--Economic aspects http://id.loc.gov/authorities/subjects/sh2008117648 | Climate change mitigation http://id.loc.gov/authorities/subjects/sh2009009655

Classification: LCC HC79.E5 P7375 2026 (print) | LCC HC79.E5 (ebook) | DDC 338.9/27--dc23/eng/20250522

LC record available at https://lccn.loc.gov/2025011210

LC ebook record available at https://lccn.loc.gov/2025011211

British Cataloguing in Publication Data
A Cataloguing in Publication record for this book is available from the British Library.

Editorial Advisory Board

Table of Contents

Detailed Table of Contents

Present study employs panel data analysis on BRICS economies data spanning from 2000 to 2023, to examine the impact of human resource development, renewable energy consumption, and carbon emissions on GDP per capita of BRICS economies, finding that while human capital strongly drives growth, renewable energy has a positive but weaker effect, and carbon emissions remain tightly linked to GDP, underscoring the persistent reliance on carbon-intensive industries, ultimately highlighting the need for policy shifts toward sustainable growth through education, clean energy transitions, and regulatory frameworks that balance industrialization with climate resilience.

While organizations struggle with the intensifying effects of climate change, incorporating sustainability into fundamental business functions has become essential. GHRM then becomes a key framework for aligning human capital with sustainable objectives, developing sustainable culture, and promoting organizational resilience. This chapter discusses the premises of GHRM, exploring its theoretical bases, essential elements, and its central role in driving climate change mitigation and resilience initiatives. Based on theoretical lenses (different theories) GHRM covers the conventional practices such as recruitment, training, performance management, employee compensation and occupational health and safety from a climate resilience approach perspective. The chapter uses Patagonia and Unilever among other good businesses to show how GHRM may drive environmental performance, employee engagement, and corporate resilience. The chapter concludes by describing the road map for GHRM in terms of cooperation, employee empowerment, and link to other CSR activities.

This chapter explores the integration of climate change mitigation and resilience strategies within organizational frameworks through strategic human resource management (HRM). By aligning HR practices with sustainability goals, organizations can foster a workforce equipped to address environmental challenges. Key strategies include developing green competencies, promoting eco-conscious behaviors, and enhancing employee engagement in sustainability initiatives. The chapter examines how HRM can drive organizational resilience by embedding climate-focused policies in recruitment, training, performance management, and leadership development. It highlights the role of HR in cultivating a culture of environmental stewardship, enabling organizations to adapt to climate-related risks while contributing to global sustainability efforts. Practical implications and future research directions are discussed to advance the intersection of HRM and climate action.

In line with sustainable development goals principles, the long-term value creation potential of businesses has started to be measured not only by economic outputs but also by environmental and social impacts (Elkington, 1997). Accordingly, it is necessary to restructure performance management and reward systems in order for organizations to effectively implement their sustainability strategies. Aligning performance management processes with sustainability goals is reported to increase employee motivation and organizational commitment, (Warhurst, 2002). In this context, rewarding sustainability-oriented performance is not limited to financial incentives, but is also supported through intangible rewards. In order for these systems to be implemented effectively, it is necessary to integrate strategic sustainability goals with corporate vision and actions, measure sustainable performance indicators, create an organizational culture conducive to stakeholder engagement, base leadership activities on sustainability principles, and establish a fair reward system.

The urgent challenge of climate change necessitates a shift in leadership and organizational culture to foster sustainability. This chapter explores Green Leadership as a transformative approach that integrates environmental responsibility into organizational goals. It examines the role of Cultural Transformation in embedding sustainability within corporate ethos through leadership-driven strategies. Theoretical frameworks, including Transformational Leadership and Ecocentric Leadership, are analyzed alongside HRM initiatives that promote sustainable practices. Case studies illustrate successful implementations, highlighting challenges such as resistance to change and greenwashing. The chapter concludes with insights on emerging sustainability trends, ESG frameworks, and HRM's role in driving innovation. By aligning leadership, culture, and sustainability, organizations can become proactive agents of environmental change and resilience.

This chapter examines how ethical leadership drives environmental sustainability in modern organizations. It explores the convergence of ethical leadership, CSR, and environmental stewardship, providing frameworks for sustainable practice implementation. Drawing from organizational behavior and stakeholder perspectives, it demonstrates how ethical leaders foster environmental responsibility cultures. The analysis covers employee engagement in sustainability initiatives and knowledge management's role in environmental performance. The chapter addresses crisis management challenges and emphasizes education's importance in developing future environmental leaders. It offers practical guidance for business leaders, organizations, and policymakers, providing a roadmap for sustainable business practices in an eco-conscious global economy.

This chapter examines how industries worldwide are adopting green policies and sustainable practices to combat climate change, reduce environmental impacts, and enhance organizational resilience. Focusing on key sectors like manufacturing,

Foreword

This book is important. Why? In its 2023 summary report, the Intergovernmental Panel on Climate Change (IPCC) noted that "*Human activities, principally through emissions of greenhouse gases, have unequivocally caused global warming, with global surface temperature reaching 1.1°C above 1850-1900 in 2011-2020*".

Many of us around the world are feeling the impacts of this global warming through increased severe weather events such as heatwaves and flooding, and in shifts in seasonal weather patterns that impact agriculture and other environmentally important ecosystems. South Asia and its populations are particularly vulnerable to the negative impacts of increasing global temperatures, although it appears that nowhere will be unaffected in the coming decades.

Global warming reflects interacting processes in physical, chemical, biological and geological systems. It is essential to better understand these systems and interactions to comprehend what is happening. It is particularly important to consider humans and their activities as part of these systems. Quoting the IPCC 2023 report again: *Global greenhouse gas emissions have continued to increase, with unequal historical and ongoing contributions arising from unsustainable energy use, land use and land-use change, lifestyles and patterns of consumption and production across regions, between and within countries, and among individuals*"

Clearly, we humans must act to reduce the impacts of global warming, in an ongoing dance of mitigation, adaptation and learning. We need to act quickly, but we also need to be informed. We especially need to understand what we don't know and be willing to continue learning. As an international researcher working at the interface of natural resource management and social science for over four decades, I know that changing human behaviours is not easy, nor linear. Supporting a move to less destructive habits and different ways of being in the world requires shared comprehension of the threat, and co-ordination of intent. Real change requires sound research, into both science and humans. This book is important then, because it brings together well evidenced research and recommendations across a range of disciplinary areas, with the intent of helping address global warming head on. The book recognises institutions and organizations as essential spaces for mitigation and adaptation activities, and that building the capacity of their processes and people cannot be ignored. By drawing attention to institutions as places for change the book highlights the importance of Human Resource Management. Crucially then, as well as advancing our understanding of the challenges of global warming the book's authors offer practical frameworks for change for managers, policymakers and scholars.

I congratulate the editors and authors on the scope and ambition of this book. While the topics addressed are often challenging, I invite readers to approach each chapter with critical curiosity. For me the challenge of a multi-disciplinary work of this type brings some satisfaction and hope, as my experience suggests such multi and transdisciplinary approaches are the only way we can make real progress in complex, dynamic situations such as learning to adapt to a warming planet.

Catherine Allan
Charles Sturt University, Australia

Preface

OVERVIEW OF THE SUBJECT MATTER

One of the most urgent challenges of the 21 st century has been climate change in the whole world. Their extensive effects can be seen in ecosystems, economies and societies and it disrupts natural balances and causes vulnerabilities that demand immediate and concerted action. Although technological developments, innovations in science and policy interventions have traditionally been the leading issues of climate discourse, there is growing awareness of the fact that sustainable transformation also needs human-driven solutions, cultural transformation and organizational resilience.

This book Promoting Climate Change Mitigation and Resilience Strategies through Human Resource Management is a book that touches upon the crucial overlap between climate science, governance, organizational responsibility, and human resource management (HRM). It examines how HRM, green leadership and sustainability-oriented practices can yield climate resilience when supplementing scientific and policy aspects of climate action. The book draws a combination of theoretical insights, empirical studies, case analysis, and practical measures to give a comprehensive picture of climate change mitigation and adaptation.

WHERE THE TOPIC FITS IN THE WORLD TODAY

Climate change is no longer something far off or something alien but a reality in the world today. Rising sea levels and extreme weather patterns, health crises and agricultural disruptions are only some of the effects of it that are being experienced in every part of the world. The urgency of cutting down on emission, investing in renewable energy and creating adaptive capacities has been highlighted by international structures like the Paris Agreement, the Sustainable Development Goals (SDGs) of the UN and country-based climate action plans.

Nevertheless, even with the high global development, meaningful climate resilience can only be attained when non-governmental and non-scientific actors are involved. Companies, businesses, and society need to ensure that they are proactive in their procedures to meet climate objectives. Human Resource Management has been on the forefront to ensure this transformation through the creation of green values, sustainable skills development, and the creation of climate-conscious behaviours among the organizational structures. HRM can position itself at the interface between the macro-level climate policies and the micro-level organizational action by introducing sustainability to the leadership, culture, and workforce practices.

This book finds its place in this new discussion and provides clues to the direction of sustainable development and climate resiliency as formed by the combination of HRM frameworks, models of governance and technological innovations..

TARGET AUDIENCE

The book is intended to be read by a wide range of audience that cuts across various fields and careers:

Scholars and Researchers: Climate science scholars, environmental scholars, business scholars, management scholars, social science scholars will be interested in extensive analyses and structures that can broaden the scholarly debate on climate resilience.

Policy Makers and Government Officials: The policy makers of climate will have a better understanding of the model of governance, adaptation strategies as well as the necessity of organizational responsibility.

Business Leaders and Managers: Corporate leaders, sustainability officers, HR professionals can use the strategies described in this book to create green organizational practices and resilience.

Students Chemistry: Undergraduate and postgraduate students enrolled in environmental management, business, and HRM programs should find the multi-facet view of science, governance and management to be resourceful.

Practitioners and NGOs: Development workers and climate activists will learn viable examples and novel approaches to community based and organizational climate resilience.

The interdisciplinary attribute of the book renders it a useful tool to any individual who aims to learn and play a role in the climate change mitigation agenda both in scientific and organizational contexts.

CHAPTER DESCRIPTIONS

Chapter 1: Understanding Climate Change and Organizational Responsibility

Climate change, driven by human activities, poses severe environmental, economic, and social challenges. This chapter examines its impacts, including rising temperatures, ecosystem disruption, and threats to food security and health, with vulnerable populations disproportionately affected. It explores the role of industries in greenhouse gas emissions and the potential for sustainable practices like renewable energy and circular economies. The chapter critiques greenwashing and emphasizes the need for stricter regulations and corporate accountability. Key policies, such as the European Green Deal and U.S. Inflation Reduction Act, are highlighted as critical steps toward climate resilience. The chapter concludes with a call for global collaboration to mitigate climate change and ensure a sustainable future.

Chapter 2: Climate Change Mitigation and Resilience Strategies: A Nexus between Governance and Sustainable Development

The critical intersection of climate change mitigation and resilience strategies with governance and sustainable development. It argues that effective climate action requires a holistic approach that integrates robust governance frameworks with sustainable development goals. The chapter examines the role of governance, from international agreements like the Paris Agreement to national policies and local community initiatives, in driving these strategies. It highlights how good governance, characterized by transparency, accountability, and inclusivity, is essential for the successful implementation of climate policies. Furthermore, the chapter delves into how sustainable development, which balances economic growth, social equity, and environmental protection, provides a crucial framework for climate action. The chapter also concludes that governance, mitigation, resilience, and sustainable development are not viewed as separate issues but as interconnected elements of a unified strategy for a sustainable and climate-secure future.

Chapter 3: Economic Growth, Human Resource Development, and Climate Resilience in BRICS- A Panel Data Analysis

Present study employs panel data analysis on BRICS economies data spanning from 2000 to 2023, to examine the impact of human resource development, renewable energy consumption, and carbon emissions on GDP per capita of BRICS economies, finding that while human capital strongly drives growth, renewable energy has a positive but weaker effect, and carbon emissions remain tightly linked to GDP, underscoring the persistent reliance on carbon-intensive industries, ultimately highlighting the need for policy shifts toward sustainable growth through education, clean energy transitions, and regulatory frameworks that balance industrialization with climate resilience.

Chapter 4: "Foundations of Green Human Resource Management (GHRM)"

While organizations struggle with the intensifying effects of climate change, incorporating sustainability into fundamental business functions has become essential. GHRM then becomes a key framework for aligning human capital with sustainable objectives, developing sustainable culture, and promoting organizational resilience. This chapter discusses the premises of GHRM, exploring its theoretical bases, essential elements, and its central role in driving climate change mitigation and resilience initiatives. Based on theoretical lenses (different theories) GHRM covers the conventional practices such as recruitment, training, performance management, employee compensation and occupational health and safety from a climate resilience approach perspective. The chapter uses Patagonia and Unilever among other good businesses to show how GHRM may drive environmental performance, employee engagement, and corporate resilience. The chapter concludes by describing the road map for GHRM in terms of cooperation, employee empowerment, and link to other CSR activities.

Chapter 5: Strategic Human Resource Management for the Advancement of Climate Change Resilience and Mitigation

This chapter explores the integration of climate change mitigation and resilience strategies within organizational frameworks through strategic human resource management (HRM). By aligning HR practices with sustainability goals, organizations can foster a workforce equipped to address environmental challenges. Key strategies include developing green competencies, promoting eco-conscious behaviors, and enhancing employee engagement in sustainability initiatives. The chapter examines how HRM can drive organizational resilience by embedding climate-focused policies in recruitment, training, performance management, and leadership development. It highlights the role of HR in cultivating a culture of environmental stewardship, enabling organizations to adapt to climate-related risks while contributing to global sustainability efforts. Practical implications and future research directions are discussed to advance the intersection of HRM and climate action.

Chapter 6: Promoting Climate Change Mitigation and Resilience Strategies Through Human Resource Management: Performance Management and Rewards for Sustainability

In line with sustainable development goals principles, the long-term value creation potential of businesses has started to be measured not only by economic outputs but also by environmental and social impacts (Elkington, 1997). Accordingly, it is necessary to restructure performance management and reward systems in order for organizations to effectively implement their sustainability strategies. Aligning performance management processes with sustainability goals is reported to increase employee motivation and organizational commitment, (Warhurst, 2002). In this context, rewarding sustainability-oriented performance is not limited to financial incentives, but is also supported through intangible rewards. In order for these systems to be implemented effectively, it is necessary to integrate strategic sustainability goals with corporate vision and actions, measure sustainable performance indicators, create an organizational culture conducive to stakeholder engagement, base leadership activities on sustainability principles, and establish a fair reward system.

Chapter 7: Green Leadership and Cultural Transformation

The urgent challenge of climate change necessitates a shift in leadership and organizational culture to foster sustainability. This chapter explores Green Leadership as a transformative approach that integrates environmental responsibility into organizational goals. It examines the role of Cultural Transformation in embedding sustainability within corporate ethos through leadership-driven strategies. Theoretical frameworks, including Transformational Leadership and Ecocentric Leadership, are analyzed alongside HRM initiatives that promote sustainable practices. Case studies illustrate successful implementations, highlighting challenges such as resistance to change and greenwashing. The chapter concludes with insights on emerging sustainability trends, ESG frameworks, and HRM's role in driving innovation. By aligning leadership, culture, and sustainability, organizations can become proactive agents of environmental change and resilience.

Chapter 8: Ethical Leadership in the Age of Green Crisis

This chapter examines how ethical leadership drives environmental sustainability in modern organizations. It explores the convergence of ethical leadership, CSR, and environmental stewardship, providing frameworks for sustainable practice implementation. Drawing from organizational behavior and stakeholder perspectives, it demonstrates how ethical leaders foster environmental responsibility cultures. The analysis covers employee engagement in sustainability initiatives and knowledge management's role in environmental performance. The chapter addresses crisis management challenges and emphasizes education's importance in developing future environmental leaders. It offers practical guidance for business leaders, organizations, and policymakers, providing a roadmap for sustainable business practices in an eco-conscious global economy.

Chapter 9: Green Policies and Practices Across Industries

This chapter examines how industries worldwide are adopting green policies and sustainable practices to combat climate change, reduce environmental impacts, and enhance organizational resilience. Focusing on key sectors like manufacturing, energy, transportation, agriculture, and services, it explores how green strategies are integrated into core operations. The role of Human Resource Management (HRM) is emphasized, highlighting its contribution in fostering green behaviors, delivering sustainability-focused training, and aligning HR practices with environmental goals. The chapter also addresses industry-specific challenges, showcases success stories, and distills lessons learned, offering actionable insights to support global climate resilience through strengthened green initiatives.

Chapter 10: A Study on Understanding Climate Change and Organizational Responsibility

Organizational responsibility, climate change, is important to their being able to tackle the global environmental crisis. A rise in temperatures, the rise of extreme weather events, disrupted ecosystems, all these are the results of human influence on the climate. In particular, organizations, both in private and public sectors, have a vast role in lessening its effects through sustainable practices, reduction of carbon footprints and advocating for policies that value environmental conservation. In context of climate change, Corporate responsibility involves taking up eco-friendly technologies, energy efficient processes and supporting climate action. Additionally, organizations are required to spread the word, teach stakeholders, and establish a sustainability based culture. Integrating climate change considerations into their operations and decision making processes has the potential of adding organization related value in the global fight against climate change and advocacy of long term environmental stewardship.

CONCLUSION

This book makes a unique contribution by bringing Human Resource Management into the climate resilience discourse. It demonstrates that while science, policy, and technology remain critical, human-driven strategies—through leadership, organizational responsibility, and HRM—are equally essential in building a sustainable future. By bridging these perspectives, the book not only advances academic scholarship but also provides actionable insights for practitioners, policymakers, and organizations worldwide.

Together, the chapters emphasize that climate resilience requires a collective effort: integrating ecological awareness, governance, innovation, and human resource strategies. The result is a comprehensive resource that both enriches the academic field and provides practical guidance for creating sustainable, climate-ready organizations and societies.

Chapter 1
Understanding Climate Change and Organizational Responsibility

Kübra Nur Uzuntaş
https://orcid.org/0009-0003-0726-5116
Istanbul Bilgi University, Turkey

ABSTRACT

Climate change, driven by human activities, poses severe environmental, economic, and social challenges. This chapter examines its impacts, including rising temperatures, ecosystem disruption, and threats to food security and health, with vulnerable populations disproportionately affected. It explores the role of industries in greenhouse gas emissions and the potential for sustainable practices like renewable energy and circular economies. The chapter critiques greenwashing and emphasizes the need for stricter regulations and corporate accountability. Key policies, such as the European Green Deal and U.S. Inflation Reduction Act, are highlighted as critical steps toward climate resilience. The chapter concludes with a call for global collaboration to mitigate climate change and ensure a sustainable future.

1. INTRODUCTION

Global climate change refers to the extensive and long-term fluctuations in the earth's weather and temperatures. Earth's climate, while naturally fluctuating throughout its existence, is of particular interest currently due to its exceptional rate and unique causation solely caused by human forces, especially after the Industrial

DOI: 10.4018/979-8-3373-2479-1.ch001

Revolution. The Intergovernmental Panel on Climate Change (IPCC) defines the climate change as "*a change in the state of the climate that can be detected by changes in the mean and/or variability of its properties and that persists for an extended period, typically decades or more*" (IPCC, 2014).

The scientific consensus is that anthropogenic climate change does exist. It is claimed that 97% of all publishing climate scientists agree that human-caused global warming is the main reason for global warming (Cook et al., 2013). This is also highlighted by the National Aeronautics and Space Administration (NASA) and the National Oceanic and Atmospheric Administration (NOAA), which have provided detailed information of evidence for anthropogenic climate change (NASA, 2021; NOAA, 2022).

Key climate change signals are rising global surface temperatures, melting ice sheets and glaciers, sea-level increase, and enhanced frequency of extreme weather events. The global mean surface temperature has increased by about 1.18°C over the late 19th century, with most of the warming happening in the last four decades, as recorded by NASA (NASA, 2021). Arctic sea ice coverage has lost approximately 12.2% every decade since 1979, reflecting the fast melting of polar ice (NASA, n.d.).

2. THE IMPACT OF CLIMATE CHANGE ON ECOSYSTEMS AND HUMAN LIFE

Climate change affects societies and natural ecosystems profoundly. Species redistribution, altered reproduction, and increased vulnerability to invasive species and disease are happening to ecosystems. For instance, many terrestrial species have experienced poleward or elevation shifts in response to higher temperatures, seeking suitable habitats (Parmesan & Yohe, 2003). These alterations disrupt ecological webs since new species interactions occur, sometimes causing competition or extinction (Putra et al., 2025).

Marine ecosystems are also threatened; coral reefs, home to a significant percentage of marine biodiversity, are experiencing widespread bleaching due to rising sea temperatures. A recent report detailed mass coral bleaching in Western Australia's Ningaloo Reef, with large expanses of reefs facing mortality within the coming weeks (The Guardian, 2025). Ocean acidification, which is caused by increased CO_2, is weakening coral skeletons and shellfish exoskeletons, impacting ocean biodiversity and undermining food webs (Putra et al., 2025). Moreover, climate change is modifying ocean zooplankton communities, which are the foundation for most oceanic food webs (Putra et al., 2025).

The human population is also vulnerable to climate change impacts. Increased and intensified extreme weather conditions, such as hurricanes, floods, and heat waves, threaten human health, safety, and infrastructure. For example, Ireland was hit by Storm Éowyn, which caused widespread destruction and left approximately 768,000 buildings in darkness (The Guardian, 2025). This accident made the country realize its need to enhance climate resilience and infrastructure to cope with extreme weather (The Times, 2025).

Climate change also threatens food and water security. Changes in precipitation patterns and increased drought frequencies affect agricultural yields, which can lead to food deficits. Heat stress and unpredictable weather patterns diminish crop production, endangering staple foods like wheat, rice, and maize (Bateman et al., 2025). Water resources are diminishing in most areas, with glaciers and snowpacks melting at record rates. These phenomena endanger livelihood and can result in mass migrations and geopolitical tensions over scarce resources. It is estimated that more than 3 billion human beings will live in places that turn into uninhabitable areas under severe climatic conditions by 2070, thus unleashing great human migrations (Natural Resources Defense Council, 2020).

Climate change-related health risks are also increasing. Rising temperatures facilitate the spread of vector-borne diseases such as malaria, dengue fever, and Lyme disease since higher temperatures boost the range of disease-spreading insects (Baker et al., 2025). Increasingly frequent heatwaves have powered the rising incidence of heatstroke and cardiovascular mortality, especially in susceptible populations like the elderly and poor communities with less access to air-conditioned facilities. In addition, increased air pollution due to wildfires and fossil fuel combustion aggravates respiratory ailments like asthma and chronic obstructive pulmonary disease (COPD) (Grossman, 2025).

The socioeconomic impacts of climate change also underscore the imperative for action to be undertaken with a sense of urgency. Climate-related disaster losses are increasing, and billions of dollars are incurred yearly in disaster recovery and infrastructure rebuilding. Increased insurance expenses, displacement of people, and disruption of sectors like agriculture and tourism impose further pressures on economies. Climate migration is already underway, with inhabitants of low-lying island states confronting existential risk from sea level rise (Bateman et al., 2025). If radical mitigation and adaptation do not occur, millions more will be displaced, posing new social and political problems for recipient and source countries.

Governments and organizations globally strive to adopt climate adaptation and mitigation measures in response to such increasing threats. Investing in climate-resilient infrastructure, establishing early warning systems, and shifting to green energy are critical aspects of climate action (De Palma et al., 2025). The growth of green technologies like solar and wind power, electric transportation, and green agriculture provides hope that greenhouse gas emissions can be curbed and further warming prevented. International agreements like the Paris Agreement emphasize the need for international collaboration in meeting climate targets and protecting the vulnerable from climate change's reverse effects (IPCC, 2014).

2.1. Historical Perspective of Industrial Climate Change

The Industrial Revolution started in the late 18th century and fundamentally transformed human history. It was a period of rapid technological advancement, urbanization, and economic development, largely driven by the utilization of fossil fuels like coal, oil, and natural gas. Before industrialization, human civilizations used biomass, wind, and water power, which had relatively small effects on atmospheric carbon. Nonetheless, the extensive utilization of coal-fired steam engines, mechanized factories, and extended transportation networks radically raised greenhouse gas emissions. The shift to fossil fuels permitted mass production, better transportation, and economic development but also sparked the intrinsic remodeling of the Earth's atmosphere, leading to today's climate crisis.

The combustion of fossil fuels also releases CO_2 and other greenhouse gases, which capture heat and intensify the natural greenhouse effect (Lüthi et al., 2008). The CO_2 increase is directly related to industrial activities, most notably the burning of coal, oil, and gas, highlighting the central role of industrialization in accelerating climate change (NOAA, 2022).

Industrialization's influence does not end with emissions. Industrial economic development has also caused deforestation and large-scale land-use changes, contributing to global warming. Forests are sinks for carbon that take up CO_2 from the atmosphere, yet industrialization has caused large-scale deforestation due to agriculture, urbanization, and mineral exploration. Deforestation lowers the ability of the planet to sequester carbon, which amplifies the effect of increasing emissions. Amazonia, the lungs of the planet, has seen record rates of deforestation and, as it were, dramatically undermined its role in regulating global carbon levels (FAO, 2021).

One of the biggest effects of industrialization is also the reorientation of patterns of global energy use. Over the 20th century, industrial economies increasingly used oil and natural gas and invested in infrastructure for fossil fuel use worldwide. Mass motor vehicle travel, the creation of internal combustion engines, and the shipping fleet further contributed to carbon emissions. The petrochemical industry that produces plastics, artificial fertilizers, and industrial chemicals was also an important indirect greenhouse gas source based on high-energy production (Lüthi et al., 2008).

The scientific community has found sufficient evidence to link industrialization with climate change. Ice core sample research shows that CO_2 levels in the atmosphere today are much higher than in the past 800,000 years (Lüthi et al., 2008). In all its reports, the Intergovernmental Panel on Climate Change (IPCC) has determined that industrial emissions are the primary cause of global warming. Observations over the ocean show that over 90% of the extra heat generated due to human activities has been absorbed by the oceans, leading to rising sea levels, ocean acidification, and severe weather (NOAA, 2024). Though industrialization has led to economic development, technological advancement, and improved living standards, it has also caused pollution, natural resource depletion, and diseases (Machalaba et al., 2025).

3. THE ROLE OF ORGANIZATIONS IN CLIMATE CHANGE

Organizations are both contributors to and potential solutions for climate change. Industrial activities, corporate supply chains, and energy consumption drive significant greenhouse gas emissions, making businesses key players in global warming. However, companies also have the resources, innovation capabilities, and financial power to implement sustainable solutions and accelerate the transition to a low-carbon economy (Sullivan & Gouldson, 2017).

While some businesses have adopted greener practices, others continue unsustainable operations, prioritizing short-term profits over environmental responsibility. The challenge lies in balancing economic growth with sustainability, requiring stronger policies, corporate accountability, and industry-wide collaboration to achieve meaningful change (Freeman et al., 2018).

3.1. Industrial Contributions to Greenhouse Gas Emissions

Industries are a vast source of global greenhouse gas (GHG) emissions, mostly from fossil fuel combustion to offer energy and other industrial processes. The manufacturing sector is the third-largest emitter of direct GHG emissions following transport and the electricity industry. Manufacturing alone in 2021 added approximately 12% to U.S. GHG emissions, such as emissions from energy consumption and chemical reactions in production processes (Congressional Budget Office, 2024). Emissions from the industrial sector arise from the direct combustion of fossil fuels and indirect emissions from electricity and heat production. As industries continue to grow to meet world demand, emissions from the sector are forecast to rise unless extreme mitigation efforts are implemented.

The production sector, in particular, emits a significant amount of CO_2 and other GHGs. This emission results from burning fossil fuels to generate energy and from processes such as cement manufacturing, steel production, and chemical processing that emit GHGs through their reactions. Cement manufacturing alone accounts for approximately 8% of worldwide CO_2 emissions, mostly due to the calcination of limestone that emits CO_2 as a waste product (Panagoda et al., 2023). Likewise, steel production notably contributes to emissions because it largely depends on coal-based blast furnaces. In 2019, the steel and iron sector alone emitted nearly 7% of global energy-related CO_2 emissions (International Energy Agency, 2020). GHG emissions in material production grew by 120% between 1995 and 2015, and 11 billion tons of CO_2-equivalent were emitted in 2015 (Hertwich, 2021). They highlight the growing difficulty of decarbonizing industrial processes and supporting economic growth.

Also, some industries contribute to GHG emissions in non-energy-related activities. The fashion industry, for example, is responsible for about 10% of all global carbon emissions annually, which is higher than the combined emissions from all global flights and ocean shipping (Maiti, 2025). The industry contributes beyond carbon emissions to significant water usage and waste production. Clothing production consumes enormous amounts of water—producing one pair of jeans requires around 7,500 liters of water, which is what an average person drinks in seven years (UN News, 2019). Fast fashion perpetuates these issues by promoting fast turnover of garments, leading to increased waste, with 85% of all textiles being sent to landfills yearly (Maiti, 2025[REMOVED HYPERLINK FIELD]).

The chemical industry also accounts for important contributions to global GHG emissions. The production of ammonia, one of the biggest industrial CO_2 emitters, is a key component of fertilizers that relies on natural gas as a hydrogen source. Global ammonia production emitted approximately 450 million metric tons of CO_2, or roughly 2% of industrial emissions overall (International Energy Agency, n.d.).

Also, another sector, aviation, relies extensively on fossil fuels, with burning jet fuel releasing enormous amounts of carbon dioxide into the environment. According to the International Air Transport Association (IATA), global aviation produced approximately 915 million tons of CO_2 in 2019 before the COVID-19 pandemic temporarily reduced demand for air transport (IATA, n.d.). The emissions will grow as air transport increases unless significant mitigation steps are undertaken.

Aviation has pledged to reduce carbon emissions by many methods, including:

- Sustainable Aviation Fuels (SAFs): SAFs, originating from bio-based materials such as algae and waste oils, can deliver CO_2 savings up to 80% below conventional jet fuel (IATA, 2024).
- Fuel-Efficient Aircraft Technology: Boeing and Airbus produce fuel-efficient aircraft models, employing lightweight material and advanced aerodynamics to reduce fuel consumption (Dolganova et al., 2022).
- Hydrogen and Electric Aircraft: Hydrogen and electric aircraft are short-term, long-term solutions for environmentally friendly aviation. ZeroAvia and Airbus enterprises are investing in hydrogen flight technology (European Commission, 2023).
- Carbon Offset Programs: Some airlines like Delta and United operate carbon offset programs where customers can compensate for flight emissions by

paying for reforestation and renewable energy projects (National Geographic, 2022).

Despite such efforts, critics argue that most aviation sustainability measures are inadequate or amount to "greenwashing" without systemic change to air travel demand and infrastructure. Legal interventions like carbon pricing and stricter emissions standards must be implemented to ensure significant reductions in aviation emissions.

Besides carbon dioxide, other greenhouse gases are released by industries, such as methane and nitrous oxide (N_2O). Methane is among the most frequent emissions in the oil and gas sector due to venting and flaring related to extraction and transportation. Methane has a 100-year global warming potential of approximately 28 times more than CO_2 and is, thus, extremely important to curb climate change mitigation (European Commission, n.d.). Nitrous oxide, released primarily from industrial processes and agriculture, is nearly 300 times stronger than CO_2 and possesses very long atmospheric residence times (Stanford Report, 2020).

A number of methods are being researched to restrict industrial emissions. Implementing CCUS technologies is one possible method with an answer: capturing emissions at the source and recycling or storing them underground. Transitioning to renewable power, such as hydrogen-powered steel production and industrial-process electrification, could also restrict emissions. Furthermore, the practices of circular economy, such as recycling content, enhancing energy efficiency, and developing alternative low-carbon production processes, are necessary to achieve long-term sustainability in industry production.

As industries will be an ongoing significant source of global emissions, policy and innovation will be the key to reducing their environmental footprint. Applying policy instruments such as carbon pricing mechanisms, emissions trading systems, and strict efficiency standards will induce industries to shift to cleaner technology. Interagency, government-industry, and academia partnerships will be required to facilitate quicker transitions toward an industrial base that is low-carbon yet economically sound and expanding.

3.2. The Importance of Organizational Responsibility

The industrial and commercial sectors are among the largest contributors to greenhouse gas emissions, primarily through energy consumption, production processes, and supply chain operations. For instance, the energy sector alone accounts for approximately 73% of total CO_2 emissions globally (D-Carbonize, 2025), which indicates the imperative impact of organizational activities on the environment.

With their substantial influence, organizations are responsible for embracing sustainable practices and reducing their environmental footprints. This is not just an ethical but also a strategic necessity; stakeholders, including consumers, investors, and regulators, are making environmental sustainability an increasing priority. Companies taking early action on climate change can enhance their reputations, address evolving regulatory demands, and potentially reap cost advantages through energy efficiency and innovation. Besides, business organizations have the potential and wherewithal to drive deep change through investing in renewable energy, resource efficiency, and circular economy approaches. Sustainable business activities minimize environmental harm and create long-term value, ensuring resilience against climate-related risks such as supply chain disruption and regulatory penalties (Sullivan & Gouldson, 2017).

Global organizations and states are also central to corporate responsibility developments. Projects such as the European Union's Green Deal and the United Nations' Sustainable Development Goals (SDGs) provide recommendations that encourage accountable business behavior. For instance, the Task Force on Climate-Related Financial Disclosures (TCFD) provides corporations with recommendations on assessing and reporting climate-related risks to develop transparency and accountability (TCFD, 2017).

3.3. Responsibility of Private and Public Sector

The debate on which public or private sector has greater responsibility in tackling climate change is complex and multifaceted. Governments create the policy and regulatory framework and implement policies, whereas corporations have the biggest role in emissions through their business operations, supply chains, and investments. Both need to work together to fight climate change but in different roles and with varying accountability based on their influence, capacity, and function.

The private sector is a vast economic activity and industrial production vehicle responsible for most of the global GHG emissions. Studies indicate that 100 firms have produced over 70% of all industrial emissions since 1988 (CDP Carbon Majors Report, 2017). Fossil fuel giants such as ExxonMobil, Shell, BP, and Chevron have significantly contributed to the world's carbon legacy (Heede, 2013). Most companies have huge supply chains in many countries, so their carbon footprint is even larger. The manufacturing, energy production, and transport industries account for enormous emissions worldwide. (IPCC, 2022) states that private sector investment in renewable energy and energy efficiency is essential to achieving global climate goals.

As a result, many companies have adopted Corporate Social Responsibility (CSR) practices, environmental, social, and governance (ESG) frameworks, and net-zero targets to make their businesses more sustainable. Google and Apple, for instance, have pledged to power their data centers and operations with 100% renewable energy (Google Sustainability Report, 2023). Amazon has introduced its Climate Pledge with the goal of reaching carbon neutrality by 2040 through reforestation efforts and investments in clean technologies (Amazon Sustainability Report, 2022). On its part, IKEA has promised to be completely circular by 2030 by focusing on using sustainable materials and reducing waste (IKEA Sustainability Report, 2021).

The government sector also has a general role in mitigating climate change through the tool of regulatory authority. Governments enact diverse policies for limiting emissions, such as carbon taxation, emissions trading systems, and subsidies for renewable energy. Sweden and Canada, for instance, have introduced carbon taxation to disincentivize the use of fossil fuels (World Bank, 2023). The EU Emissions Trading System (EU ETS), the largest carbon market globally, limits the emissions of industry and power plants. Likewise, the U.S. Inflation Reduction Act (2022) offers tax credits amounting to billions of dollars for renewable energy projects, electric vehicles, and green infrastructure (U.S. Department of Energy, 2023).

Besides policy-making, governments promote green infrastructure and low-carbon research to propel low-carbon options. Green buildings, low-carbon public transport, and renewable energy generation are at the leading edge of lowering a country's carbon footprint. The public sector can accelerate systems change by incorporating sustainability into urban planning, transport policy, and industrial regulation (IPCC, 2023).

In addition, governments need to establish corporate responsibility by enacting environmental policies, penalties, and transparency obligations. For example, the U.S. Securities and Exchange Commission (SEC) has proposed rules mandating corporations to reveal their climate risk and emissions (SEC, 2022). Likewise, the EU Corporate Sustainability Due Diligence Directive (2023) obliges large companies to detect and mitigate environmental hazards in their value chains (European Commission, 2023).

Organizations, governments, and international organizations are the major stakeholders in combating climate change; they have different ways of cutting carbon emissions, switching to clean energy, and innovating sustainable business models.

4. REGULATORY FRAMEWORKS AND CLIMATE POLICIES

Climate legislation and policies determine how individuals, industries, and governments respond to climate change. Environmental legislation, carbon pricing, and climate agreements are all designed to reduce emissions and promote sustainability. The effectiveness of the policies, however, depends on enforcement, international cooperation, and compliance by the industries (IPCC, 2023).

Countries are adopting different strategies, from the European Union's strict carbon market regulations to incentive-based policies like the U.S. Inflation Reduction

Act. While these moves promote cleaner production, they also face economic and political resistance, a sign of the existing dilemma between environmental conservation and economic competitiveness (Ockenfels & Pahle, 2025).

4.1 The EU Green Deal and Its Implications for Businesses

The European Green Deal (2019) is the most ambitious EU environmental policy blueprint to transform Europe into the world's first climate-neutral continent by 2050. The policy approach is intended to steer the EU economy to sustainable development and social equity. The Green Deal is a multi-sector policy, including energy, industry, agriculture, nature and biodiversity, and transport, and is in harmony with the United Nations Sustainable Development Goals (SDGs) (Paleari, 2022).

Among the most significant tools of the Green Deal is the EU Emissions Trading System (ETS), charging industry with a restriction on total greenhouse gas emissions and permitting the trading of emission credits. The cap-and-trade mechanism has reduced emissions from power plants and industrial installations in the European Union in order. The EU further extends the application of the ETS to other sectors, such as shipping and aviation, which are typically difficult to control in the case of emissions (Ockenfels & Pahle, 2025).

The second enabling pillar of the Green Deal is the Carbon Border Adjustment Mechanism (CBAM), a carbon tax on imports of products from less climate-ambitious nations. CBAM would not render European businesses unfairly competitive relative to foreign rivals with weaker environmental regulations. CBAM has, nonetheless, been denounced by developing nations and trading partners as generating economic asymmetries and distorting global trade flows.

For companies, the Green Deal represents an opportunity and a threat. On the negative side, additional emissions goals and compliance expenses are a cost to companies, particularly those with fossil fuel exposure. On the positive side, the policy creates new markets for clean technology, drives green innovation in sustainable industries, and encourages investment in clean energy, circular economy technologies, and energy efficiency. European businesses that adapt swiftly to such policies will certainly be advantageous in the world's shift towards sustainability.

Although substantial challenges hinder its ambitious target, the Green Deal. The high cost of implementation, opposition from some EU member states, and the challenge of making the transition fair for workers in carbon-intensive sectors are still major impediments. Nevertheless, as the EU advances with binding legal obligations, the Green Deal continues to influence global climate policy and act as a template for other economies (Ockenfels & Pahle, 2025).

4.2 The U.S. Inflation Reduction Act and Climate Targets

The Inflation Reduction Act (IRA), signed into law in 2022, represents the largest climate investment in U.S. history, allocating $369 billion for clean energy incentives, carbon reduction programs, and environmental justice initiatives (Thom, 2025). The IRA aims to reduce U.S. greenhouse gas emissions by 40% by 2030 while promoting economic growth and job creation in renewable energy sectors. One of the major provisions of the IRA is extensive tax credits and subsidies for solar, wind, and nuclear energy projects, as well as battery storage and electric vehicles (EVs). The law provides:

- Production tax credits (PTCs) and investment tax credits (ITCs) for businesses that develop clean energy projects.
- Incentives for consumers to purchase electric vehicles (EVs) and install energy-efficient home systems.
- Financial support for carbon capture and storage (CCS) technologies aims to reduce emissions from fossil fuel industries.

Beyond clean energy incentives, the IRA also includes measures to address environmental justice, ensuring that historically marginalized communities disproportionately affected by climate change and pollution receive financial and technical support. Billions of dollars are allocated to help disadvantaged communities access clean energy technologies and improve climate resilience.

Despite its ambitious objectives, the IRA faces political and economic challenges.

- Political resistance: Some lawmakers argue that the IRA imposes unnecessary regulations on businesses and increases federal spending.
- Legal uncertainties: Changes in administration or judicial rulings could limit the act's long-term effectiveness.
- Supply chain concerns: The transition to clean energy requires critical minerals, such as lithium and cobalt, which are heavily concentrated in a few countries. Securing these resources poses a challenge to large-scale implementation.

Nevertheless, the IRA has already accelerated the deployment of renewable energy projects across the U.S., boosted investment in green industries, and reinforced American leadership in climate action. Its success will depend on continued funding, regulatory support, and bipartisan commitment to sustainability.

The IRA's success will depend on government implementation and how businesses and organizations align their strategies with these new climate incentives. Companies investing in renewable energy, carbon capture technologies, and sustainable infrastructure stand to benefit significantly, highlighting the intersection of government action and corporate responsibility in tackling climate change.

4.3. Carbon Taxation and Emission Trading Systems

Carbon Taxation

A carbon tax is a popular market-based climate policy aimed at reducing carbon emissions by imposing a direct price on carbon dioxide (CO_2) emissions. By raising the cost of pollution, carbon taxes provide a financial incentive for businesses and individuals to switch to cleaner alternatives. Countries such as Sweden, Canada, and Germany have successfully implemented carbon taxes, leading to significant reductions in emissions.

- Sweden introduced its carbon tax in 1991, now set at around $130 per ton of CO_2—one of the highest rates globally. Since its introduction, Sweden has reduced emissions by over 25% while maintaining strong economic growth (Nadeem, 2025).
- Canada implemented a nationwide carbon pricing system in 2019, requiring provinces to meet minimum pricing standards. Revenue from the tax is returned to citizens through rebates, ensuring that low-income households are not disproportionately affected.
- Germany launched its carbon pricing scheme in 2021, with plans to gradually increase the tax rate to accelerate coal phase-out and renewable energy investment.

While carbon taxes are effective in reducing emissions, they are politically controversial. Opponents argue that carbon pricing raises energy costs for consumers and industries, potentially harming economic competitiveness. Additionally, without global coordination, carbon taxes may lead to carbon leakage, where businesses relocate to countries with less stringent regulations.

Emission Trading Systems (ETS)

The Emissions Trading System (ETS), or cap-and-trade, is yet another market mechanism for curbing greenhouse gas emissions. Under ETS, governments place a cap (limit) on emissions and sell or auction permits for emissions to industries.

Industries that lower their emissions can sell surplus permits, while those who need more buy additional permits.

The EU ETS is the world's largest and most successful cap-and-trade system. Since its introduction in 2005, it has led to substantial emissions cuts by the power and industrial sectors while keeping the economy expanding (Rahman et al., 2025). Some other prominent ETS programs are:

- California's Cap-and-Trade Program is among the strongest in the United States.
- China's National Carbon Market, which became the world's largest ETS in 2021.
- The Northeastern US Regional Greenhouse Gas Initiative (RGGI) has aimed to lower emissions while generating funds for clean energy initiatives.

ETS is a more responsive strategy than carbon taxes, with firms permitted to sell and buy permits and respond more promptly. However, efficiency relies on firm regulatory enforcement because loose caps or over-issued permits can diminish carbon prices and efficiency.

5. CORPORATE SOCIAL RESPONSIBILITY

CSR is a company's commitment to business ethics, conservation, and social welfare beyond the legal minimum (Carroll & Shabana, 2010). CSR originated from philanthropy and evolved to a strategic model wherein companies incorporate environmental and social concerns in the company's business purpose.

One of the most pressing CSR issues today is climate action. Corporates are one of the highest carbon emitters in the world, and their climate change mitigation can have a significant impact on them. Corporates tackle climate action by making carbon reduction pledges, investing in renewable energy, and sustainable supply chain management (Porter & Kramer, 2011). Some MNCs have made net-zero pledges, committing to reduce greenhouse gas emissions to a specific date.For instance, Microsoft has committed to being carbon-negative by 2030, i.e., it will have more carbon offsets than it emits (Microsoft, 2020). Similarly, Unilever integrates sustainability in its supply chain by using palm oil from sustainable plantations, which reduces deforestation and biodiversity loss (Unilever, 2022).

Despite such attempts, there is skepticism about the effectiveness of voluntary corporate climate commitments. Companies use CSR initiatives as PR tools rather than taking real change, as argued by critics (Delmas & Burbano, 2011). This necessitates having regulatory mechanisms and independent verification systems to ensure that corporate climate commitments are delivered in terms of real impact.

5.1. Triple Bottom Line: People, Planet, Profit

The Triple Bottom Line (TBL) approach, first taken to the limelight by John Elkington in 1994, incorporates social, environmental, and economic dimensions to conventional parameters for business success (Elkington, 1999). Companies employing this approach assess their performance in terms of profit and long-term value to people and the planet. This view acknowledges that a firm's success can be gauged not just by its economic performance but also by its effects on nature and society (Savitz & Weber, 2013).

The TBL principle is based on the aspect that businesses depend on natural capital, human capital, and economic infrastructure to do business successfully. If they degrade these elements, they decrease their ability to function over time (Bansal & DesJardine, 2014). For instance, companies that exploit workers or damage the environment can face regulatory penalties, loss of customer trust, and damage to reputation that erodes profitability (Dyllick & Hockerts, 2002). On the other hand, firms that pursue sustainability aim to build stronger brands, more loyal customer bases, and long-term financial health (Freeman et al., 2018).

Figure 1. Triple Bottomline

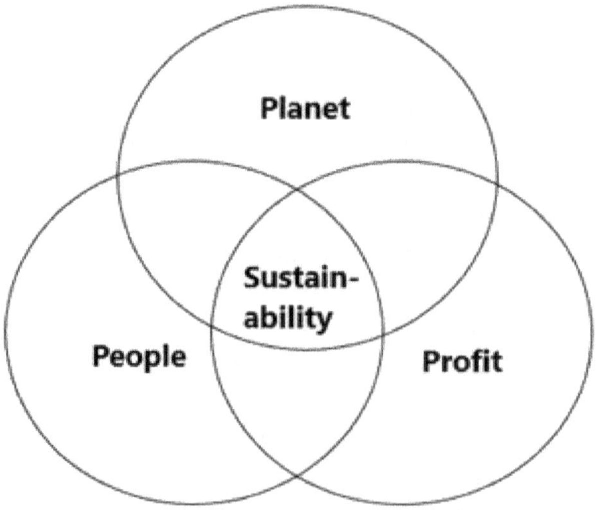

People: The Social Dimension

The initial pillar of the TBL approach is people and entails the social responsibility of businesses to their workers, customers, and society. A socially responsible business values good salaries, secure workplaces, diversity, inclusivity, and employees' well-being (Carroll & Shabana, 2010). Social sustainability incorporates businesses' contribution to education, health, and community development (Crane et al., 2014).

Most firms are aware that treating employees with kindness enhances productivity, job satisfaction, and innovation (Porter & Kramer, 2011). For instance, firms offering employee training, mental health support, and family-friendly policies have lower employee turnover and greater employee engagement (Kim et al., 2018). It also helps when studies demonstrate that diverse and inclusive organizations outperform their rivals since they foster creativity and offer various problem-solving perspectives (Harrison & Klein, 2007).

Planet: The Environmental Dimension

The second column of the TBL model is a planet that addresses businesses' impact on the physical world. Greenhouse gas emissions, deforestation, pollution, and loss of biodiversity are issues brought about by most industries (Rockström et al., 2009). As awareness of these issues grows, companies are compelled to lower their carbon footprint, reduce waste, and use resources efficiently (Lacy & Rutqvist, 2015).

Environmental sustainability generally means using renewable energy, circular economy, and green manufacturing (Geissdofer et al., 2017). For instance, most businesses have moved to utilize solar and wind power to fuel their activities, lowering fossil fuel use (Gielen et al., 2019).

Profit: The Economic Dimension

The TBL model contends that profitability must not come at the cost of people or the planet (Savitz & Weber, 2013). The more conventional business models are generally fixated on short-term financial returns, sometimes at the cost of exploitative working conditions, destruction of the environment, and economic imbalances (Stiglitz, 2012). However, studies have established that companies that adopt ethical, long-term financial policies perform better in the long run (Freeman et al., 2018).

Firms that integrate sustainability into their business strategies are likely to achieve cost efficiency, enhanced operation, and enhanced brand equity (Eccles et al., 2014). For instance, energy-saving factories reduce power costs, while waste reduction plans reduce the material cost (Hart & Milstein, 2003). Green bonds and ethical funds, which are types of sustainable investment, are increasingly prevalent as investors look for good governance and long-term-oriented firms (Friede et al., 2015).

Economic sustainability also includes the manner in which firms engage with local economies. Firms that invest in ethical trade enable small suppliers and invest in local development to build stable markets and long-term customers (Sachs, 2015). Governments and banks also increasingly favor sustainable firms through tax relief, grants, and low-interest loans (OECD, 2021).

5.2. GreenWashing

The term greenwashing was first used by environmentalist Jay Westerveld in 1986. It describes corporate marketing tactics that create a false impression of environmental responsibility without real, substantial action to reduce environmental harm (Pearson, 2010). Greenwashing has become widespread, with companies across industries making misleading claims about their sustainability efforts.

Greenwashing involves selective disclosure, misleading labels, vague claims, or exaggerated sustainability achievements. It allows companies to capitalize on consumer demand for eco-friendly products while avoiding the costs associated with true sustainability initiatives (Jamil & Wahyuni, 2024). Greenwashing can manifest in multiple forms, from misleading advertising to exaggerated sustainability commitments. Below are some of the most common types:

I. Hidden Trade-offs: Companies may highlight a single green feature of their product and ignore other characteristics that are harmful to the environment. For example, a water plastic bottle with "50% recycled plastic" may still be a top source of plastic waste and have intense carbon emissions through production and transport.
II. Deceptive Labels and Certifications: Most companies use false sustainability labels or misleading eco-certifications to convince consumers of their green status. Others create their own "green labels" without third-party verification, which misleads consumers into thinking they are purchasing environmentally friendly products (Peng et al., 2024).
III. Ambiguity and Vagueness: Companies tend to use imprecise terms such as "green," "eco-friendly," "natural," or "sustainable" without respective evidence or properly defining them. These words can mislead consumers into perceiving a product or service as more sustainable than it actually is (Nahari, 2024).
IV. Irrelevant Claims: Some companies promote environmental claims that are technically true but irrelevant. For instance, advertising a "CFC-free" aerosol spray as an environmental benefit is misleading because CFCs (chlorofluorocarbons) have already been banned worldwide since the 1980s.

V. Changed Information: In extreme cases, companies falsify data or manipulate scientific evidence to appear more environmentally friendly than they actually are. The Volkswagen Dieselgate scandal is a notorious example, where the company installed software to cheat emissions tests while falsely claiming its vehicles were low-emission (Ewing, 2017).

As more cases of greenwashing emerge, consumer skepticism about corporate sustainability claims increases. A 2023 survey found that 58% of consumers do not trust companies' environmental claims due to widespread deceptive marketing (Obeid et al., 2024). Trust erosion negatively impacts legitimate, sustainable businesses and consumer engagement in environmental issues.

Greenwashing delays real climate action by preventing companies from making meaningful sustainability commitments. False sustainability claims encourage overconsumption, leading to greater environmental harm rather than meaningful reductions in waste and emissions (Nobanee, 2024).

6. CLIMATE CHANGE AND SOCIAL JUSTICE

Climate change is not just an environmental issue but also a grave social justice crisis. The most vulnerable populations—poor people, Indigenous peoples, and poorer nations—are those that contribute the least to greenhouse gas emissions but get the worst of climate change's damaging effects (IPCC, 2023). This disparity highlights the urgent need for climate justice, with a promise that the benefits and burdens of climate policy are fairly shared so that the interests of the most affected communities are heard in the policy-making process.

Climate change aggravates existing social and economic disparities. The more frequent and intense extreme weather events increase, the poorer and vulnerable communities cannot deal with them since they lack economic resources and institutional support. For example, protracted droughts and erratic rainfall patterns in Africa and South Asia threaten food security, uprooting millions of people living in agriculture (Bateman et al., 2025). United Nations projections estimate that more than 3 billion people will be forced to migrate by 2070 as the climate becomes increasingly uninhabitable (NRDC, 2020).

The most egregious form of climate injustice is the burden of responsibility for historical emissions. Industrialized nations have emitted the lion's share of greenhouse gases since the Industrial Revolution, yet the developing world will suffer the greatest impacts (Lüthi et al., 2008). Bangladesh suffers from catastrophic flooding and loss of territory due to sea-level rise, while the small island states face complete flooding (IPCC, 2021). The concept of "climate debt" is that industrial nations, having gained from using fossil fuels in the past, have both economic and moral obligations to pay for climate adaptation in the developing world.

Environmental racism is another feature that marks climate injustice. Minority and low-income communities carry the concentration of industrial pollutants in them, and they experience more air and water pollution. In the US, research indicates that Black and Latino neighborhoods are more apt to be circled by toxic waste plants and factories, with more resultant health risks (Bullard, 2018). Likewise, the indigenous peoples of the Amazonian rainforest experience land disappropriation and forest destruction due to illegal mining and logging. Such a structural imbalance underscores the necessity of firmer environmental policy and legislation acting in the favor of marginalized peoples.

Climate change is disproportionately affecting women, particularly in the developing world. Women have the primary role of securing food, water, and fuel for their families, subjecting them to climate disturbance. UN Women (2023) argues that climate disasters increase gender-based violence, economic marginalization, and forced displacement of women. The underrepresentation of women in political and economic decision-making also denies them a say in influencing climate decisions.

Climate change fuels social injustice, and it affects the most vulnerable individuals the most. Climate justice is achievable through climate change reduction policies and structural inequality tackling. The developed world needs to own its past responsibility and boost climate adaptation finance in vulnerable countries. Governments also need to make sure that climate policies defend workers, enable marginalized groups, and mainstream gender equality concerns. Sustainable development was achievable through a combined strategy for environmental, economic, and social justice.

7. CONCLUSION

Climate change is arguably the most characteristic issue of our era, with wide-ranging environmental, economic, and social impacts. Anthropogenic sources—mainly fossil fuel combustion, industrial activities, land clearing, and intensive farming—have been major drivers of the rise in greenhouse gas emissions in the environment, causing a rapid global temperature (IPCC, 2023). This warming trend has set off a chain of interrelated crises, such as rising ocean levels, intense weather conditions, loss of biodiversity, and interruptions in food and water security (NASA, 2021). The intensity of these effects requires that there be collective global action to limit emissions, shift to shifting climate patterns, and provide future generations with a sustainable future.

Climate change impacts are unevenly distributed but disproportionately affect poor groups, small island states, and low-capacity developing countries (UN Women, 2023). Unpredictable weather and global warming have resulted in more frequent droughts, desertification, and lower agricultural output, compromising food security across most of the world's regions (Bateman et al., 2025). Simultaneously, sea-level increases put coastal communities at risk of extinction, displacing millions to date and risking billions more by the end of the century (NRDC, 2020). Climate migration, wars over resources, and economic disruption add to current inequalities, making it clear that climate change is not only an environmental issue but a humanitarian and geopolitical one.

On the business side, industries contribute to climate change and solutions. The energy industry alone is responsible for almost 73% of global CO_2 emissions, followed by transport, industry, and agriculture (D-Carbonize, 2025). While some businesses have gone above and beyond to minimize their carbon footprint, greenwashing is an ongoing issue where companies mislead others by saying that they are doing their share for the planet but not necessarily making a tangible difference (Bowen & Aragon, 2014). Regulatory policies should be tightened to facilitate corporate responsibility by requiring transparency in sustainability reporting and measurable commitments for emissions reduction. Governments must take the initiative to enact enforceable standards precluding misleading sustainability declarations and promoting genuine corporate responsibility.

At a policy level, international agreements like the Paris Agreement have set frameworks for international climate action, yet implementation is sought unevenly between countries (IPCC, 2023). Some countries have launched ambitious policies to limit emissions, like the European Green Deal, which seeks to make Europe the first carbon-neutral continent in the world by 2050 (European Commission, 2023). Likewise, the U.S. Inflation Reduction Act allocated billions of dollars in tax credits to renewable energy, carbon capture technologies, and green infrastructure projects (U.S. Department of Energy, 2023). Although these policies are a huge step in the right direction, their effectiveness relies on long-term commitment, bipartisanship, and the capacity to balance economic growth with environmental sustainability. Carbon pricing instruments, emissions trading schemes, and imposing stricter regulation on high-emission sectors are instrumental to accelerating the shift towards a low-carbon economy.

In addition to government and corporate intervention, individuals also play a significant role in the fight against climate change. Consumer behavior directly affects carbon emissions, as demand for low-carbon products, vegetarianism, and sustainable products all influence market demand (Ellen MacArthur Foundation,

2020). Personal changes in lifestyle, including energy efficiency, minimizing wastage, and patronizing eco-friendly brands, all sum up to emission savings. For all that, systemic change must follow individual action if the effect is to be sustained. Investing in green infrastructure and public transportation systems and making affordable access to clean energy are fundamental steps toward putting sustainable alternatives into practice at the societal level.

The second critical ingredient of climate action is climate justice, acknowledging that those who have least contributed to climate change are the ones who suffer its most catastrophic impacts. With minimal historical per capita emissions, the developing world faces apocalyptic climate-related catastrophes without the financial capital to adapt (Lüthi et al., 2008). Richer countries, having contributed the most to industrial-era emissions, have both an economic and moral duty to assist with climate adaptation and mitigation finance and technology in those vulnerable populations (World Bank, 2023). Second, being custodians of the natural ecosystems for centuries, Indigenous peoples must be represented in climate governance to facilitate traditional ecological knowledge and support sustainable solutions. To strengthen climate action, several key recommendations must be prioritized:

Speeding Up the Shift to Renewable Energy

The fast replacement of fossil fuels with renewable energy sources like solar, wind, and green hydrogen is key to carbon neutrality. Governments must raise subsidies for renewable energy initiatives while lowering subsidies for fossil fuel companies (Gielen et al., 2019). Investing in energy storage and smart grid technology can make renewable energy systems more efficient and dependable.

Strengthening Corporate Accountability and Sustainable Business Models

Industries should be made responsible for their ecological footprint through regulation, open reporting, and third-party verification of sustainability claims. Businesses must embrace circular economy strategies by designing durable products, minimizing waste, and sourcing materials sustainably (Ellen MacArthur Foundation, 2020).

Climate Adaptation Solutions for Vulnerable Communities

Governments and international agencies must offer financial support and technical aid to vulnerable climate areas. Investing in flood defenses, drought-tolerant agriculture, and disaster preparedness systems is essential to build resilience to climate shocks (UN Women, 2023).

Developing Credits-Based Climate Policies

Carbon pricing instruments, including carbon taxes and cap-and-trade systems, must be scaled up to offer economic incentives for reducing emissions. The policies should be formulated to reduce adverse socio-economic effects on poor communities while enhancing compliance by industries with high emissions (World Bank, 2023).

Encouraging Public Involvement and Education

Climate education must be integrated into school curriculums to prepare future generations with the knowledge and skills to achieve sustainability. Public awareness, community-based climate actions, and grassroots movements are important in compelling governments and corporations to be accountable for acting on climate.

Driving Technological Innovations in Climate Solutions

Clean technology research and development, such as carbon capture and storage (CCS), electric mobility, and sustainable agriculture, must be fast-tracked to speed up the transition towards a low-carbon economy (International Energy Agency, 2020). Green technology startup investment and collaboration among governments, industry, and academia can lead to innovation in climate solutions.

Finally, a holistic approach is needed to combat climate change that balances environmental sustainability with economic development and social equity. Despite the progress that has been seen in recent years, the rate of change remains insufficient to meet global climate objectives. Definitive, science-driven policy, corporate responsibility, and personal accountability must merge to limit the rise in the world's temperature and avert its far-reaching consequences. Without immediate intervention, the world risks irreversible environmental destruction, mass displacement, and economic volatility. However, with the appropriate policies, innovations, and international collaboration, a sustainable and equal future is still within reach.

REFERENCES

Baker, L., Jacobson, H., McGrosky, A., Hinz, E., & Associates. (2025). *Ambient temperature and wet bulb globe temperature outperform heat index in predicting hydration status and heat perception in a semi-arid environment.* Annals of Human Biology. Retrieved from https://www.tandfonline.com/doi/full/10.1080/03014460.2025.2456152

Bansal, P., & DesJardine, M. R. (2014). Business sustainability: It is about time. *Strategic Organization, 12*(1), 70–78. DOI: 10.1177/1476127013520265

Bateman, K. M., McCausland, J., & Walsh, N. (2025). *Exploring human impact on the Hudson River over time and space.* Justice-Oriented Science Education. Retrieved from https://books.google.com/books?id=68xIEQAAQBAJ

Bowen, F., & Aragon-Correa, J. A. (2014). Greenwashing in corporate environmentalism research and practice: The importance of what we say and do. *Organization & Environment, 27*(2), 107–112.

Bullard, R. D. (2018). *Dumping in Dixie: Race, Class, and Environmental Quality*. Westview Press.

Carbon Majors Report, C. D. P. (2017). *The 100 Companies Responsible for 70% of Global Carbon Emissions.* https://www.cdp.net/en

Carroll, A. B., & Shabana, K. M. (2010). The business case for corporate social responsibility: A review of concepts, research and practice. *International Journal of Management Reviews, 12*(1), 85–105. DOI: 10.1111/j.1468-2370.2009.00275.x

Center for International Environmental Law. (2019). *Plastic & climate: The hidden costs of a plastic planet.* Retrieved from https://www.ciel.org/plasticandclimate

Center for Sustainable Systems. (2021). Carbon Footprint Factsheet. University of Michigan. https://css.umich.edu/publications/factsheets/sustainability-indicators/carbon-footprint-factsheet

Congressional Budget Office. (2024, February 15). The budget and economic outlook: 2023 to 2033. Retrieved March 5, 2025, from https://www.cbo.gov/publication/60030

Cook, J., Oreskes, N., Doran, P. T., Anderegg, W. R. L., Verheggen, B., Maibach, E. W., & Rice, K. (2013). Quantifying the consensus on anthropogenic global warming in the scientific literature. *Environmental Research Letters, 8*(2), 024024. DOI: 10.1088/1748-9326/8/2/024024

Crane, A., Matten, D., Glozer, S., & Spence, L. J. (2014). *Business ethics: Managing corporate citizenship and sustainability in the age of globalization*. Oxford University Press., DOI: 10.1093/hebz/9780198810070.001.0001

D-CARBONIZE. (2025, February 22). *Comparison of GHG emissions by sector*. D-CARBONIZE. Retrieved March 5, 2025, from https://d-carbonize.eu/greenhouse-gases/comparison-ghg-emissions-sector/#:~:text=The%20energy%20sector%20is%20responsible,significantly%20contributing%20to%20global%20warming

De Palma, A., Lindsey, R., Proost, S., Riou, Y., & Trannoy, A. (2025). *Why combating climate change is so challenging*. Université de Cergy-Pontoise. Retrieved from https://thema.u-cergy.fr/IMG/pdf/2025-02.pdf

Delmas, M. A., & Burbano, V. C. (2011). The drivers of greenwashing. *California Management Review*, *54*(1), 64–87. DOI: 10.1525/cmr.2011.54.1.64

Dolganova, I., Bach, V., Rödl, A., Kaltschmitt, M., & Finkbeiner, M. (2022). Assessment of critical resource use in aircraft manufacturing. *Circular Economy and Sustainability*, *2*(3), 1193–1212.

Dyllick, T., & Hockerts, K. (2002). Beyond the business case for corporate sustainability. *Business Strategy and the Environment*, *11*(2), 130–141. DOI: 10.1002/bse.323

Eccles, R. G., Ioannou, I., & Serafeim, G. (2014). The impact of corporate sustainability on organizational processes and performance. *Management Science*, *60*(11), 2835–2857. DOI: 10.1287/mnsc.2014.1984

Elkington, J. (1999). *Cannibals with forks: The triple bottom line of 21st century business*. Capstone.

Ellen MacArthur Foundation. (2020). *Circular economy in detail*. https://www.ellenmacarthurfoundation.org/circular-economy/concept

European Commission. (2023). The European Green Deal: A Just Transition. Retrieved from https://ec.europa.eu/info/strategy/priorities-2019-2024/european-green-deal_en

European Commission. (2023). *Corporate Sustainability Due Diligence Directive*. Retrieved from https://ec.europa.eu/commission/presscorner/detail/en/ip_23_1104

European Commission. (2023). Hydrogen-powered aviation: Future developments. Retrieved from https://ec.europa.eu

European Commission. (2023, June). *Progress report* (v.1.0). European Commission. Retrieved March 5, 2025, from https://defence-industry-space.ec.europa.eu/system/files/2023-06/Progress%20report%20v.1.0.pdf

European Commission. (n.d.). *Methane emissions*. European Commission. Retrieved March 5, 2025, from https://energy.ec.europa.eu/topics/carbon-management-and-fossil-fuels/methane-emissions_en

Ewing, J. (2017). *Faster, higher, farther: The inside story of the Volkswagen scandal*. Random House.

Food and Agriculture Organization (FAO). (2021). *State of the World's Forests 2021*. Rome: FAO. Retrieved from https://www.fao.org/state-of-forests

Freeman, R. E., Harrison, J. S., & Zyglidopoulos, S. (2018). *Stakeholder theory: Concepts and strategies*. Cambridge University Press.

Friede, G., Busch, T., & Bassen, A. (2015). ESG and financial performance: Aggregated evidence from more than 2000 empirical studies. *Journal of Sustainable Finance & Investment*, *5*(4), 210–233. DOI: 10.1080/20430795.2015.1118917

Gielen, D., Boshell, F., Saygin, D., Bazilian, M. D., Wagner, N., & Gorini, R. (2019). The role of renewable energy in the global energy transformation. *Energy Strategy Reviews*, *24*, 38–50. DOI: 10.1016/j.esr.2019.01.006

Grossman, K. R. (2025). *The role of mindfulness practice in warding off despair and sustaining ecological action in the context of existential threat*. ProQuest Dissertations. Retrieved from https://search.proquest.com/openview/a8b48976572cf9b286b67a83cf19e7d4/1?pq-origsite=gscholar

Harrison, D. A., & Klein, K. J. (2007). What's the difference? Diversity constructs are separation, variety, or disparity in organizations. *Academy of Management Review*, *32*(4), 1199–1228.

Hart, S. L., & Milstein, M. B. (2003). Creating sustainable value. *The Academy of Management Perspectives*, *17*(2), 56–67. DOI: 10.5465/ame.2003.10025194

Heede, R. (2013). Tracing anthropogenic carbon dioxide and methane emissions to fossil fuel and cement producers, 1854–2010. *Climatic Change*, *122*(1–2), 229–241. DOI: 10.1007/s10584-013-0986-y

Hertwich, E. G. (2021). Increased carbon footprint of materials production driven by rise in investments. *Nature Geoscience*, *14*, 151–155. DOI: 10.1038/s41561-021-00690-8

IKEA. (2021). *Sustainability Report*. Retrieved from https://about.ikea.com/en/sustainability

Intergovernmental Panel on Climate Change (IPCC). (2014). *Climate change 2014: Synthesis report*. Geneva, Switzerland: IPCC. https://doi.org/DOI: 10.1017/CBO9781107415324

Intergovernmental Panel on Climate Change (IPCC). (2021). *Climate change 2021: The physical science basis*. Retrieved from https://www.ipcc.ch/report/ar6/wg1

Intergovernmental Panel on Climate Change (IPCC). (2022). *Sixth Assessment Report: Mitigation of Climate Change*. https://www.ipcc.ch/ar6/

Intergovernmental Panel on Climate Change (IPCC). (2023). *AR6 synthesis report: Climate change 2023: Synthesis report*. Intergovernmental Panel on Climate Change. Retrieved March 5, 2025, from https://www.ipcc.ch/report/ar6/syr/downloads/report/IPCC_AR6_SYR_LongerReport.pdf

Intergovernmental Panel on Climate Change (IPCC). (2023). *Climate Change 2023: Synthesis Report*. Geneva: IPCC. DOI: https://doi.org/DOI: 10.1016/j.scitotenv.2023.164942

International Air Transport Association (IATA). (2024). *Fact sheet: Sustainable aviation fuels*. Retrieved March 5, 2025, from https://www.iata.org/en/iata-repository/pressroom/fact-sheets/fact-sheet-sustainable-aviation-fuels/

International Air Transport Association (IATA). (n.d.). *Fact sheet on climate change*. Retrieved March 5, 2025, from https://www.iata.org/contentassets/d13875e9ed784f75bac90f000760e998/fact_sheet_on_climate_change.pdf

International Energy Agency. (2020, October). Iron and steel technology roadmap. Retrieved March 5, 2025, from https://www.iea.org/reports/iron-and-steel-technology-roadmap

International Energy Agency. (n.d.). *Ammonia technology roadmap: Executive summary*. Retrieved March 5, 2025, from https://www.iea.org/reports/ammonia-technology-roadmap/executive-summary

IPCC. (2014). [*Synthesis Report*. Intergovernmental Panel on Climate Change.]. *Climatic Change*, •••, 2014.

IPCC. (2021). Climate Change 2021: The Physical Science Basis. Retrieved from https://www.ipcc.ch/report/ar6/wg1

IPCC. (2023). Climate Change 2023: Synthesis Report. Intergovernmental Panel on Climate Change. Retrieved from https://www.ipcc.ch/report/ar6/syr/

Jamil, N. N., & Wahyuni, E. T. (2024). Do Some Sustainability Reporting Measures Enable ESG Greenwashing: A Case Study of Oil and Gas Industry in Malaysia. *Global Business and Management Research, 16*(3s), 123–128.

Kim, H., Woo, E., Uysal, M., & Kwon, N. (2018). The effects of corporate social responsibility (CSR) on employee well-being in the hospitality industry. *International Journal of Contemporary Hospitality Management, 30*(3), 1584–1600.

Kramer, M. R., & Porter, M. (2011). *Creating shared value* (Vol. 17). FSG.

Lacy, P., & Rutqvist, J. (2015). *Waste to wealth: The circular economy advantage.* Palgrave Macmillan.

Lüthi, D., Le Floch, M., Bereiter, B., Blunier, T., Barnola, J. M., & Siegenthaler, U. (2008). High-resolution carbon dioxide concentration record 650,000–800,000 years before present. *Nature, 453*(7193), 379–382. DOI: 10.1038/nature06949 PMID: 18480821

Machalaba, C., Cybulska, L., & Fogarty, F.. (2025). Application of a One Health Approach to Translate Policies to Action. *CABI One Health.* Advance online publication. DOI: 10.1079/cabionehealth.2025.0005

Maiti. (2025, January 20). *Fast fashion's detrimental effect on the environment.* Retrieved March 5, 2025, from https://earth.org/fast-fashions-detrimental-effect-on-the-environment/

Microsoft. (2020, January 16). *Microsoft will be carbon negative by 2030.* Microsoft. https://blogs.microsoft.com/blog/2020/01/16/microsoft-will-be-carbon-negative-by-2030/

Nadeem, S. (2025). *Selecting Green Wall Adoption Best Solutions Using Optimization.* American University in Cairo. PDF: https://fount.aucegypt.edu/cgi/viewcontent.cgi?article=3542&context=etds

Nahari, M. (2024). Vague sustainability terms and consumer confusion. *The Journal of Consumer Research, 46*(2), 215–231.

NASA. (2021). *Global climate change: Vital signs of the planet.* https://climate.nasa.gov/

NASA. (n.d.). *Arctic sea ice: Vital signs of the planet.* NASA Climate. Retrieved March 5, 2025, from https://climate.nasa.gov/vital-signs/arctic-sea-ice/?intent=121

National Geographic. (2022, January 21). *Should you buy carbon offsets for your air travel?* National Geographic. Retrieved March 5, 2025, from https://www.nationalgeographic.com/travel/article/should-you-buy-carbon-offsets-for-your-air-travel

National Oceanic and Atmospheric Administration (NOAA). (2022). *Sea level rise and coastal flooding impacts.* https://www.noaa.gov/

Nobanee, H. (2024). Digital Entrepreneurship and Greenwashing: Addressing Ethical Challenges in Sustainable Business. *Available at SSRN* 5073927.

NRDC. (2020). By 2070, more than 3 billion people may live outside the human climate niche. Retrieved from https://www.nrdc.org/stories/2070-more-3-billion-people-may-live-outside-human-climate-niche

Obeid, R., Smith, A., & Gonzalez, A. (2024). Consumer skepticism and trust erosion in corporate sustainability claims. *Sustainability Science, 19*(4), 42–59.

Ockenfels, A., & Pahle, M. (2025). *Das Pariser Abkommen droht zu scheitern. Was jetzt?* Zeitschrift für Wirtschaftspolitik., DOI: 10.1515/zfwp-2025-2002

OECD. (2021). The role of government in sustainable business. Retrieved from https://www.oecd.org

Paleari, S. (2022). The impact of the European Green Deal on EU environmental policy. *Journal of Environment & Development, 31*(2), 196–220.

Panagoda, S. S., Ranasinghe, H., Perera, V., Sandunika, I., Tilanka, G., Alwis, S., & Dilka, S. (2023). Cement manufacturing process and its environmental impact. *J. Res. Technol. Eng, 4*(3), 161–168.

Parmesan, C., & Yohe, G. (2003). A globally coherent fingerprint of climate change impacts across natural systems. *Nature, 421*(6918), 37–42. PMID: 12511946

Pearson, J. (2010). Turning point. Are we doing the right thing? Leadership and prioritisation for public beneft. *Journal of Corporate Citizenship, 2010*(37), 37–40. DOI: 10.9774/gleaf.4700.2010.sp.00006

Peng, Y., Lu, X., & Zhou, L. (2024). False labels: A growing problem in corporate sustainability. *International Journal of Business Ethics, 27*(3), 301–315.

Putra, A., Aini, S., Maulida, C., Qhadaffi, L. A. J., & Associates. (2025). *Impact of global climate shifts on the biodiversity and functionality of marine zooplankton communities.* Journal of Tropical Biology. Retrieved from https://journal.ugm.ac.id/v3/jtbb/article/view/14575

Rahman, M. J., Rana, T., & Zhu, H. (2025). *The Relationship between Carbon Accounting, Carbon Dioxide Emissions, and Climate Change.* Google Books: https://books.google.com/books?hl=en&id=TfNGEQAAQBAJ

Report, S. (2020, October 7). *Laughing gas: A growing climate problem.* Stanford News. Retrieved March 5, 2025, from https://news.stanford.edu/stories/2020/10/laughing-gas-growing-climate-problem#:~:text=In%20the%20industrial%20era%2C%20carbon,over%20a%20100%2Dyear%20period

Rockström, J., Steffen, W., Noone, K., Persson, Å., Chapin, F. S., Lambin, E. F., & Foley, J. A. (2009). Planetary boundaries: Exploring the safe operating space for humanity. *Ecology and Society, 14*(2), 32. DOI: 10.5751/ES-03180-140232

Sachs, J. D. (2015). *The age of sustainable development.* Columbia University Press.

Savitz, A. W., & Weber, K. (2013). *The triple bottom line: How today's best-run companies are achieving economic, social, and environmental success—and how you can too.* Wiley.

Stiglitz, J. E. (2012). *The price of inequality: How today's divided society endangers our future.* W.W. Norton & Company.

Sullivan, R., & Gouldson, A. (2017). The governance of corporate responses to climate change: An international comparison. *Business Strategy and the Environment, 26*(4), 413–425.

Task Force on Climate-related Financial Disclosures. TCFD. (2017). *Final report: Recommendations of the task force on climate-related financial disclosures.* https://www.fsb-tcfd.org/recommendations

The Guardian. (2025). Storm Éowyn leaves 768,000 buildings without power in Ireland. Retrieved from https://www.theguardian.com/world/2025/jan/24/storm-eowyn-homes-without-power-ireland-record-gusts

The Guardian. (2025). *Coral bleaching spreads in Ningaloo Reef as marine heatwave intensifies.* Retrieved from https://www.theguardian.com/environment/2025/feb/26/coral-bleaching-spreads-in-ningaloo-reef

The Guardian. (2025, January 24). *Storm Éowyn leaves homes without power in Ireland as record gusts recorded.* Retrieved March 5, 2025, from https://www.theguardian.com/world/2025/jan/24/storm-eowyn-homes-without-power-ireland-record-gusts

The Times. (2025). *Storm Éowyn leaves 768,000 buildings without power in Ireland.* Retrieved from https://www.thetimes.co.uk/article/storm-eowyn-damage-ireland

Thom, F. (2025). *Fundamental Changes in Food: Improving the Toolbox of Simulation Modeling*. University of Göttingen. Online Link: https://ediss.uni-goettingen.de/handle/11858/15823

United Nations. (UN Women). (2023). Gender and Climate Change: Women's Role in Achieving a Sustainable Future. Retrieved from https://www.unwomen.org/en

United Nations Women (UN Women). (2023). *Gender and Climate Change: Women's Role in Achieving a Sustainable Future*. Retrieved from https://www.unwomen.org/en

U.S. Department of Energy. (2023). *Inflation Reduction Act of 2022*. U.S. Department of Energy. Retrieved March 5, 2025, from https://www.energy.gov/lpo/inflation-reduction-act-2022

U.S. Securities and Exchange Commission. SEC. (2022). *Proposed Rule on Climate-Related Disclosures*. Retrieved from https://www.sec.gov/news/press-release/2022-46

World Bank. (2023). *Carbon Pricing Dashboard*. Retrieved from https://carbonpricingdashboard.worldbank.org/

Chapter 2
Climate Change Mitigation and Resilience Strategies:
A Nexus Between Governance and Sustainable Development

Amrik Singh
https://orcid.org/0000-0003-3598-8787
Lovely Professional University, India

ABSTRACT

The critical intersection of climate change mitigation and resilience strategies with governance and sustainable development. It argues that effective climate action requires a holistic approach that integrates robust governance frameworks with sustainable development goals. The chapter examines the role of governance, from international agreements like the Paris Agreement to national policies and local community initiatives, in driving these strategies. It highlights how good governance, characterized by transparency, accountability, and inclusivity, is essential for the successful implementation of climate policies. Furthermore, the chapter delves into how sustainable development, which balances economic growth, social equity, and environmental protection, provides a crucial framework for climate action. The chapter also concludes that governance, mitigation, resilience, and sustainable development are not viewed as separate issues but as interconnected elements of a unified strategy for a sustainable and climate-secure future.

DOI: 10.4018/979-8-3373-2479-1.ch002

INTRODUCTION

Climate change represents one of the most profound and pressing challenges confronting humanity in the twenty-first century, acting as both a catalyst and a magnifier of environmental, economic, social, and political vulnerabilities, thereby making it impossible to address in isolation from the frameworks of governance and sustainable development. Since the problem and its solutions are deeply interwoven with policy coherence, institutional capacity, societal participation, and technological innovation, and as global average temperatures continue to rise due to the anthropogenic greenhouse gas emissions primarily carbon dioxide, methane, and nitrous oxide from fossil fuel combustion, deforestation, industrial processes, and agricultural activities (Alibašić, 2018). The cascading impacts such as melting polar ice caps, sea level rise, intensified tropical cyclones, prolonged droughts, shifting precipitation patterns, biodiversity loss, ocean acidification, and the destabilization of food and water systems are not merely environmental concerns but existential threats to human well-being and planetary stability (Al-Romeedy & Singh, 2026). Prompting urgent calls for transformative approaches that combine mitigation efforts to reduce or prevent the emission of greenhouse gases with resilience strategies that strengthen the capacity of communities, economies, and ecosystems to anticipate, absorb, adapt to, and recover from climate-related shocks, all within the broader agenda of sustainable development as envisioned by the United Nations Sustainable Development Goals (SDGs), particularly SDG 13 on climate action, SDG 7 on affordable and clean energy, SDG 11 on sustainable cities and communities, and SDG 15 on life on land, where governance acts as the central mechanism to orchestrate the balance between environmental protection, economic growth, and social equity (Al Romeedy et al., 2025). The effective climate governance entails multilevel coordination global treaties like the Paris Agreement, national policy frameworks, subnational climate action plans, and local community initiatives while also requiring transparency, accountability, inclusiveness, and adaptive capacity to respond to new scientific evidence and socio-economic realities (Ashipala & Singh, 2025). In this nexus, mitigation strategies include transitioning to renewable energy sources such as solar, wind, hydropower, and geothermal to decarbonize electricity generation; improving energy efficiency in buildings, transportation, and industry through advanced materials, smart technologies, and circular economy practices; promoting sustainable land use and forest conservation to enhance carbon sinks; investing in carbon capture, utilization, and storage (CCUS) technologies (Bahadur et al., 2013). Reforming the agricultural practices to reduce methane emissions from livestock and nitrous oxide emissions from fertilizers through precision farming, regenerative agriculture, and alternative protein sources that phasing out fossil fuel subsidies and implementing carbon pricing mechanisms such as carbon taxes or cap-and-trade systems to internalize the environmental costs of emissions; fostering behavioral changes through public awareness campaigns (Bangar et al., 2020). Education, and incentives for low-carbon lifestyles; and supporting research and innovation for breakthrough solutions in clean technology and climate-friendly infrastructure, while resilience strategies, which are equally critical, encompass strengthening critical infrastructure against climate hazards such as flood defenses, stormwater management systems, and resilient road and bridge designs improving early warning systems and disaster preparedness (Beg et al., 2002). To reduce loss of life and property, enhancing ecosystem-based adaptation by restoring mangroves, wetlands, and forests that act as natural buffers against storms and floods, diversifying livelihoods to reduce dependence on climate-sensitive sectors, integrating climate risk assessments into urban planning to prevent maladaptive development in high-risk zones, ensuring access to climate-resilient seeds and irrigation methods for farmers, building social safety nets to protect vulnerable populations from climate-induced economic shocks, and embedding resilience considerations into public and private investment decisions to avoid lock-in of high-risk assets, and all of these strategies, to be effective and equitable, depend on governance systems that are capable of aligning short-term political cycles with long-term climate objectives (Das et al., 2025). Moreover, harmonizing policies across sectors such as energy, transport, agriculture, water, and health, enabling stakeholder participation including marginalized and indigenous communities whose traditional knowledge often offers valuable adaptive insights, ensuring climate justice so that the burdens and benefits of climate action are fairly distributed between and within countries (Ferguson & Wollersheim, 2023). Fostering international cooperation through finance, technology transfer, and capacity building, particularly for developing countries that have contributed the least to historical emissions but face some of the most severe impacts, and in this light, sustainable development provides the ethical and strategic framework to integrate climate change responses into economic and social planning, recognizing that poverty eradication, gender equality, education, health, and inclusive economic growth are not separate from climate goals but mutually reinforcing, as for example, investments in clean energy can create green jobs, reduce air pollution, and improve public health while cutting emissions, and climate-resilient infrastructure can safeguard economic productivity and social stability (Galderisi, 2016). The nexus between governance, climate mitigation, resilience, and sustainable development becomes a dynamic interplay where the effectiveness of climate policies is determined not only by their technical soundness but also by political will, institutional design, societal trust, and the capacity to navigate trade-offs and synergies across goals. This requires adaptive governance approaches that embrace systems thinking, continuous learning, scenario planning, and iterative policy adjustments based on monitoring, evaluation, and feedback, because climate change is a moving target with complex feedback loops and uncertainties, and it is in this space that innovations such as climate budgeting, mainstreaming climate considerations into public financial management, establishing independent climate advisory bodies, leveraging digital technologies for monitoring emissions and climate impacts, and forging public-private partnerships for low-carbon and climate-resilient investments can significantly enhance governance effectiveness, and as the scientific consensus underscores the urgency of limiting global warming to well below 2°C above pre-industrial levels—ideally to 1.5°C—to avoid the most catastrophic consequences, the scale and speed of transformation required is unprecedented, necessitating not incremental adjustments but systemic shifts in energy systems, consumption patterns, industrial processes, urban design, and ecosystem management, and here governance plays the role of both catalyst and steward, creating enabling environments for innovation, managing social transitions to ensure fairness, and safeguarding public goods against short-term exploitation (Hasan et al., 2020). While sustainable development ensures that these transformations enhance rather than undermine human development, leaving no one behind, and if nations fail to act decisively, the escalating costs of inaction manifested in climate-induced disasters, displacement, conflict over scarce resources, loss of biodiversity, and irreversible damage to critical Earth systems will far exceed the investments required for mitigation and resilience today, making the moral, economic, and environmental case for urgent, coordinated, and sustained action indisputable, and thus the path forward lies in embedding climate change mitigation and resilience strategies (Islam & Singh, 2025). The governance systems that are transparent, participatory, accountable, and adaptive, guided by the vision of sustainable development that recognizes the inseparable links between environmental integrity, economic prosperity, and social well-being, for only through such an integrated and forward-looking approach can humanity hope to navigate the profound challenges of the climate crisis while building a more equitable, resilient, and sustainable future for generations to come (Kaur & Sharma, 2024).

Climate Change Governance Sustainable Development Nexus

The global challenge of climate change is inextricably linked with how we govern ourselves and how we pursue development. Addressing this challenge effectively requires a holistic understanding of the intricate relationships between climate change, governance, and sustainable development (Kaur & Singh, 2025). This "nexus" perspective recognizes that climate action cannot be isolated; it must be integrated into the very fabric of our social, economic, and political systems (Kaur et al., 2025). A failure to acknowledge these connections risks siloed, ineffective, and potentially counterproductive policies.

Climate Change Mitigation and Resilience

Climate action is broadly divided into two core strategies: mitigation and resilience. Mitigation refers to actions aimed at reducing the sources or enhancing the sinks of greenhouse gases (GHGs) that cause climate change. The primary goal of mitigation is to stabilize or reduce the concentration of GHGs in the atmosphere, thereby limiting the magnitude of future warming (Kaur et al., 2025a). This includes transitioning from fossil fuels to renewable energy sources, improving energy efficiency, and protecting and restoring forests. Resilience, on the other hand, involves preparing for and adapting to the unavoidable impacts of climate change that are already locked in or will occur in the future. This is often referred to as adaptation. Resilience strategies focus on building the capacity of social, economic, and ecological systems to cope with climate-related shocks and stresses, such as more frequent extreme weather events, sea-level rise, and shifts in agricultural productivity (Venema & Rehman, 2007). Examples include developing early warning systems for floods, building sea walls, and cultivating drought-resistant crops. While mitigation seeks to prevent the problem from worsening, resilience is about living with the consequences. Both are essential and complementary components of a comprehensive climate strategy.

Principles of Governance in Climate Action

Effective governance is the engine that drives successful climate action. It provides the institutional framework and mechanisms for developing, implementing, and enforcing climate policies. Good governance in this context is built on several key principles. Transparency is crucial, as it ensures that climate policies and financial flows are open to public scrutiny, building trust and accountability. This means clear communication about goals, progress, and setbacks. Accountability holds governments and other actors responsible for their climate commitments, whether they are international pledges or domestic targets. This can be achieved through legal frameworks, independent oversight bodies, and robust monitoring and reporting systems (Suprayitno et al., 2024). Inclusivity ensures that a diverse range of stakeholders, including vulnerable communities, women, youth, and indigenous peoples are part of the decision-making process. This not only makes policies more equitable but also more effective, as it draws on local knowledge and ensures broader public buy-in. Finally, coordination across different levels of government (local, national, and international) and across different sectors (energy, agriculture, transport) is vital to avoid conflicting policies and to create a coherent and unified approach.

Dimensions of Sustainable Development

Sustainable development provides a framework for addressing human needs while preserving the planet's life support systems. It is often conceptualized as a balancing act between three interconnected dimensions: social, economic, and environmental. The environmental dimension focuses on protecting natural resources and ecosystems. This includes a stable climate, clean air and water, and biodiversity. The economic dimension is about creating prosperous and equitable economies. This involves promoting sustainable growth, creating jobs, and ensuring that economic benefits are widely shared (Kaur & Singh, 2025). The social dimension is concerned with human well-being and equity. It includes poverty eradication, access to education and healthcare, gender equality, and social justice. The challenge and the opportunity lies in finding policies that advance all three dimensions simultaneously. For example, investing in public transport improves air quality (environmental), creates jobs (economic), and provides affordable mobility for all citizens (social).

InterlinkagesBetween Governance and Sustainability in Climate Policy

The governance-sustainability nexus is where climate policy truly takes shape. Strong governance is not just about enacting policies; it's about embedding them within a sustainable development framework. When governance is effective, climate policies can be designed to deliver co-benefits across the social, economic, and environmental dimensions. For example, a government with a strong, transparent governance structure can implement a carbon tax. This policy not only mitigates climate change by making polluting activities more expensive (environmental) but can also generate revenue that is then invested in social programs or used to reduce taxes on low-income households (social and economic). Similarly, governance that promotes public participation can lead to the successful implementation of local renewable energy projects, where communities own and benefit from their energy production (economic), reduce emissions (environmental), and strengthen social cohesion (social) (Pinkse & Kolk, 2012). Conversely, weak governance can undermine sustainable development and climate action. Corruption can divert funds intended for climate projects, leading to poorly built infrastructure or greenwashing. Lack of inclusivity can lead to policies that disproportionately harm marginalized communities, such as the displacement of indigenous peoples for large-scale renewable energy projects. Therefore, strengthening governance is not a separate goal but an essential prerequisite for achieving both climate stability and sustainable development. The success of international agreements like the Paris Agreement and the Sustainable Development Goals (SDGs) depends on countries' ability to translate these global commitments into effective national and local governance that integrates mitigation, resilience, and all three dimensions of sustainable development.

DISCUSSION

Climate change represents one of the most pressing and multifaceted challenges of our time, with far-reaching implications for global systems and human society. The issue is not just an environmental one; it's a complex web of interconnected crises spanning socioeconomic, political, and ecological domains.

Global Climate Change Challenges

Understanding the scope of these challenges is crucial for developing effective and equitable solutions. This involves a clear-eyed assessment of current climate trends, a thorough analysis of their impacts, a recognition of who is most vulnerable, and a careful study of past efforts to learn from our successes and failures.

Current Climate Change Trends and Projections

The Earth's climate is changing at an unprecedented rate, a phenomenon unequivocally linked to human activities, particularly the emission of greenhouse gases. The global average temperature has already risen by approximately 1.4°C since the pre-industrial era. This seemingly small increase has significant consequences. Recent analyses project that global temperatures will continue at or near record levels in the next five years (Rai & Singh, 2026).The World Meteorological Organization (WMO) predicts an 86% chance that at least one of the next five years will be more than 1.5°C above the pre-industrial average.While the long-term, decade-averaged warming remains below 1.5°C, this increasing probability of exceeding the threshold in individual years is a stark reminder of the accelerating pace of change.These rising temperatures are not uniform. The Arctic, for example, is warming at a rate two to three times the global average, leading to the rapid melting of glaciers and sea ice. This has a cascade of effects, including rising sea levels and the potential disruption of global ocean currents.Precipitation patterns are also becoming more erratic and extreme, with some regions experiencing more intense rainfall and flooding, while others face prolonged and severe droughts.These shifts, coupled with the warming climate, are intensifying extreme weather events such as heatwaves, hurricanes, and wildfires, which are becoming more frequent and destructive.

Socioeconomic and Ecological Impacts

The impacts of these climate trends are already being felt across the globe, with profound socioeconomic and ecological consequences.Economically, climate change is a significant threat to global stability. Extreme weather events destroy infrastructure, disrupt supply chains, and cause billions of dollars in damage. Sectors heavily reliant on natural resources, such as agriculture, fishing, and tourism, are particularly vulnerable.For instance, shifting weather patterns and water scarcity threaten crop yields, leading to food insecurity and higher prices. The World Bank warns that climate change could push more than 100 million people into poverty by 2030, highlighting the direct link between environmental and economic well-being. Ecologically, the impacts are equally severe (Rai & Singh, 2026; Rasul & Sharma, 2016). Rising temperatures and ocean acidification are devastating marine ecosystems, leading to the widespread bleaching of coral reefs and threatening fisheries. On land, ecosystems are struggling to adapt to the rapid changes. Many plant and animal species are being pushed out of their natural habitats, leading to biodiversity loss and, in some cases, extinction. Forest degradation, driven by a combination of drought, pest outbreaks, and wildfires, further exacerbates the problem by reducing the planet's capacity to absorb carbon dioxide. The stability of entire ecosystems, from tropical rainforests to mountain glaciers, is at risk, with consequences that ripple through global ecological systems.

Vulnerable Regions and Populations

While climate change affects everyone, its impacts are disproportionately felt by the world's most vulnerable regions and populations.Low-income countries and small island developing states are often the most exposed to climate hazards and have the least capacity to adapt.Their economies are frequently dependent on climate-sensitive sectors like agriculture and fisheries, and they lack the financial resources and robust infrastructure to cope with natural disasters.Within and across countries, certain populations face heightened risks.Indigenous peoples and local communities, whose livelihoods and cultural identities are deeply intertwined with the natural environment, are on the front lines of climate change. For example, indigenous communities in the Arctic are seeing their traditional hunting and fishing grounds disappear due to melting ice (Sarkodie et al., 2023). Women and children, particularly in developing countries, are often more vulnerable due to existing social inequalities.They may have limited access to resources and decision-making processes and are at greater risk from climate-related health issues and displacement. The elderly and those with pre-existing health conditions also face greater risks from extreme heat and air pollution.

Lessons from Global Case Studies

Global case studies offer invaluable insights into effective climate action and highlight the importance of context-specific solutions. In cities like Barcelona, Spain, the focus has been on adapting to urban heat island effects and water scarcity. The city's Green Infrastructure and Biodiversity Plan promotes the expansion of green spaces and the use of drought-resistant trees, which not only cool the city but also help manage stormwater runoff. This demonstrates the co-benefits of integrated urban planning.Another powerful example comes from Bangladesh, a country highly vulnerable to cyclones and flooding. The country has invested in early warning systems and the construction of cyclone shelters, which have significantly reduced the loss of life from natural disasters (Sarkodie et al., 2023). This shows the importance of building resilience through proactive, community-based adaptation measures. In the Netherlands, a country with extensive low-lying land, innovative water management strategies like the "Room for the River" program are being used to create space for rivers to flood naturally, rather than relying solely on hard infrastructure like dikes. This approach combines engineering with natural processes to manage flood risk more effectively.These case studies teach us that successful climate action is not a one-size-fits-all solution. It requires a combination of mitigation and adaptation, strong governance, and a deep understanding of local social and ecological contexts. The lessons from these global examples underscore the need for flexible, multi-pronged strategies that empower communities and build resilience from the ground up, while simultaneously working to reduce global emissions.

Governance Frameworks for Climate Action

Effective climate action is built upon a multi-layered system of governance that spans international, national, and local levels (Singh, 2025). This interconnected framework ensures that global ambitions are translated into tangible policies and projects on the ground. A comprehensive understanding of this governance structure is crucial for anyone seeking to engage with or influence climate policy.

International Climate Agreements and Frameworks

At the global level, governance for climate action is primarily shaped by the United Nations Framework Convention on Climate Change (UNFCCC). This international treaty, established in 1992, provides the overarching framework for global climate negotiations. Its ultimate objective is to stabilize greenhouse gas concentrations "at a level that would prevent dangerous anthropogenic interference with the climate system." The Conference of the Parties (COP), the supreme decision-making body of the UNFCCC, meets annually to review progress.A major milestone under the UNFCCC was the Paris Agreement, adopted in 2015. Unlike previous agreements, it is a legally binding international treaty applicable to all Parties (Singh & Bathla, 2023). The central goal of the Paris Agreement is to limit global warming to well below 2°C, and preferably to 1.5°C, compared to pre-industrial levels. A key feature is the Nationally Determined Contributions (NDCs), which are voluntary, country-specific climate action plans. This bottom-up approach allows each country to set its own targets, fostering a sense of ownership and collective responsibility. The Paris Agreement also mandates a five-year cycle for countries to review and enhance their NDCs, ensuring a continuous progression of ambition.

National and Regional Climate Policies

International agreements like the Paris Agreement are only effective when they are translated into national and regional climate policies. These policies are the primary vehicles for implementing a country's NDC. They can take various forms, including legislative acts, executive orders, and regulatory frameworks (Singh & Kumar, 2025). For example, many countries have adopted national climate laws that set legally binding emissions reduction targets and establish carbon pricing mechanisms, such as a carbon tax or an emissions trading system (ETS). Regional policies, such as the European Union's Green Deal, integrate climate action into economic policy, promoting sustainable industries, energy efficiency, and a circular economy. The effectiveness of these policies often depends on the political will, economic capacity, and institutional strength of a given country or region.

Role of Local Governments and Decentralized Governance

The implementation of climate policy often rests on the shoulders of local governments and decentralized governance. While national governments set the broad strokes, cities, municipalities, and regional authorities are on the front lines of climate change, both in terms of mitigation and adaptation. They have direct control over key sectors like urban planning, transportation, waste management, and local energy systems (Singh & Sihmar, 2024). Many cities have set their own ambitious climate targets, sometimes even exceeding their national commitments. This decentralized governance approach allows for policies to be tailored to local needs and contexts, fostering community engagement and innovation. For instance, a city might invest in green infrastructure to manage stormwater, promote public transit to reduce emissions, or establish local building codes that mandate energy efficiency. The success of these local initiatives demonstrates that climate action is not just a top-down process, but a multi-level effort that requires collaboration and empowerment at all scales.

Public–Private Partnerships in Climate Mitigation

Addressing the climate crisis requires massive investment and innovation, which the public sector cannot achieve alone. This has led to an increasing reliance on public–private partnerships (PPPs)in climate mitigation. In these arrangements, governments collaborate with private sector companies to design, finance, and operate climate-friendly projects. Examples include building large-scale renewable energy facilities, developing and deploying new green technologies, and modernizing public infrastructure. A government may provide the policy framework and financial incentives, such as tax credits or subsidies, while a private company brings in the capital, technological expertise, and operational efficiency. This collaboration not only helps to mobilize the necessary resources for the energy transition but also shares the risks and rewards (Singh & Singh, 2025). Successful PPPs in countries like India have been instrumental in the rapid expansion of solar and wind energy, accelerating the pace of decarbonization beyond what either sector could achieve on its own.

Climate Change Mitigation Strategies

Addressing climate change requires a diverse portfolio of mitigation strategies aimed at reducing greenhouse gas emissions across all sectors of the economy. These strategies work in tandem to shift our global systems away from a fossil-fuel-dependent model toward a more sustainable, low-carbon future.A fundamental component of climate mitigation is the rapid transition to renewable energy. This involves replacing fossil fuels (coal, oil, and natural gas) with clean, sustainable sources such as solar, wind, hydropower, and geothermal energy.This shift is critical for decarbonizing the electricity sector, which is one of the largest sources of global emissions. Governments and private sector companies are increasingly investing in large-scale renewable energy projects, while policies like feed-in tariffs and tax credits encourage their adoption at a smaller, decentralized level (Suprayitno et al., 2024). The falling cost of renewable technologies has made this transition not only an environmental imperative but also an economic opportunity, creating new jobs and industries.Beyond the source of energy, improving energy efficiency is a powerful mitigation strategy. This means using less energy to achieve the same or a better outcome. This can be achieved through various measures, such as upgrading to more efficient appliances and lighting, improving building insulation, and optimizing industrial processes. Complementing this are low-carbon technologies, which include a range of innovations designed to reduce emissions. Examples include carbon capture and storage (CCS), which captures CO_2 from industrial sources and stores it underground, and the development of new batteries for electric vehicles and energy storage (Kaur & Singh, 2025; Singh & Sihmar, 2025). These technologies play a crucial role in sectors that are difficult to decarbonize, like heavy industry. The transportation sector is a significant contributor to greenhouse gas emissions. Sustainable transportation systems are key to reducing this footprint. This involves a multi-pronged approach that includes promoting electric vehicles (EVs), which are powered by clean electricity, and investing in high-quality public transit systems. Encouraging the use of bicycles and promoting walkability in urban planning also plays a vital role. Policies that support this transition include regulations on fuel efficiency, incentives for purchasing EVs, and investments in charging infrastructure. To incentivize emissions reductions, governments are increasingly using carbon pricing mechanisms (Singh & Sihmar, 2025). A carbon tax directly charges emitters for each ton of CO_2 they release, making polluting activities more expensive. This encourages businesses and consumers to find cleaner alternatives. An Emissions Trading Scheme (ETS), or "cap-and-trade" system, sets a limit (cap) on the total amount of emissions allowed and then issues permits that can be bought and sold. Companies that reduce their emissions can sell their unused permits, creating a market-based incentive to innovate and decarbonize. These economic tools are powerful drivers for market-led climate action.Agriculture and land use practices are both sources and sinks of greenhouse gases. Climate-smart agriculture and forestry focus on practices that increase productivity sustainably, enhance resilience to climate change, and reduce emissions. This includes techniques like agroforestry, which integrates trees into farming systems, and improved soil management to increase carbon sequestration. Sustainable forestry practices, such as preventing deforestation and promoting reforestation, are essential for preserving the planet's natural carbon sinks.The circular economy is a systemic approach to economic development designed to benefit businesses, society, and the environment (Singh & Singh, 2025). It involves minimizing waste and keeping products and materials in use for as long as possible. Instead of a "take, make, dispose" linear model, it promotes recycling, reusing, and remanufacturing. This not only reduces the need for raw materials, but also cuts down on the energy and emissions associated with their extraction and processing. Waste reduction, therefore, is a key mitigation strategy that also offers significant economic and social benefits.

Role of Governance in Driving Climate Action

Governance is the linchpin of effective climate action, providing the essential framework for translating goals into reality. Without robust governance, international agreements remain mere promises, and local initiatives lack the necessary support and coherence to scale. Good governance in this context is characterized by its multi-level and multi-actor nature, encompassing a complex web of interactions between international bodies, national governments, local authorities, civil society, and the private sector. The journey of a climate policy often begins at the international level with agreements like the Paris Agreement, which sets a global ambition. However, the true test of this ambition is its translation into Nationally Determined Contributions (NDCs), which are specific climate action plans set by each country.At the national level, governance involves creating the legal and regulatory frameworks needed to meet these NDCs (Kaur & Singh, 2025). This includes setting emissions reduction targets, implementing carbon pricing mechanisms like a carbon tax or an emissions trading scheme, and creating policies to promote renewable energy and energy efficiency. The effectiveness of these national policies, however, is heavily influenced by principles of transparency, accountability, and inclusivity. Transparent governance ensures that the public can scrutinize how climate funds are spent and whether targets are being met, fostering trust and discouraging corruption. Accountability mechanisms, such as independent auditing bodies, hold governments and corporations responsible for their climate commitments. Inclusivity ensures that climate policy is shaped by a diverse range of voices, including marginalized communities who are often most vulnerable to climate impacts.The role of local governments is particularly critical, as they are on the front lines of both mitigation and adaptation. They can implement policies on urban planning, waste management, and public transportation that have a direct impact on emissions. Decentralized governance allows for flexible, context-specific solutions that are more likely to be successful because they are tailored to local needs and realities. Furthermore, the private sector's role is amplified through public-private partnerships (Singh & Sihmar, 2024). By creating a predictable and supportive policy environment, governments can mobilize private capital and innovation to scale up climate solutions like renewable energy projects and green infrastructure. Ultimately, effective climate governance is not just about making rules; it is about building a coherent, collaborative, and accountable system that empowers all actors to drive meaningful and lasting change.

Case Studies and Best Practices

To address climate change effectively, we can learn from successful case studies and best practices from around the world. These examples showcase how different regions, with varying levels of resources and vulnerabilities, are implementing creative and impactful solutions for mitigation and resilience.

Mitigation and Adaptation in Urban Areas

Cities are at the forefront of the climate crisis, facing both significant emissions from dense populations and infrastructure and a heightened risk from climate impacts. Barcelona, Spain, provides an excellent example of urban adaptation. Facing rising temperatures and water scarcity, the city developed a Green Infrastructure and Biodiversity Plan. This plan focuses on expanding urban green spaces and planting drought-resistant trees (Singh, 2025; Singh & Bathla, 2023). These initiatives not only combat the urban heat island effect and improve air quality but also help manage stormwater runoff, showcasing the co-benefits of integrated urban planning. Similarly, Basel, Switzerland, has pioneered an incentive program for green roofs. The city's building law mandates that all new and renovated flat roofs be greened. These roofs act as a mitigation measure by improving energy efficiency in buildings, while also serving as an adaptation strategy by absorbing rainwater to reduce flood risk and providing habitat for urban biodiversity.

Building Resilience in Vulnerable Communities

Case studies from vulnerable regions highlight the importance of proactive, community-based adaptation. Bangladesh, a country highly susceptible to cyclones and coastal flooding, has significantly reduced loss of life by investing in robust early warning systems and building thousands of cyclone shelters. This strategy of preparedness and community empowerment has proven more effective than reactive measures alone (Singh & Kumar, 2025).In the Netherlands, facing the threat of rising sea levels, the country's “Room for the River” program offers a powerful lesson in living with water rather than fighting it. Instead of just building higher dikes, the program created space for rivers to flood naturally by widening them and relocating levees (Kaur & Singh, 2025). This natural infrastructure approach has successfully increased flood safety while also creating new recreational areas.

Public-Private Partnerships and National Policies

Climate action requires significant investment and collaboration, and Public-Private Partnerships (PPPs) have proven to be a key best practice.India's railways present a compelling case, with a goal of achieving net-zero emissions by 2030. The project involves installing solar power units at thousands of railway stations, electrifying tracks, and adopting energy-efficient technologies. This initiative demonstrates how a major public utility can leverage a mix of government policies and private sector innovation to drive large-scale mitigation.Finally, Finland provides a long-standing example of effective national policy through its pioneering carbon tax, which was first introduced in 1990. This economic tool places a price on emissions, providing a clear incentive for businesses and consumers to reduce their carbon footprint. Finland's experience demonstrates that well-designed carbon pricing can be a powerful and durable policy for driving market-led climate action. These global examples underscore that successful climate strategies are often a blend of innovation, strong governance, and a commitment to integrating climate goals with broader societal and economic objectives.

Barriers and Challenges in Climate Change

Effectively addressing climate change is hindered by a complex array of barriers and challenges that span political, economic, social, and institutional domains. Overcoming these obstacles is as crucial as developing the mitigation and resilience strategies themselves.

Political and Institutional Barriers

One of the most significant hurdles is the presence of political and institutional barriers. Political gridlock and short-term electoral cycles often prevent governments from adopting long-term climate policies that require sustained commitment and investment. Vested interests from the fossil fuel industry and other high-carbon sectors exert considerable influence, lobbying against climate regulations and promoting a status quo that benefits them financially (Singh & Sihmar, 2024). Furthermore, the lack of coordination across different levels of government from local municipalities to national bodies can lead to conflicting policies and inefficient use of resources. Institutional inertia within government bureaucracies can also slow the pace of change, as old systems and ways of thinking are difficult to overhaul.

Economic and Technological Constraints

Economic and technological constraints also pose major challenges. While the cost of renewable energy is falling, the initial investment required for a large-scale energy transition remains a significant barrier for many countries, especially developing nations. The necessary infrastructure for a low-carbon economy, such as smart grids, electric vehicle charging stations, and public transit systems, requires immense capital.Additionally, some hard-to-abate sectors, like aviation and heavy industry, still lack commercially viable, low-carbon technologies, making their transition particularly difficult and costly (Singh, 2025; Singh & Bathla, 2023). The economic reliance of many countries on fossil fuel exports also creates a strong disincentive to transition away from these sources.

Social and Behavioral Challenges

Beyond policy and economics, social and behavioral challenges can impede climate action. Public awareness and concern about climate change are not always sufficient to drive significant behavioral change. The "tragedy of the commons" principle, where individuals prioritize their own short-term interests over the collective good, is a pervasive issue. People may resist new regulations, like carbon taxes or restrictions on car use, that they perceive as a direct threat to their personal freedom or financial well-being (Venema & Rehman, 2007). Additionally, misinformation and climate change denial, often amplified by social media, can sow doubt and erode public support for climate policies. Shifting deeply ingrained social norms and consumption patterns requires more than just policy it requires a profound cultural change.

Gaps in Policy Implementation

Finally, there are significant gaps in policy implementation. Even when good policies are in place, they may not be effectively enforced or adequately funded. In many cases, ambitious climate goals are set without a clear and realistic plan for their execution. This can be due to a lack of technical capacity within government agencies, insufficient financial resources, or a failure to coordinate with key stakeholders (Singh & Sihmar, 2024). For example, a national mandate for renewable energy may falter if local governments lack the authority or expertise to approve and connect new projects to the grid. These implementation gaps create a chasm between the intentions of policy and the realities on the ground, ultimately slowing the progress of climate action.

Future Directions and Recommendations

Looking ahead, a sustainable and resilient future requires a new, integrated approach to climate action (Singh & Sihmar, 2025). The path forward must move beyond siloed thinking to connect climate policy with broader societal goals, ensuring that our response to the crisis is not only effective but also just and equitable. This means prioritizing several key future directions and recommendations.

Strengthening the Governance–Sustainability Link

To make lasting progress, we need to fundamentally strengthen the governance–sustainability link. This involves integrating climate considerations into the core of all government functions, from finance and trade to agriculture and urban planning.

It's about more than just having a dedicated climate ministry; it's about making climate a central part of every policy decision. This requires a shift towards adaptive governance, where frameworks are flexible enough to respond to new information and changing climate risks. We must foster greater transparency and accountability in climate finance, ensuring that funds are allocated efficiently and reach those who need them most (Sarkodie et al., 2023; Singh, 2025). Moreover, there needs to be a clear mechanism for holding governments and the private sector accountable for their commitments, whether through legal frameworks or public scrutiny.

Integrating Climate Policies into Development Planning

A critical step is to integrate climate policies into development planning. Historically, climate action and development have been viewed as separate, and sometimes competing, objectives. This must change. Climate policies should be seen as a tool to achieve sustainable development goals (SDGs), and vice versa (Suprayitno et al., 2024). For example, investing in renewable energy not only reduces emissions (mitigation) but also improves energy access, creates jobs, and reduces air pollution (development). Similarly, building resilient infrastructure protects against climate impacts (resilience) while also safeguarding economic gains and improving quality of life. This requires a "two-in-one" approach where development strategies are screened for climate risks and opportunities, and climate policies are designed to deliver co-benefits for human well-being and economic growth.

Promoting Inclusive and Participatory Climate Action

Effective climate solutions are those that are built from the ground up, with the input and participation of diverse stakeholders. We must promote inclusive and participatory climate action. This means ensuring that vulnerable populations, indigenous peoples, women, and youth are not just consulted but are at the table, empowered to shape the policies that affect them. Their traditional knowledge and lived experiences are invaluable for designing effective and equitable adaptation and mitigation strategies (Singh & Sihmar, 2025). This can be achieved through mechanisms like citizens' assemblies, participatory budgeting for climate projects, and formal representation of these groups in decision-making bodies. True climate justice requires that those most affected by climate change are the ones leading the charge for a better future.

Advancing Research, Innovation, and Technology Deployment

Finally, we must continuously advance research, innovation, and technology deployment. While we have many of the solutions needed to tackle climate change, we also need to invest in a broad portfolio of new and emerging technologies, especially for hard-to-abate sectors. This includes R&D in green hydrogen, advanced battery storage, and carbon capture technologies (Singh & Sihmar, 2024; Singh & Sihmar, 2025). Equally important is the rapid deployment and diffusion of these technologies, particularly to developing countries, through international cooperation and technology transfer mechanisms. We also need to leverage innovative financial models and public-private partnerships to de-risk investment in new technologies, ensuring that promising innovations move from the lab to the market quickly and at scale.

CONCLUSION

The global response to climate change requires a fundamental shift from a fragmented, siloed approach to an integrated, holistic strategy. The nexus between climate change, governance, and sustainable development is the key to unlocking a more effective and equitable future. Mitigation and resilience strategies, while distinct in their focus, must be viewed as two sides of the same coin, each essential for navigating the challenges ahead. Mitigation aims to tackle the root cause by reducing greenhouse gas emissions, while resilience prepares us to adapt to the unavoidable impacts already in motion. Neither can succeed in isolation.The linchpin connecting these strategies to tangible results is robust and inclusive governance. We've seen how effective governance, from international agreements like the Paris Agreement to local-level policies and public-private partnerships, provides the institutional backbone for climate action. Principles of transparency, accountability, and inclusivity are not mere buzzwords; they are the bedrock of policies that are not only effective but also just. Without these principles, climate initiatives risk becoming corrupt, exclusionary, and ultimately ineffective.Furthermore, the pursuit of climate action must be inextricably linked with the three dimensions of sustainable development: social, economic, and environmental. The most successful strategies are those that create co-benefits, where policies designed to mitigate climate change also lead to economic prosperity, social equity, and improved environmental health. For example, investing in renewable energy creates jobs, improves air quality, and provides energy access to underserved communities. Conversely, weak governance and a failure to consider sustainable development can lead to policies that disproportionately harm the most vulnerable, exacerbating existing inequalities.The path forward is not without its challenges. Political inertia, economic constraints, and social barriers all stand in the way of progress. However, the global case studies and best practices highlighted in this book offer a beacon of hope, demonstrating that a transformative approach is possible. By strengthening the governance sustainability link, integrating climate policies into all aspects of development planning, and promoting truly inclusive and participatory action, we can build a more resilient and sustainable world. The task ahead is immense, but by working together within a framework that recognizes the deep connections between climate, governance, and development, we can create a future that is not only climate-secure but also more just and prosperous for all.

REFERENCES

Al-Romeedy, B. S., Hussein, H., & Singh, A. (2025). Peaks and Valleys: Navigating Over-Tourism in Mountain Destinations. In *Balancing Mountain Tourism, Cultural Heritage, and Environmental Stability* (pp. 431-444). IGI Global Scientific Publishing.

Al-Romeedy, B. S., & Singh, A. (2026). Emerging Green Entrepreneurship: A Path to Green Business Models. In *Incentives and Benefits for Adopting Green Entrepreneurship Practices* (pp. 79-104). IGI Global Scientific Publishing.

Alibašić, H. (2018). The nexus of sustainability and climate resilience planning: Embedding climate resilience policies in local governments. *The International Journal of Climate Change: Impacts and Responses*, *10*(2), 21–33.

Ashipala, D. O., & Singh, A. (2025). Successful Local Experience in Waste Management of Urban Destination. In *Solid Waste Management for Rural Regions* (pp. 257–274). IGI Global Scientific Publishing.

Bahadur, A. V., Ibrahim, M., & Tanner, T. (2013). Characterising resilience: Unpacking the concept for tackling climate change and development. *Climate and Development*, *5*(1), 55–65.

Bangar, V., Goyal, R., & Pandey, R. (2020). Climate change responses and sustainable development: integration of mitigation and adaptation. In *Sustainable Development Goals: An Indian Perspective* (pp. 203–214). Springer International Publishing.

Beg, N., Morlot, J. C., Davidson, O., Afrane-Okesse, Y., Tyani, L., Denton, F., & Rahman, A. A. (2002). Linkages between climate change and sustainable development. *Climate Policy*, *2*(2-3), 129–144.

Das, R., Singh, A., & Roy, K. (2025). Homestay Initiative: Building a Resilient Pathway for Garden Communities in Darjeeling's Tea Tourism. In *Global Practices and Innovations in Sustainable Homestay Tourism* (pp. 229-258). IGI Global Scientific Publishing.

Ferguson, P., & Wollersheim, L. (2023). From sustainable development to resilience?(Dis) continuities in climate and development policy governance discourse. *Sustainable Development*, *31*(1), 67–77.

Galderisi, A. (2016). Nexus approach to disaster risk reduction, climate adaptation and ecosystems' management: New paths for a sustainable and resilient urban development. In *Peri-urban areas and food-energy-water nexus: Sustainability and resilience strategies in the age of climate change* (pp. 11–21). Springer International Publishing.

Hasan, M. A., Abubakar, I. R., Rahman, S. M., Aina, Y. A., Chowdhury, M. M. I., & Khondaker, A. N. (2020). The synergy between climate change policies and national development goals: Implications for sustainability. *Journal of Cleaner Production, 249*, 119369.

Islam, N., & Singh, A. (2025). Solid Waste Management in Rural Regions. In *Solid Waste Management for Rural Regions* (pp. 239–256). IGI Global Scientific Publishing.

Kaur, A., & Sharma, V. (2024). Nexus between climate change and mitigation approaches for sustainable development: A bibliometric review. *Global Knowledge, Memory and Communication*.

Kaur, M., Singh, A., & Kaur, A. (2025). Eco-Friendly Skies: Sustainable Marketing Strategies for the Aviation Industry. In *Sustainable Marketing Practices in the Aviation Industry* (pp. 151-182). IGI Global Scientific Publishing.

Kaur, M., Singh, A., & Kaur, A. (2025a). Navigating the environmental pitfalls of overtourism and finding sustainable solutions. In *Solutions for Managing Overtourism in Popular Destinations* (pp. 61–84). IGI Global Scientific Publishing.

Kaur, S., & Singh, A. (2025). India's Peaks Under Pressure: The Impact of Climate Change and Gender Issues on Mountain Tourism. In *Balancing Mountain Tourism, Cultural Heritage, and Environmental Stability* (pp. 309-320). IGI Global Scientific Publishing.

Pinkse, J., & Kolk, A. (2012). Addressing the climate change—sustainable development nexus: The role of multistakeholder partnerships. *Business & Society, 51*(1), 176–210.

Rai, V., & Singh, A. (2026). The Digital Renaissance of Hospitality: Workforce Mastery in the Smart Revolution. *Smart Operations and Enhancing Guest Experience in the Hospitality Industry*, 125-144.

Rasul, G., & Sharma, B. (2016). The nexus approach to water–energy–food security: An option for adaptation to climate change. *Climate Policy, 16*(6), 682–702.

Sarkodie, S. A., Ahmed, M. Y., & Owusu, P. A. (2023). Advancing COP26 climate goals: Leveraging energy innovation, governance readiness, and socio-economic factors for enhanced climate resilience and sustainability. *Journal of Cleaner Production, 431*, 139757.

Singh, A. (2025). India's Peaks Under Pressure: The Impact of Climate Change and. *Balancing Mountain Tourism, Cultural Heritage, and Environmental Stability*, 309.

Singh, A., & Bathla, G. (2023). Environmental, social, and governance (ESG) measures and their impact on insurance industry: A global perspective. *The impact of climate change and sustainability standards on the insurance market*, 417-427.

Singh, A., & Kumar, V. (2025). Sustainable Waste Management in Tourism and Hospitality Sectors. In *Sustainable Waste Management in the Tourism and Hospitality Sectors* (pp. 27–44). IGI Global Scientific Publishing.

Singh, A., & Sihmar, P. (2024). Resource Management. *Green HRM Awareness and Training in Higher Education Institutions*, 1.

Singh, A., & Sihmar, P. (2025). Higher education institution sustainability: Role of green human resource management. In *Green HRM awareness and training in higher education institutions* (pp. 1–14). IGI Global.

Singh, M., & Singh, A. (2025). The Role of Artificial Intelligence in Driving Sustainability in the Tourism Industry: A Marketing Perspective. In *Marketing Technology-Infused Hospitality: Upskilling Frontline Employees for Competitiveness* (pp. 115-132). IGI Global Scientific Publishing.

Sudmeier-Rieux, K., Fernández, M., Gaillard, J. C., Guadagno, L., & Jaboyedoff, M. (2016). Introduction: exploring linkages between disaster risk reduction, climate change adaptation, migration and sustainable development. In *Identifying Emerging Issues in Disaster Risk Reduction, Migration, Climate Change and Sustainable Development: Shaping Debates and Policies* (pp. 1–11). Springer International Publishing.

Suprayitno, D., Iskandar, S., Dahurandi, K., Hendarto, T., & Rumambi, F. J. (2024). Public policy in the era of climate change: Adapting strategies for sustainable futures. *Migration Letters : An International Journal of Migration Studies*, *21*(S6), 945–958.

Venema, H. D., & Rehman, I. H. (2007). Decentralized renewable energy and the climate change mitigation-adaptation nexus. *Mitigation and Adaptation Strategies for Global Change*, *12*(5), 875–900.

Chapter 3

Economic Growth, Human Resource Development, and Climate Resilience in BRICS: A Panel Data Analysis

Varsha Sharma
https://orcid.org/0000-0001-5462-4693
Chandigarh University, Mohali, India

Pragti Dixit
https://orcid.org/0009-0002-4034-4572
Chandigarh University, Mohali, India

Mukul Bhatnagar
https://orcid.org/0000-0002-7773-5641
Graphic Era University, Uttarakhand, India

Nikhil Varghese
Christ University, Bengaluru, India

Anshuman Singh
Graphic Era University, India

ABSTRACT

Present study employs panel data analysis on BRICS economies data spanning from 2000 to 2023, to examine the impact of human resource development, renewable energy consumption, and carbon emissions on GDP per capita of BRICS economies,

DOI: 10.4018/979-8-3373-2479-1.ch003

finding that while human capital strongly drives growth, renewable energy has a positive but weaker effect, and carbon emissions remain tightly linked to GDP, underscoring the persistent reliance on carbon-intensive industries, ultimately highlighting the need for policy shifts toward sustainable growth through education, clean energy transitions, and regulatory frameworks that balance industrialization with climate resilience.

INTRODUCTION

The BRICS economies—comprising Brazil, Russia, India, China, and South Africa—serve as a formidable engine of global economic dynamism, collectively accounting for a substantial share of worldwide GDP, trade, and investment flows, while exhibiting diverse yet complementary developmental trajectories that underscore their strategic significance in reshaping the geopolitical and economic landscape, driving industrialization (Lavrik & Saurombe, 2025), technological innovation, and infrastructural expansion, all while grappling with the intricate challenges of socio-economic disparities, climate vulnerabilities, and sustainable development imperatives, thereby positioning them as pivotal actors in fostering a multipolar world order, advancing South-South cooperation, and spearheading policy frameworks that balance rapid economic growth with long-term environmental resilience and human capital enhancement (Leal-Arcas et al., 2022).

The interplay between economic growth, human capital, and climate resilience in BRICS economies manifests as a multidimensional nexus wherein sustained economic expansion is inherently dependent on strategic investments in human resource development—encompassing education, workforce upskilling, technological innovation, and institutional capacity-building—while simultaneously necessitating robust climate resilience frameworks that mitigate environmental vulnerabilities through adaptive policies, green energy transitions, and sustainable resource management, thereby ensuring that economic progress does not come at the expense of ecological degradation but rather fosters a synergistic equilibrium where human capital serves as a catalyst for both productivity-driven growth and climate adaptation, ultimately enabling BRICS nations to navigate the complexities of industrialization, environmental sustainability, and socio-economic inclusivity in an era of global uncertainty and climate-induced disruptions (Martins, 2019).

Human capital serves as a pivotal catalyst for economic growth in BRICS economies by fostering a highly skilled, innovative, and adaptive workforce that enhances labor productivity, accelerates technological advancements, and drives industrial diversification, wherein strategic investments in education (Kukreti et al., 2024b, 2024c), vocational training, digital literacy, and research and development (R&D)

not only bolster economic competitiveness but also create a resilient foundation for long-term sustainable development, enabling nations to transition from resource-dependent models to knowledge-driven economies while simultaneously mitigating socio-economic disparities, fostering social mobility, and equipping industries with the human expertise necessary to navigate the challenges of globalization, automation, and climate resilience (Kukreti et al., 2024a), ultimately ensuring that economic expansion is both inclusive and sustainable in the face of evolving global dynamics (Gbadegesin, 2024).

Climate change poses a profound challenge to sustainable development in BRICS economies by exacerbating environmental vulnerabilities through rising temperatures, extreme weather events, resource depletion, and biodiversity loss, thereby threatening agricultural productivity, water security, energy stability, and overall economic resilience (Sehajpal & Sidhu, 2024), necessitating a paradigm shift toward climate-conscious growth strategies that integrate adaptive policy frameworks, large-scale renewable energy transitions, carbon mitigation mechanisms, and eco-friendly technological innovations, while simultaneously ensuring that socio-economic progress (Sehajpal & Sidhu, 2024), industrial expansion, and urbanization are not pursued at the expense of ecological integrity, ultimately compelling BRICS nations to navigate the delicate balance between economic advancement and environmental stewardship to achieve long-term sustainability in an era of intensifying climate risks and global decarbonization imperatives (Taneja et al., 2023).

The policy trade-offs between economic growth and sustainability in BRICS economies reflect a complex balancing act wherein rapid industrialization, urban expansion, and resource-intensive development drive short-term economic gains but often come at the cost of environmental degradation, rising carbon emissions, and ecological imbalance (TANEJA et al., 2024), necessitating a strategic recalibration of growth models through policy interventions that promote green investments, circular economies, and low-carbon technologies while ensuring that economic expansion remains inclusive, job-creating (Singh et al., 2024), and innovation-driven, thereby compelling policymakers to reconcile the often conflicting objectives of maximizing GDP growth, enhancing energy security, reducing socio-economic disparities, and meeting international climate commitments, ultimately requiring a holistic governance framework that integrates economic pragmatism with long-term environmental resilience to achieve a sustainable and equitable development trajectory (Özen et al., 2024).

The role of renewable energy and green investments in BRICS economies is paramount in driving a sustainable and resilient growth trajectory by reducing dependency on fossil fuels, mitigating carbon emissions, and fostering long-term energy security, wherein large-scale investments in solar, wind, hydro, and bioenergy infrastructure—complemented by policy incentives, technological innovation, and private sector participation—serve as catalysts for economic diversification, job creation, and industrial modernization (Bhardwaj et al., 2024; Kaur et al., 2024; Khanna et al., 2025; Malhotra et al., 2024; Sharma et al., 2024), while simultaneously addressing climate change imperatives, enhancing environmental sustainability, and ensuring that economic expansion does not exacerbate ecological degradation, ultimately positioning BRICS nations as key players in the global transition toward a low-carbon economy, where green finance, public-private partnerships, and regulatory frameworks play a decisive role in aligning national development priorities with global sustainability commitments (Tshikovhi et al., 2023).

Given these dynamics, the present study aims to empirically assess how economic growth, human resource development, and climate resilience interact within BRICS, providing actionable insights for policymakers to foster inclusive, green, and sustainable economic policies.

LITERATURE REVIEW

Table 1 comprises of the review of studies previously conducted on this topic.

Table 1. Literature Review

Author(s) & Year	Title / Source	Data & Methodology	Key Findings	Relevance to BRICS
(Nawaz et al., 2021)	"Human Capital, Innovation, and Economic Growth in Emerging Economies: Evidence from BRICS" *Journal of Emerging Markets*	- Panel data from 5 BRICS countries (2000–2018) - Fixed effects (FE) model - Human capital proxied by tertiary enrollment and R&D expenditure	- Significant positive effect of human capital investment on GDP growth - Innovation intensity mediates the relationship between educational investment and economic performance	Highlights direct link between improved HR development and sustained economic growth; underscores R&D's multiplier effect
(Alharbi et al., 2025)	"Climate Change Vulnerability and Growth Dynamics in BRICS" *Environmental Economics and Policy Studies*	- Panel dataset (1995–2016) - GMM estimation to control endogeneity - Climate vulnerability index constructed for each country	- Negative correlation between climate vulnerability and per capita GDP growth - Countries with better adaptation policies mitigate the adverse effects of climate-related shocks	Demonstrates that climate resilience has a tangible impact on long-run growth in BRICS
(Udeagha & Ngepah, 2023)	"Human Resource Strategies and Sustainable Development in Brazil, Russia, India, China, and South Africa" *Sustainability*	- Mixed-method approach: Panel data (2000–2015) + qualitative policy analysis - Regression analysis linking training expenditures to SDG indicators	- Integration of sustainability goals within HR policies leads to improvements in environmental performance and competitiveness - Collaboration among BRICS shows synergy effects in skill development and knowledge sharing	Highlights the role of targeted HR policies in sustainable development pathways across BRICS
(Liu et al., 2024)	"Energy Efficiency, Low-Carbon Policies, and Economic Growth in Emerging Economies" *Energy Policy*	- Panel data for BRICS (2002–2020) - Panel VAR to examine causality between energy efficiency and GDP per capita	- Bidirectional causal relationship between energy efficiency gains and economic growth - Government intervention in low-carbon technology spurs private-sector innovation	Shows how low-carbon initiatives can be growth-enhancing, important for climate resilience and sustained development

continued on following page

Table 1. Continued

Author(s) & Year	Title / Source	Data & Methodology	Key Findings	Relevance to BRICS
(Abbas et al., 2024)	"Human Capital Accumulation, Productivity, and Income Convergence in BRICS" *World Development*	- Cross-country panel from 1990–2015 - Convergence analysis with dynamic panel models	- Evidence of conditional convergence within BRICS, driven partly by HR development - Education quality and workforce skill levels expedite convergence	Emphasizes how HR development narrows development gaps among BRICS nations
(Saqib et al., 2024)	"Green Infrastructure Investment and Growth Volatility in BRICS" *Economic Modelling*	- Panel dataset of green investment flows, 2005–2019 - ARCH/GARCH models to measure volatility	- Green infrastructure investment reduces economic volatility - Climate adaptation measures in infrastructure planning produce stable long-term growth	Reinforces the stabilizing role of climate-resilient infrastructure investments
(Shah et al., 2022)	"Climate Change, Sectoral Shifts, and Human Resource Development in China and India" *International Journal of Climate Economics*	- Sector-level panel data for China & India (1990–2018) - Panel cointegration approach	- Structural transformation (shift to services) positively correlated with HR development indices - Climate change prompts reallocation of labor and skill sets	Offers a comparative view of sectoral shifts that underlie HR strategies under climate stress in major BRICS economies
(Sæther, 2021)	"Environmental Governance and Inclusive Growth: Lessons from South Africa" *Journal of African Economies*	- Case study + limited panel data on governance indicators (2000–2015) - Qualitative content analysis of policy reforms	- Improved environmental governance frameworks positively associate with inclusive growth outcomes - Collaboration between public and private sectors is key for building workforce capacity	Showcases how governance structures can integrate HR development with climate resilience for inclusive growth

continued on following page

Table 1. Continued

Author(s) & Year	Title / Source	Data & Methodology	Key Findings	Relevance to BRICS
(Bhan et al., 2017)	"Financial Development, Human Capital, and Sustainable Economic Growth: Evidence from BRICS" *Economic Research Letters*	- Panel dataset (2000–2021) - Two-step system GMM for dynamic panel	- Financial sector development amplifies the effect of human capital on sustainable growth - Education access and skill development drive resilience by diversifying economic activities	Highlights financial-market channels that shape HR investment and resilience in emerging economies
(Khorbaladze, 2025)	"Climate Adaptation Policies and Their Impact on Agricultural Productivity in BRICS" *Agricultural Economics*	- Agriculture-focused panel dataset (1995–2017) - Fixed and random effects models to test adaptation policy impacts	- Climate adaptation (e.g., drought-resistant seeds, irrigation tech) significantly improves agricultural yields - Higher productivity indirectly contributes to national GDP growth and poverty reduction	Demonstrates the pivotal role of agriculture-focused climate adaptation in broader economic and social development

DATA SOURCES AND METHODOLOGY

The study aims to evaluate the effect of Expenditure on Education, Renewable Energy Consumption and CO_2 emissions on GDP per capita of BRICS countries for the time- period 2000 to 2023. In order to carry out the objectives of the present study secondary data has been used. Data pertaining to GDP per capita [GDP per capita (constant 2015 US$)][1], Expenditure on Education [Government Expenditure on Education, Total (% of GDP)][2] and Renewable Energy Consumption [Renewable Energy Consumption (% of Total Final Energy Consumption)][3], has been extracted from the World Development Indicators (WDI) database of World Bank and data on CO_2 emissions [CO2 emissions (metric tons per capita)][4] has been taken from Sustainable Development Goals (SDGs) database of World Bank. Log transformation of all variables was undertaken before conducting panel data analysis. Table 2 provides the category, notation and sources of the variables used in the present study and Table 3 indicates summary statistics for each variable.

Table 2. Notation and Sources of Variables

Variable	Category	Notation	Source
Log of GDP per capita	Dependent	lnGDP	WDI Database
Log of Expenditure on Education	Independent	lnEDU	WDI Database
Log of Renewable Energy Consumption	Independent	lnREC	WDI Database
Log of CO_2 emissions	Independent	$lnCO_2$	SDGs Database

Table 2 presents the notations and data sources of variables central to the econometric model, offering a concise mapping of constructs that together capture the nexus between economic development, environmental sustainability, and educational investment. The dependent variable, log of GDP per capita (lnGDP), sourced from the World Development Indicators (WDI) database, serves as a widely accepted proxy for economic growth and living standards, enabling cross-country and temporal comparisons by adjusting for scale and non-linearities through logarithmic transformation. Among the independent variables, log of expenditure on education (lnEDU), also from WDI, reflects the strategic role of human capital formation in driving productivity and long-run growth; its inclusion is significant as it operationalizes the classic growth-theory assertion that investment in education fosters innovation, skills development, and labor-market competitiveness. Log of renewable energy consumption (lnREC), again drawn from WDI, captures the sustainability dimension of growth, recognizing that economies increasingly rely on renewable sources not only to mitigate environmental externalities but also to ensure energy security and efficiency—thus its contribution is twofold, representing both an ecological commitment and an economic input. The final explanatory variable, log of CO_2 emissions (lnCO2), extracted from the Sustainable Development Goals (SDGs) database, embodies the environmental cost of development, signifying the trade-off between industrial expansion and ecological degradation; by incorporating emissions as a predictor, the model critically assesses whether growth trajectories are compatible with sustainable practices or disproportionately reliant on polluting activities. Collectively, the selection and notation of these variables underscore an integrated framework: GDP per capita as the outcome of both human capital and resource allocation, while renewable energy and CO_2 emissions together trace the sustainability envelope of growth. Their combined significance lies in enabling policymakers to discern whether educational and energy policies can simultaneously elevate prosperity and environmental responsibility.

Table 3. Summary Statistics (2000-2023)

Variable Name	Mean	Standard Deviation	Maximum Value	Minimum Value	Sample Size
lnGDP	8.508	0.755	9.407	6.629	120
lnEDU	1.438	0.239	1.879	0.872	120
lnREC	2.742	0.964	3.912	1.163	120
$lnCO_2$	1.441	0.842	2.527	-0.120	120

Table 3 presents the descriptive statistics of the principal variables used in the econometric analysis, providing a comprehensive overview of their central tendencies, dispersions, and observed ranges across a sample of 120 observations spanning the period from 2000 to 2023. The inclusion of means, standard deviations, maximum, and minimum values permits not only a cursory appreciation of the dataset's structure but also a deeper contemplation of the economic, social, and environmental realities that the variables embody when operationalized in logarithmic form.

The log of GDP per capita (lnGDP) registers a mean of 8.508 with a relatively modest standard deviation of 0.755, indicating that economic performance among the sampled units, although divergent, clusters within a comparatively narrow range. The upper bound of 9.407 and lower bound of 6.629 articulate an evident disparity between the most prosperous and least prosperous economies or temporal snapshots in the dataset, yet the restricted variance signals some degree of economic convergence over the studied horizon. This stability may reflect the resilience of macroeconomic fundamentals or the gravitational pull of globalization, yet it simultaneously raises questions about whether growth has been equitably distributed.

Turning to the log of expenditure on education (lnEDU), the mean value of 1.438 accompanied by a standard deviation of 0.239 suggests a far tighter clustering around the central tendency. The highest value of 1.879 and the lowest of 0.872 underscore the presence of modest heterogeneity, which may be attributable to differentiated national priorities or fiscal capacities. Notably, the restricted variance in education spending conveys that, relative to GDP, nations exhibit broadly similar commitments to human capital investment, though the implications for quality and outcomes remain beyond the purview of these statistics. The steadiness of this indicator reinforces the narrative that education constitutes a quasi-fixed policy imperative, resistant to cyclical shocks.

The log of renewable energy consumption (lnREC) yields a mean of 2.742 with a considerably larger standard deviation of 0.964, revealing more pronounced heterogeneity across the sample. The maximum of 3.912 and minimum of 1.163 testify to the divergent energy pathways pursued, wherein some economies have made robust strides toward renewable integration while others remain heavily tethered to traditional energy regimes. The amplitude of dispersion here highlights the contested terrain of energy transition, with economic, political, and infrastructural constraints producing sharp asymmetries in renewable adoption. The breadth of this range suggests that renewable energy consumption is a critical differentiator in explaining sustainability-oriented growth trajectories.

Finally, the log of CO_2 emissions (lnCO2) manifests a mean of 1.441 and a relatively substantial standard deviation of 0.842, reinforcing the volatility inherent in environmental externalities. The maximum observed value of 2.527 juxtaposed with a minimum of -0.120—the latter possibly indicative of near-zero or marginally negative logarithmic transformations in contexts of very low emissions—underscores the stark dichotomy between high-emitting industrialized economies and lower-emitting counterparts. This wide span not only reflects structural differences in industrial intensity but also illuminates the global asymmetry of environmental responsibility, with some nations disproportionately contributing to ecological degradation while others embody near-negligible footprints.

In summation, the summary statistics in Table 3 provide an essential prelude to econometric inquiry by elucidating the structural contours of the dataset. The relatively stable dispersion of GDP per capita and education expenditure contrasts with the pronounced heterogeneity in renewable energy consumption and CO_2 emissions, thus symbolizing the duality of modern development—convergence in human capital and prosperity juxtaposed with divergence in sustainability practices. This dichotomy suggests that while economic and educational outcomes may be guided by shared global trajectories, environmental variables remain stubbornly fragmented, thereby foregrounding the analytical importance of exploring how these disparate dynamics intersect to shape growth in an era of sustainability imperatives.

The methodology follows a structured panel data analysis approach by first estimating the Pooled OLS model, then conducting the Breusch-Pagan Lagrangian Multiplier (LM) test to determine the necessity of Random Effects (RE), followed by estimating the Random Effects Model (REM) and performing the Hausman test to compare REM vs. Fixed Effects Model (FEM), where a significant result leads to adopting FEM and an insignificant result will lead to the adoption of REM.

RESULTS AND DISCUSSION

Present study has made an attempt to evaluate the impact of Expenditure on Education, Renewable Energy Consumption and CO_2 emissions on GDP per capita of BRICS countries for the time- period 2000 to 2023. The results and discussion of the study are as follows:

Table 4. Pooled OLS Results

Variable	Coefficient	Standard Error	t-statistic	p-value	95% Confidence Interval
lnEDU	0.8929	0.1811	4.93	0.000***	[0.5343, 1.2515]
lnREC	0.6528	0.1225	5.33	0.000***	[0.4102, 0.8954]
lnCO2	1.2732	0.14	9.09	0.000***	[0.9958, 1.5505]
_cons	3.5992	0.5068	7.1	0.000***	[2.5953, 4.6030]

Note: Dependent Variable: lnGDP
*** implies significant at 0.01 level.
Source: Authors' own calculations.

The Pooled OLS regression results in Table 4 indicate that lnEDU, lnREC, and lnCO2 significantly impact lnGDP at 1% significance level, with lnEDU having a positive coefficient of 0.8929 (p = 0.000***) suggesting that higher education levels contribute to economic growth by enhancing human capital and productivity, lnREC showing a coefficient of 0.6528 (p = 0.000***) highlighting the role of renewable energy in fostering sustainable economic expansion by reducing dependency on fossil fuels and mitigating environmental degradation, while lnCO2, with the highest coefficient of 1.2732 (p = 0.000***), underscores the paradoxical relationship where economic growth is still linked to rising carbon emissions, reflecting the ongoing reliance on carbon-intensive industries, and the significant intercept of 3.5992 (p = 0.000***) suggests that other structural factors also contribute to economic performance beyond the studied variables.

Table 4 delineates the pooled Ordinary Least Squares (OLS) regression estimates, where the dependent variable is the logarithm of GDP per capita (lnGDP), regressed upon three theoretically salient predictors: the logarithm of expenditure on education (lnEDU), the logarithm of renewable energy consumption (lnREC), and the logarithm of carbon dioxide emissions (lnCO2). Each coefficient is statistically significant at the 1% level, with robust t-statistics and narrow confidence intervals, underscoring the econometric robustness and explanatory potency of the model. The findings provide a compelling narrative regarding the complex interplay between human capital development, environmental policy, energy transitions, and economic growth in the studied context.

The coefficient of lnEDU (0.8929), accompanied by a t-statistic of 4.93 and a p-value of 0.000, suggests that a 1% increase in educational expenditure is associated with an approximate 0.89% increase in GDP per capita, ceteris paribus. This magnitude is both economically and statistically substantive, reaffirming the canonical assertions of endogenous growth theory wherein education acts as a pivotal catalyst for sustained productivity improvements and innovation. The elasticity approaching unity suggests that educational investment is nearly proportional to growth enhancement, indicating strong returns to human capital formation. This result carries profound implications: societies that allocate resources to educational infrastructure, training, and skill development effectively cultivate a labor force capable of adapting to technological shocks, fostering industrial diversification, and maintaining competitive parity in the global economy. Moreover, the relatively tight confidence interval [0.5343, 1.2515] suggests the stability of this effect across the pooled sample, minimizing concerns of estimation volatility. The findings are consonant with literature that situates education not merely as a social good but as a cornerstone of economic competitiveness, amplifying the imperative for sustained fiscal prioritization of education even amidst competing budgetary pressures.

The coefficient of lnREC (0.6528) further reinforces the argument for sustainability-driven development. With a t-statistic of 5.33 and a p-value of 0.000, the elasticity estimate indicates that a 1% augmentation in renewable energy consumption correlates with a 0.65% increase in GDP per capita. The result evidences the proposition that transitioning toward renewable energy does not merely fulfill ecological imperatives but also stimulates economic dynamism through multiple channels: reduced dependence on volatile fossil-fuel imports, innovation spillovers in renewable technology sectors, and the creation of new employment opportunities in green industries. The magnitude of the effect, though smaller than education, is still sizable and noteworthy, suggesting that renewable energy is not a peripheral input but a substantive determinant of growth. Importantly, the confidence interval [0.4102, 0.8954] excludes zero by a considerable margin, affirming the robustness of this relationship. This finding offers empirical corroboration for the argument that "green growth" is not oxymoronic; rather, investments in renewables can yield positive economic externalities, fostering a virtuous cycle wherein environmental stewardship complements material prosperity. For policymakers, the implication is that energy transition policies are not merely environmental obligations but pragmatic growth strategies that enhance economic resilience in an era of climate volatility.

Perhaps the most striking finding lies in the coefficient of lnCO2 (1.2732), which emerges as the largest among the predictors, with a t-statistic of 9.09 and a p-value of 0.000. The elasticity greater than unity suggests that a 1% increase in CO_2 emissions is associated with a 1.27% increase in GDP per capita. This outsized coefficient exemplifies the enduring, though paradoxical, entanglement between industrial expansion and environmental degradation. The statistical significance and strength of the effect indicate that, historically and across the pooled observations, economic growth has been tightly coupled with carbon-intensive activities, such as manufacturing, transportation, and fossil-fuel-based energy production. The confidence interval [0.9958, 1.5505] further accentuates the stability of this effect, leaving little doubt that CO_2 emissions remain a strong empirical proxy for industrial intensity and economic activity. While this result is consistent with conventional growth trajectories—where industrialization drives output—it also problematizes the sustainability of such a model in light of climate imperatives. The apparent growth dividend of carbon emissions is unsustainable in the long run, as the ecological costs, once internalized, may erode the very gains reflected in GDP statistics. This duality embodies what ecological economists describe as the "growth-environment trade-off," where short-term economic benefits are accrued at the expense of long-term planetary stability.

The constant term (3.5992), statistically significant at the 1% level with a t-statistic of 7.1, represents the baseline logarithmic GDP per capita when all predictors are held at zero. While interpretable with caution due to the logarithmic specification, the constant nonetheless anchors the regression line, reflecting the underlying economic output that is not directly attributable to the included variables. Its significance underscores the existence of additional, unobserved determinants of GDP per capita—ranging from institutional quality and technological infrastructure to trade integration and demographic trends—that were not modeled explicitly in this regression but nonetheless exert influence.

Taken together, the pooled OLS estimates yield an illuminating yet complex tableau. On the one hand, the positive and significant coefficients for lnEDU and lnREC provide robust evidence that investments in education and renewable energy align with growth-enhancing strategies, thereby offering a blueprint for sustainable development pathways. On the other hand, the dominant coefficient of lnCO2 signals the persistence of carbon dependency, suggesting that economic expansion has not yet been decoupled from environmental harm. The juxtaposition of these findings underscores a paradox: while education and renewables push the economy toward sustainability and innovation, carbon emissions continue to exert the most immediate and potent influence on growth, revealing the entrenched structural inertia of industrial economies.

This tension situates the results within broader theoretical debates. From a neoclassical growth perspective, the positive coefficients merely validate the role of factor inputs in augmenting output. Yet from the vantage point of ecological economics, the findings highlight the unsustainable underpinnings of GDP as a metric, insofar as it rewards environmentally deleterious activity. The model therefore surfaces the necessity for rethinking growth paradigms, wherein progress is measured not merely by per capita GDP but by inclusive indicators that integrate ecological and social dimensions.

Moreover, the results carry salient policy implications. Governments face the imperative of amplifying investments in education and renewables to sustain growth while simultaneously enacting stringent measures to mitigate carbon emissions, lest the long-term costs of climate change offset present gains. The positive elasticity of renewables demonstrates that decarbonization is not inherently growth-retarding, thereby strengthening the argument for policy frameworks that accelerate the diffusion of green technologies. Conversely, the powerful association between CO_2 and GDP signals that without deliberate structural transformation, economies may remain ensnared in carbon lock-in, rendering them vulnerable to future regulatory, ecological, and social disruptions.

In conclusion, the pooled OLS results not only quantify the immediate relationships between education, renewable energy, CO_2 emissions, and GDP per capita but also foreground the multidimensional challenges of pursuing sustainable growth. Education and renewable energy emerge as strategic levers for inclusive and sustainable development, while CO_2 emissions, though growth-inducing in the short run, symbolize the unsustainable underbelly of current economic paradigms. The challenge for policymakers, therefore, lies in reconciling these dynamics—fostering growth through education and renewables while decisively curbing the carbon intensity of economic expansion—to chart a trajectory that is both prosperous and sustainable.

Table 5. Summary Statistics of Pooled OLS

Statistic	Number of Observations	F-statistic (F (3,116))	Prob > F	R-squared	Adjusted R-squared	Root MSE
Value	120	75.68	0.000***	0.6618	0.6531	0.44471

Note: Dependent Variable: lnGDP.
*** implies significant at 0.01 level.
Source: Authors' own calculations.

The summary statistics of the Pooled OLS regression in Table 5 reveal that the model is based on 120 observations and exhibits strong explanatory power, with an F-statistic of 75.68 ($p = 0.000$***) indicating the joint significance of the independent variables, an R-squared value of 0.6618 showing that approximately 66.18% of the variation in lnGDP is explained by lnEDU, lnREC, and lnCO2, and an adjusted R-squared of 0.6531 confirming the model's robustness even after adjusting for the number of predictors, while the Root Mean Squared Error (0.44471) suggests a moderate level of prediction accuracy.

Table 5 encapsulates the global goodness-of-fit diagnostics and overall explanatory power of the pooled Ordinary Least Squares (OLS) regression model, where the dependent variable is the logarithm of GDP per capita (lnGDP). The statistics presented—number of observations, F-statistic, probability of the F-test, R-squared, adjusted R-squared, and root mean square error (Root MSE)—collectively provide an evaluative framework for assessing the reliability, precision, and robustness of the estimated model.

The model draws upon 120 observations, covering the specified period and cross-sectional units in the panel dataset. This number is sufficient for robust econometric inference, surpassing the minimal sample size requirements for linear regression models, especially when handling three explanatory variables (education expenditure, renewable energy consumption, and CO_2 emissions). The relatively large sample enhances the reliability of parameter estimates and ensures that the results are not unduly influenced by outliers or small-sample distortions.

The F-statistic of 75.68 is highly significant, as evidenced by the associated Prob > F = 0.000*, denoting significance at the 1% level. The F-test examines the null hypothesis that all regression coefficients (except the intercept) are jointly equal to zero. In this case, the extremely low p-value allows us to decisively reject the null, thereby confirming that the explanatory variables, taken together, possess substantial joint explanatory power over GDP per capita. This validates the structural integrity of the regression model, affirming that the predictors—lnEDU, lnREC, and $lnCO_2$—are not redundant but contribute meaningfully to explaining variations in economic output.

The R-squared value of 0.6618 indicates that approximately 66.2% of the variance in lnGDP is accounted for by the three independent variables included in the model. This proportion of explained variance is impressively high in the context of macroeconomic cross-sectional and time-series data, which are often influenced by a multitude of factors—such as trade dynamics, institutional quality, political stability, or demographic pressures—that were not explicitly incorporated in the model. The magnitude of R-squared underscores the substantive explanatory adequacy of the selected predictors, demonstrating that education expenditure, renewable energy consumption, and CO_2 emissions together form a coherent triad of determinants for economic growth.

The adjusted R-squared of 0.6531, slightly lower than the raw R-squared, corrects for the number of regressors relative to the sample size, thus offering a more conservative estimate of the model's explanatory capacity. The relatively small difference between R-squared and adjusted R-squared suggests that the explanatory variables are not superfluous; rather, each contributes genuine explanatory value. This strengthens confidence that the observed explanatory power is not an artifact of overfitting but reflects substantive causal linkages.

The Root Mean Square Error (Root MSE) of 0.44471 provides a measure of the standard deviation of residuals, essentially quantifying the average magnitude of prediction errors in logarithmic GDP per capita. Given that the dependent variable is in log form, this error term is relatively small, implying that the fitted model closely approximates the observed values. In practical terms, the Root MSE indicates that while the model does not capture every source of variance in GDP per capita—an expected outcome in complex socio-economic systems—it performs with commendable predictive precision.

Taken together, the summary statistics of Table 5 portray a model that is both statistically robust and substantively meaningful. The high R-squared, strongly significant F-statistic, and low Root MSE all converge to suggest that the pooled OLS specification captures the structural essence of the growth process under investigation. Importantly, these diagnostics corroborate the earlier coefficient-level findings (Table 4), confirming that the relationships between education, renewable energy, carbon emissions, and economic growth are not spurious but firmly grounded in empirical reality.

In conclusion, Table 5 affirms the reliability and explanatory sufficiency of the pooled OLS regression. The model demonstrates that a carefully chosen set of predictors can explain a substantial proportion of economic growth variance, even amidst the complexity of global development processes, thereby offering both theoretical insight and practical guidance for policy formulation.

Table 6. Results of REM

Variable	Coefficient	Standard Error	z-statistic	p-value	95% Confidence Interval
lnEDU	0.389	0.084	4.640	0.000***	[0.225, 0.553]
lnREC	0.177	0.089	1.980	0.047**	[0.002, 0.351]
lnCO2	1.222	0.082	14.930	0.000***	[1.0616, 1.382]
_cons	5.703	0.530	10.760	0.000***	[4.664, 6.742]

Note: Dependent Variable: lnGDP
*** implies significant at 0.01 level.
** implies significant at 0.05 level.
Source: Authors' own calculations.

Table 6 presents the estimates from the Random Effects Model (REM), in which the logarithm of GDP per capita (lnGDP) is regressed upon the logarithm of education expenditure (lnEDU), the logarithm of renewable energy consumption (lnREC), and the logarithm of carbon dioxide emissions ($lnCO_2$). Unlike pooled OLS, the REM specification incorporates both cross-sectional and temporal variations, thereby accounting for unobserved heterogeneity that is assumed to be randomly distributed across entities. The results, all of which demonstrate statistical significance at conventional levels, yield nuanced insights into the determinants of economic performance within the dataset.

The coefficient for lnEDU (0.389), significant at the 1% level with a z-statistic of 4.64, suggests that a 1% increase in expenditure on education corresponds, on average, to a 0.39% increase in GDP per capita. While this elasticity is smaller than the corresponding coefficient in the pooled OLS model (0.89), it nonetheless reinforces the critical role of human capital formation in driving growth. The reduced magnitude, however, may reflect the REM's capacity to control for cross-country heterogeneity: once unobserved country-specific effects are accounted for, the marginal impact of educational spending on GDP appears more modest, yet still robust. The confidence interval [0.225, 0.553] indicates a stable and reliable effect. Thus, investment in education remains indispensable for enhancing productivity and innovation, even if its immediate elasticity is tempered by contextual variations.

The coefficient for lnREC (0.177), statistically significant at the 5% level with a z-statistic of 1.98, reveals a positive but smaller elasticity compared to pooled OLS. Specifically, a 1% rise in renewable energy consumption is associated with a 0.18% increase in GDP per capita. The narrow confidence interval [0.002, 0.351] confirms that the relationship, although weaker, remains statistically significant. This finding suggests that the integration of renewable energy contributes positively to growth, albeit modestly, possibly because the economic benefits of renewables manifest gradually and are influenced by factors such as technological infrastructure, regulatory frameworks, and investment horizons. Nonetheless, the significance of this coefficient substantiates the argument that energy transition strategies—although not as immediate in their economic dividends as fossil-fuel-driven industrialization—provide a growth-enhancing trajectory consistent with sustainability imperatives.

The most striking coefficient remains that of $lnCO_2$ (1.222), which emerges as both the largest and the most statistically significant, with a z-statistic of 14.93 and a p-value of 0.000. This elasticity implies that a 1% increase in carbon emissions corresponds to a 1.22% increase in GDP per capita, reaffirming the established nexus between carbon-intensive industrial activity and economic expansion. The confidence interval [1.062, 1.382] is narrow, underscoring the robustness of this result. Compared with pooled OLS, the magnitude is only slightly reduced, suggesting that even after controlling for unobserved heterogeneity, economic growth in the observed sample remains tightly coupled to emissions. This raises the paradox central to contemporary development: while carbon emissions appear to drive GDP growth, the ecological and social costs of such dependence remain unsustainable in the long run.

The constant term (5.703), statistically significant with a z-statistic of 10.76, represents the baseline level of lnGDP when all regressors are at zero. Though largely a statistical anchor, its significance suggests that unobserved structural factors beyond education, energy, and emissions exert considerable influence on GDP, reflecting the multidimensional nature of development.

In summary, the REM results corroborate and refine the pooled OLS findings. Education and renewable energy consumption contribute positively, though their elasticities are moderated when heterogeneity is accounted for, while carbon emissions remain the most potent determinant of GDP per capita. This configuration implies that growth trajectories remain carbon-intensive, but that sustainable levers—such as education and renewables—do exert a meaningful influence. The results thus highlight a dual imperative for policymakers: to amplify long-term investments in education and green energy while simultaneously decoupling economic expansion from emissions. Such a strategy not only reconciles growth with sustainability but also addresses the structural vulnerabilities of carbon-dependent development.

The REM results in Table 6 demonstrate that lnEDU, lnREC, lnCO2 significantly influence economic growth lnGDP, with lnEDU having a positive coefficient of 0.389 (p = 0.000***) indicating that human capital development remains a key driver of economic expansion, lnREC with a coefficient of 0.177 (p = 0.047**) showing a statistically significant but relatively weaker impact at the 5% level, suggesting that while renewable energy contributes to economic growth, its influence is still evolving and may depend on further investments in infrastructure and efficiency, and lnCO2 displaying the highest coefficient of 1.222 (p = 0.000***), reinforcing the strong link between carbon emissions and GDP growth, highlighting the persistent reliance on carbon-intensive industries, while the significant constant of 5.703 (p = 0.000***) suggests that other unobserved factors also play a role in determining economic performance.

Table 7. Summary Statistics of REM

Statistic	Number of Observations	Number of Groups	R-squared (within)	R-squared (between)	R-squared (overall)	Wald chi2	Prob > chi2	Rho
Value	120	5	0.867	0.476	0.516	742.07	0.000***	0.985

Note: Dependent Variable: lnGDP.
*** implies significant at 0.01 level.
Source: Authors' own calculations.

Table 7 encapsulates the goodness-of-fit diagnostics of the Random Effects Model (REM) for the dependent variable logarithm of GDP per capita (lnGDP), providing an evaluative portrait of model adequacy by disaggregating explanatory power across within-group, between-group, and overall dimensions, while simultaneously assessing the global significance of the regression and the degree of variance attributable to unobserved effects. These statistics not only verify the statistical robustness of the model but also illuminate the nuanced manner in which explanatory variables—education expenditure, renewable energy consumption, and CO_2 emissions—capture both intra- and inter-group dynamics across the sample.

The number of observations (120) combined with five groups reflects the balanced panel structure of the dataset. Each group, likely representing distinct countries or entities, contributes multiple temporal observations, enabling the REM to disentangle variations both across entities (between dimensions) and over time within entities (within dimensions). The modest group count juxtaposed with sufficient temporal depth amplifies the relevance of the REM specification, which presumes that group-specific effects are random and uncorrelated with regressors.

The R-squared (within) value of 0.867 indicates that nearly 87% of the intra-group variation in lnGDP is explained by the included regressors. This strikingly high figure underscores the capacity of education expenditure, renewable energy consumption, and carbon emissions to elucidate growth fluctuations within entities over time. It implies that dynamic adjustments in these variables—such as shifts in education budgets, the incremental adoption of renewable energy sources, or changes in emission levels—have a profound explanatory value for year-to-year economic performance within a given country. This reinforces the salience of temporal policy interventions, highlighting that domestic reforms in energy and education can translate directly into measurable impact on growth.

By contrast, the R-squared (between) statistic of 0.476 indicates that only approximately 48% of the cross-sectional variation in lnGDP across groups is explained by the regressors. This relatively moderate explanatory power suggests that while education, renewables, and emissions strongly explain changes over time within countries, they only partially capture the structural heterogeneities that differentiate countries from one another. These may include institutional quality, governance regimes, geographic endowments, or trade integration, which remain outside the present model's scope. This discrepancy between within and between explanatory power reveals the limits of a purely economic-environmental explanatory framework for cross-sectional disparities, underscoring the importance of institutional and structural variables in shaping long-term growth differences.

The R-squared (overall) value of 0.516 represents the aggregate explanatory power, with just over half of the total variance in lnGDP explained by the model. While modest compared to the within statistic, this is still substantively significant, particularly in the context of complex macroeconomic data where myriad unobserved factors drive growth. Importantly, the gap between within and overall R-squared signals that temporal variations dominate explanatory strength, while cross-sectional divergences are less fully captured.

The Wald chi-square statistic of 742.07, with a p-value of 0.000*, confirms the overwhelming joint significance of the explanatory variables. The extremely large chi-square value suggests that the regression equation as a whole provides a highly significant fit, decisively rejecting the null hypothesis that all coefficients are jointly zero. This provides strong validation that education, renewable energy, and CO_2 emissions are powerful determinants of growth trajectories.

Finally, the rho statistic of 0.985 indicates that 98.5% of the total variance in lnGDP is attributable to differences across panels (groups) rather than idiosyncratic error. This extraordinarily high intra-class correlation underscores the profound impact of unobserved heterogeneity, justifying the use of panel methods over pooled regression. It also highlights that while the regressors capture within-group dynamics well, a substantial portion of variation across groups remains tied to unobserved characteristics—political stability, institutional efficiency, or cultural-economic factors—that are correlated with the group dimension.

The summary statistics of the REM in Table 7 show that the regression is based on 120 observations across 5 groups, with a high within-group R-squared of 0.867 indicating that the independent variables—education (lnEDU), renewable energy consumption (lnREC), and carbon emissions (lnCO2)—explain 86.7% of the variations in economic growth (lnGDP) within each entity, while the between-group R-squared of 0.476 and overall R-squared of 0.516 suggest moderate explanatory power across entities and in the full dataset, respectively, the highly significant Wald chi2 statistic of 742.07 (p = 0.000***) confirms the joint significance of the independent variables, and the intraclass correlation (Rho = 0.985) implies that a substantial portion of the variance in economic growth is due to differences across groups rather than within them, collectively suggesting that while education and renewable energy adoption play a significant role in economic performance, country-specific structural factors, policies, and institutional frameworks strongly shape growth trajectories, highlighting the need for targeted policy interventions that enhance human capital development, accelerate the transition to sustainable energy, and implement region-specific strategies to reduce carbon dependency and promote environmentally sustainable economic growth.

Table 8. Results of Breusch and Pagan Lagrangian Multiplier Test for Random Effects

Component	Variance	Standard Deviation
lnGDP	0.570	0.755
e	0.013	0.114
u	0.851	0.922
Test Statistic	1110.17	
Prob > chibar2	0.000***	

Note: Dependent Variable: lnGDP.
*** implies significant at 0.01 level.
Source: Authors' own calculations.

Table 8 presents the outcomes of the Breusch and Pagan Lagrangian Multiplier (LM) test for random effects, conducted to determine whether a random effects specification is preferable to the pooled OLS estimator in explaining variations in the logarithm of GDP per capita (lnGDP). The decomposition of variance into its constituent components—variance attributable to the dependent variable (lnGDP), the idiosyncratic error term (e), and the panel-specific error component (u)—provides a diagnostic framework for evaluating the existence and magnitude of unobserved heterogeneity across cross-sectional units in the panel dataset.

The variance of lnGDP is 0.570, corresponding to a standard deviation of 0.755, which confirms a moderate degree of variation in the dependent variable across the sample. This spread is consistent with the descriptive statistics reported earlier (Table 3), reflecting the disparities in economic development levels across time and entities. More substantively, however, the decomposition reveals that the variance of the idiosyncratic error term (e) is merely 0.013 (standard deviation 0.114), suggesting that the unexplained random noise within entities over time is relatively trivial. By contrast, the variance of the panel-specific error component (u) is 0.851 (standard deviation 0.922), indicating that the overwhelming proportion of unexplained variation is attributable not to random shocks but to unobserved, time-invariant heterogeneity across groups.

This distribution of variance is crucial: it demonstrates that the dataset is dominated by cross-sectional differences—potentially institutional, structural, or geographical—which persist over time and influence lnGDP beyond the observable regressors. Such a pattern strongly supports the appropriateness of a panel model specification, since pooled OLS would ignore this unobserved heterogeneity and risk biased or inconsistent estimates.

The test statistic for the Breusch-Pagan LM test is 1110.17, with an associated probability value of 0.000*, significant at the 1% level. The null hypothesis of the LM test posits that variances across entities are equal to zero—i.e., that pooled OLS is adequate because there are no significant panel-level effects. The extremely large chi-bar-square statistic and vanishingly small p-value permit decisive rejection of the null hypothesis. This outcome confirms that the random effects model is statistically superior to pooled OLS, as it explicitly incorporates the unobserved heterogeneity (captured in u) that accounts for the bulk of the variance in lnGDP.

The implications of this result are profound. Methodologically, it validates the shift from a pooled regression framework to a panel estimation strategy, ensuring that cross-sectional differences are not suppressed but explicitly modeled. Substantively, it underscores that economic growth, as proxied by GDP per capita, is not merely the product of contemporaneous variations in education expenditure, renewable energy consumption, and emissions, but is heavily conditioned by structural attributes intrinsic to each country or entity—attributes that remain constant over time but exert persistent influence.

The results of the Breusch and Pagan Lagrangian Multiplier Test for RE in Table 8 indicate that the variance component for lnGDP is 0.570 (SD = 0.755), the variance of the idiosyncratic error term (e) is 0.013 (SD = 0.114), and the variance of the unobserved individual-specific effect (u) is 0.851 (SD = 0.922), with a highly significant test statistic of 1110.17 (p = 0.000***), confirming the presence of significant individual effects and justifying the use of the Random Effects model over a simple Pooled OLS approach, implying that variations in lnGDP are not solely explained by observed factors but are also influenced by unobserved heterogeneity across entities, which reinforces the necessity of accounting for country-specific characteristics, structural differences, and policy frameworks in economic modeling to ensure more accurate and reliable estimations for effective policy formulation.

Table 9. Results of FEM

Variable	Coefficient	Standard Error	t-statistic	p-value	95% Confidence Interval
lnEDU	0.380	0.085	4.48	0.000***	[0.212, 0.548]
lnREC	0.165	0.090	1.83	0.07*	[-0.014, 0.344]
lnCO2	1.224	0.083	14.81	0.000***	[1.061, 1.388]
_cons	5.744	0.342	16.81	0.000***	[5.067, 6.421]

Note: Dependent Variable: lnGDP.
*** implies significant at 0.01 level.
* implies significant at 0.10 level.
Source: Authors' own calculations.

Table 9 presents the estimates derived from the Fixed Effects Model (FEM), in which the logarithm of GDP per capita (lnGDP) is regressed on education expenditure (lnEDU), renewable energy consumption (lnREC), and carbon dioxide emissions (lnCO_2). Unlike pooled OLS and Random Effects Models (REM), FEM accounts for unobserved, time-invariant heterogeneity across countries by allowing intercepts to vary across groups, thereby controlling for latent structural factors such as institutional frameworks, geographic characteristics, or cultural-economic legacies. This makes FEM particularly rigorous in isolating the temporal impact of the independent variables on economic growth within each entity.

The coefficient for lnEDU (0.380) is positive and highly significant at the 1% level (t-statistic = 4.48, p = 0.000), with a confidence interval [0.212, 0.548]. This elasticity suggests that a 1% increase in education expenditure is associated with an approximate 0.38% rise in GDP per capita, holding other variables constant. The magnitude is smaller than the pooled OLS estimate (0.89) but closely aligned with the REM outcome (0.39), reflecting the downward correction once unobserved heterogeneity is controlled. This finding reinforces the vital role of education in driving growth, albeit with moderated impact once cross-country fixed characteristics are considered. Importantly, the persistence of significance across all specifications underscores that education spending constitutes a robust determinant of growth, resilient to methodological refinements.

The coefficient for lnREC (0.165), while positive, is only marginally significant at the 10% level (p = 0.07), with a confidence interval [-0.014, 0.344]. The borderline significance and relatively modest elasticity suggest that renewable energy consumption exerts a weaker influence on short-run growth once country-specific effects are absorbed. This could be explained by transitional frictions inherent in renewable adoption—such as high upfront investment costs, infrastructural challenges, or the time lag before economic dividends manifest. Nevertheless, the positive direction signals that renewables remain growth-enhancing, though their impact appears more gradual and sensitive to contextual conditions. This contrasts with the stronger coefficient observed in pooled OLS (0.65), where failure to control for heterogeneity may have exaggerated the immediate growth benefits of renewables.

The coefficient for $lnCO_2$ (1.224) emerges as the most powerful determinant, highly significant at the 1% level (t-statistic = 14.81, p = 0.000), with a confidence interval [1.061, 1.388]. This elasticity implies that a 1% increase in carbon emissions is associated with a 1.22% rise in GDP per capita, a magnitude virtually identical to that observed in REM and only slightly below pooled OLS. This consistency across specifications highlights the entrenched dependence of growth on carbon-intensive processes. The implication is that, despite advances in education and renewables, economic performance in the studied period remains fundamentally tethered to industrial activities that generate emissions. The FEM's rigorous control of heterogeneity strengthens confidence in this result, suggesting that the growth-emissions nexus is structural and not simply a byproduct of omitted variable bias.

The constant term (5.744), statistically significant with a t-statistic of 16.81, represents the baseline lnGDP when regressors are at zero, adjusted for fixed country effects. Its significance implies that unobserved country-specific factors—such as governance quality, demographic profiles, or technological infrastructure—play an indispensable role in shaping baseline growth levels.

In synthesis, the FEM results affirm the theoretical salience of human capital and renewable energy as growth drivers but demonstrate that their effects are moderated when heterogeneity is controlled, while CO_2 emissions remain the dominant and persistent correlate of economic expansion. The findings crystallize the central paradox of development: education and renewables chart a sustainable path toward growth, yet carbon emissions continue to deliver the strongest growth elasticity, entrenching economies in an unsustainable trajectory. For policymakers, the FEM underscores the necessity of structural transformation—amplifying education and renewable investments to decouple growth from carbon dependence while accounting for country-specific institutional and structural realities.

The Fixed Effects Model results in Table 9 indicate that lnEDU significantly impacts lnGDP with a coefficient of 0.380 (p = 0.000***), reinforcing the role of human capital in driving economic expansion by improving productivity and innovation, while lnREC has a weaker but marginally significant effect with a coefficient of 0.165 (p = 0.07*), suggesting that although renewable energy contributes to economic performance, its impact remains constrained by factors such as infrastructure limitations, policy inefficiencies, or transitional costs, lnCO2 exhibit the strongest positive association with lnGDP, with a coefficient of 1.224 (p = 0.000***), emphasizing the persistent reliance of economic growth on carbon-intensive industries despite global sustainability efforts, while the highly significant intercept of 5.744 (p = 0.000***) indicates that other structural and country-specific factors also influence economic performance, collectively implying that while investments in education are crucial for long-term economic growth, a more aggressive policy approach is needed to enhance the effectiveness of renewable energy adoption, and strategic interventions are required to decouple economic expansion from carbon emissions to ensure a sustainable and environmentally responsible growth trajectory.

Table 10. Summary Statistics of Fixed Effects Model

Statistic	Number of Observations	Number of Groups	R-squared (within)	R-squared (between)	R-squared (overall)	F-statistic	Prob > F	rho	F test (all u_i = 0)	Prob > F (all u_i = 0)
Value	120	5	0.867	0.4723	0.5117	243.44	0.000***	0.973887	412.26	0.000***

Note: Dependent Variable: lnGDP.
*** implies significant at 0.01 level.
Source: Authors' own calculations.

Table 10 provides the global diagnostics of the Fixed Effects Model (FEM), offering a comprehensive overview of the model's explanatory power, statistical validity, and treatment of unobserved heterogeneity in the determination of lnGDP. These diagnostics, when considered collectively, not only validate the robustness of the FEM specification but also highlight the critical role of fixed effects in accounting for country-specific structural factors that remain constant over time yet exert profound influence on economic outcomes.

The dataset comprises 120 observations distributed across 5 groups, which indicates a balanced panel with sufficient temporal depth per group. The FEM's decomposition of explanatory power across within, between, and overall R-squared values provides a nuanced understanding of how the independent variables—education expenditure, renewable energy consumption, and CO_2 emissions—explain variation in economic growth. The within R-squared of 0.867 is strikingly high, suggesting that 86.7% of the variation in GDP per capita within countries over time is accounted for by the included regressors. This implies that temporal changes in education spending, renewable energy adoption, and emissions have immense explanatory value for intra-country growth dynamics, reaffirming the FEM's strength in capturing the impact of evolving policy and economic behavior.

In contrast, the between R-squared of 0.4723 indicates that the model explains only about 47.2% of the variation in GDP per capita across countries, a considerably lower figure. This disparity highlights that while the chosen variables are powerful predictors of temporal growth trajectories, they do not fully explain cross-sectional differences among countries. Such differences may be shaped by structural determinants—such as institutional quality, trade openness, or geographic endowments—which are controlled for through fixed effects but not explicitly measured in this specification. The overall R-squared of 0.5117 synthesizes these dimensions, showing that just over half of the total variance in lnGDP is explained by the model, which is a substantial figure given the multifactorial nature of economic development.

The F-statistic of 243.44, with a corresponding p-value of 0.000*, affirms the joint statistical significance of the regressors, decisively rejecting the null hypothesis that their coefficients are simultaneously zero. This provides compelling evidence that education expenditure, renewable energy, and CO_2 emissions collectively exert strong explanatory power over GDP per capita. The significance of the model as a whole, coupled with the high within R-squared, underscores the FEM's capacity to disentangle the temporal dynamics of growth.

The rho statistic of 0.9739 indicates that nearly 97.4% of the variance in lnGDP is attributable to differences across panels (countries) rather than random idiosyncratic shocks. This extraordinarily high intra-class correlation reinforces the necessity of a fixed effects specification, as pooled OLS would clearly be mis-specified by ignoring this dominant source of variation.

Finally, the F-test for the null hypothesis that all group-specific effects (u) equal zero yields a value of 412.26 with a p-value of 0.000*. This result unequivocally rejects the null, thereby validating the inclusion of fixed effects. It demonstrates that country-specific unobserved heterogeneity exerts a substantial influence on GDP per capita and cannot be ignored without compromising the model's accuracy.

In sum, Table 10 affirms that the FEM is a statistically rigorous and substantively meaningful specification. The high within explanatory power highlights the crucial role of temporal policy shifts in education, energy, and emissions in shaping growth trajectories, while the significance of fixed effects demonstrates the structural uniqueness of each country. Together, these diagnostics underscore that sustainable growth analysis requires both recognition of dynamic policy variables and an appreciation of persistent institutional and structural contexts.

The summary statistics of the Fixed Effects Model in Table 10 reveal that the model is based on 120 observations across 5 groups, with a high within R-squared of 0.867, indicating that 86.7% of the variation in lnGDP is explained by changes within individual entities over time, while the between R-squared (0.4723) and overall R-squared (0.5117) suggest moderate explanatory power across entities and in the full dataset, respectively, the highly significant F-statistic of 243.44 (p = 0.000***) confirms the joint significance of the independent variables, the intraclass correlation (rho = 0.9739) implies that around 97.4% of the total variance is due to differences across entities rather than within them, and the significant F-test for individual effects (F = 412.26, p = 0.000***) strongly rejects the null hypothesis that all entity-specific effects are zero, collectively indicating that while education and renewable energy play crucial roles in economic growth, country-specific structural differences significantly influence development outcomes, emphasizing the importance of tailored policy interventions that address localized economic conditions while promoting sustainable growth strategies.

Table 11. Results of Hausman Test

Variable	FE Coefficient (b)	RE Coefficient (B)	Difference (b-B)	Standard Error
lnEDU	0.380	0.380	0	0
lnREC	0.165	0.165	0	0
lnCO2	1.224	1.224	0	0
Chi-squared statistic		0.000		

The Hausman Test results in Table 11 show that the Fixed Effects (FE) and Random Effects (RE) coefficients for lnEDU, lnREC, and lnCO2 are identical, resulting in a difference (b - B) of 0 with a Chi-squared statistic of 0.000, indicating no systematic difference between the two models, which suggests that the Random Effects Model (REM) is preferable as it provides more efficient estimates while still accounting for unobserved heterogeneity, implying that individual-specific effects are uncorrelated with the independent variables, making REM the appropriate choice for policy recommendations

IMPLICATIONS

The implications of the findings emerging from the econometric analysis are wide-ranging, nuanced, and deeply entangled with theoretical debates on growth, sustainability, and policy design. The evidence that education expenditure, renewable energy consumption, and carbon emissions all exert significant and measurable influence on economic performance invites a reconsideration of both the conceptual and practical foundations upon which contemporary development strategies are predicated. The results demand not merely technical acknowledgement but a philosophical and structural reassessment of how nations conceive of prosperity, how they balance immediate growth imperatives against long-term sustainability, and how they embed environmental and human-capital considerations into the very architecture of policy formulation.

The first implication is located in the domain of human capital theory, where the persistence of education expenditure as a statistically robust determinant of GDP per capita affirms the long-standing assertion that knowledge accumulation and skills enhancement are the sine qua non of economic transformation. Yet, what is particularly striking is the moderated elasticity observed once heterogeneity is controlled for: education's influence on growth, while positive and durable, is revealed as less dramatic than pooled OLS initially suggested. This refinement underscores the necessity of contextual sensitivity in policymaking: while education is universally beneficial, its potency is conditioned by institutional quality, labor market absorptive capacity, and technological complementarities. The implication is that governments cannot assume education alone is sufficient; rather, they must construct ecosystems where educational investments translate into innovation, productive employment, and industrial upgrading. The complexity of this dynamic further suggests that policy interventions must transcend the quantitative expansion of education budgets and instead prioritize curricular relevance, alignment with technological shifts, and equitable access, without which the returns to education may stagnate or accrue disproportionately to elites.

The second set of implications arises from the positive, albeit modest, role of renewable energy consumption. The evidence that renewables enhance growth supports the proposition that ecological responsibility need not be antithetical to economic dynamism. Yet the attenuated coefficients observed in both random and fixed effects estimations point to the transitional nature of these gains. Renewable energy, while beneficial, appears to contribute incrementally rather than dramatically, perhaps reflecting infrastructural lags, high capital costs, and the entrenched dominance of carbon-intensive systems. This finding carries a dual policy message: on the one hand, states should be emboldened to pursue aggressive renewable adoption strategies, knowing that such investments yield growth dividends in addition to environmental co-benefits; on the other hand, they must recognize that renewables alone are insufficient to displace carbon dependence in the short run. Consequently, complementary measures—such as subsidies for green innovation, infrastructural modernization, and institutional mechanisms to facilitate technology transfer—are indispensable. The broader implication is that the so-called green growth paradigm must be conceived as an iterative, cumulative process requiring sustained political will rather than as a rapid, self-fulfilling trajectory.

The most disquieting implication emanates from the enduring dominance of carbon emissions as the strongest predictor of GDP per capita across all specifications. The elasticity greater than unity reveals the uncomfortable truth that, at least within the studied temporal horizon, economic expansion remains structurally wedded to carbon-intensive processes. This finding resonates with the environmental Kuznets curve debate but complicates its optimistic arc: the persistence of carbon's explanatory strength suggests that the decoupling of growth from emissions remains aspirational rather than empirically validated. For policymakers, the implication is stark: unless structural transformations are deliberately engineered, the trajectory of growth will remain environmentally unsustainable, with escalating ecological and social costs that may eventually undermine the very prosperity reflected in GDP metrics. This recognition necessitates a profound rethinking of how growth is measured and pursued. GDP, while capturing economic output, does not internalize environmental externalities; thus, reliance on GDP as the lodestar of policy risks institutionalizing perverse incentives that valorize emission-heavy expansion. The broader implication is the urgent need to embed sustainability indicators into national accounts, so that the trade-offs between prosperity and planetary stability are neither obfuscated nor postponed.

The diagnostic tests further enrich the landscape of implications. The Breusch-Pagan test results, which decisively reject the pooled OLS model in favor of random effects, confirm that cross-sectional heterogeneity is a central feature of economic performance. Similarly, the fixed effects diagnostics reveal that nearly all variance in GDP is attributable to country-specific characteristics, thereby affirming that growth is not merely a function of universal economic laws but is deeply conditioned by structural, institutional, and historical factors. This insight compels policymakers to eschew one-size-fits-all strategies and instead pursue context-sensitive development pathways that align with their unique socio-political configurations. The implication is that global policy templates—however well-intentioned—must be indigenized if they are to be effective, recognizing that growth drivers manifest differently across diverse environments.

Another critical implication pertains to the tension between short-term and long-term horizons. The results demonstrate that carbon emissions yield immediate and powerful economic dividends, whereas renewable energy and education produce smaller, albeit positive, effects. This divergence invites policymakers into a temporal dilemma: the temptation of short-term expansion through carbon-intensive pathways versus the gradualist, patient accumulation of long-term sustainability via renewables and human capital. The implication here is not merely economic but profoundly political: states must negotiate domestic constituencies, international pressures, and intergenerational justice when charting their growth trajectories. The findings thus sharpen the debate between political expediency and sustainability imperatives, revealing the difficulty of aligning electoral cycles with the slow gestation of structural transformation.

At the societal level, the implications stretch even further. The persistence of emissions as a growth determinant suggests that societies are effectively subsidizing present prosperity with future ecological degradation, thereby raising ethical questions about distributive justice across generations. The weaker immediate impact of renewables and education implies that social patience and collective vision are prerequisites for sustainable transformation, yet these are often in short supply in political economies driven by immediacy. The implication is that without cultural shifts toward valuing long-term resilience, the empirical tendencies revealed in the data will continue to reinforce unsustainable equilibria.

From a theoretical perspective, the findings implicate the limits of neoclassical growth paradigms and point toward the necessity of integrating insights from ecological economics, institutional theory, and political economy. The divergence between within-country explanatory power and between-country explanatory weakness underscores that growth dynamics are embedded in institutional and historical matrices that defy purely econometric abstraction. The implication for scholars is that interdisciplinary approaches must be embraced to more fully apprehend the complex causal web linking education, energy, emissions, and prosperity.

In totality, the implications of this study are as expansive as they are sobering. They suggest that while education and renewables represent genuine pathways toward inclusive and sustainable growth, the gravitational pull of carbon dependence continues to define the trajectory of economic expansion. Policymakers are thereby confronted with the dual task of amplifying sustainable levers while dismantling the structural entrenchment of carbon-intensive activity. At the same time, the high explanatory power of fixed effects demonstrates that context matters deeply, necessitating localized strategies and rejecting simplistic universality. The findings thus illuminate the profound complexity of contemporary development: a domain where prosperity, sustainability, and equity are inextricably intertwined, and where the path forward requires not incremental adjustments but structural reimaginings of how growth itself is conceptualized and pursued.

The choice of the REM, supported by the Hausman test results, implies that lnEDU, lnREC, and lnCO2 significantly influence lnGDP while unobserved country-specific effects are random and uncorrelated with the explanatory variables, meaning that structural, institutional, and policy-driven differences across entities do not systematically bias the estimated relationships, suggesting that investments in education remain a crucial long-term driver of economic expansion by enhancing human capital and productivity, renewable energy adoption—though statistically significant—still faces challenges in fully translating into economic growth due to infrastructure constraints and transitional costs, and the strong positive relationship between carbon emissions and GDP highlights the continued reliance on carbon-intensive industries despite global sustainability efforts, emphasizing the urgent need for a balanced approach that fosters green energy investments, strengthens policy incentives for sustainable development, and implements gradual decarbonization strategies to achieve long-term economic resilience while ensuring environmental sustainability in line with global climate goals.

CONCLUSION

This study conclusively elucidates that economic expansion within the analyzed framework is inexorably contingent upon a confluence of human capital proliferation, sustainable energy integration, and the entrenched structural dependencies on carbon-intensive economic paradigms, wherein the empirical validation through the REM substantiates the hypothesis that while education manifests as an unequivocal catalyst for long-term macroeconomic fortification, the ostensibly positive yet relatively subdued influence of renewable energy consumption delineates systemic impediments—ranging from infrastructural inadequacies to policy inertia—that stymie its full-scale assimilation into the growth architecture, concomitantly underscoring the paradox wherein carbon emissions remain inextricably intertwined with GDP augmentation, thereby necessitating a paradigmatic recalibration of developmental strategies toward a synergistic equilibrium that reconciles economic dynamism with ecological exigencies through judicious policy orchestration, incentivization of green technological innovations, and the gradual decoupling of industrial productivity from environmental degradation, ensuring a resilient, equitable, and sustainable economic trajectory aligned with global climate imperatives and long-term prosperity objectives.

REFERENCES

Abbas, Q., HongXing, Y., Ramzan, M., & Fatima, S. (2024). BRICS and the climate challenge: Navigating the role of factor productivity and institutional quality in CO2 emissions. *Environmental Science and Pollution Research International, 31*(3), 4348–4364. DOI: 10.1007/s11356-023-31321-x PMID: 38102427

Alharbi, S. S., Tabash, M. I., Farooq, U., & Issa, S. S. (2025). How does climate policy uncertainty determine green innovation adoption? New perspectives from the BRICS. *Journal of Economic Asymmetries*. Advance online publication. DOI: 10.1016/j.jeca.2025.e00411

Bhan, M., Sharma, D., Ashwin, A. S., & Mehra, S. (2017). Policy forum: Nationally-determined climate commitments of the BRICS: At the forefront of forestry-based climate change mitigation. *Forest Policy and Economics, 85*, 172–175. DOI: 10.1016/j.forpol.2017.09.013

Bhardwaj, M., Prakash, N., & Malhotra, R. K. (2024). Energy Sector and Carbon Trading in the Future: A Reference to India. *2024 International Conference on Smart Devices, ICSD 2024*. https://doi.org/DOI: 10.1109/ICSD60021.2024.10751388

Gbadegesin, T. K. (2024). Climate change and energy dynamics: Evidence from BRICS and G7 countries. *Energy & Environment*. Advance online publication. DOI: 10.1177/0958305X241302692

Kaur, J., Khanna, R., Kumar, R., & Sunil, G. (2024). Role of Blockchain Technologies in Goods and Services Tax. *Proceedings - 2024 3rd International Conference on Sentiment Analysis and Deep Learning, ICSADL 2024*, 607–612. https://doi.org/DOI: 10.1109/ICSADL61749.2024.00104

Khanna, R., Sharma, S., Czeh, T., & Adhana, K. (2025). Challenges, Interventions, and Strategies for Higher Education Transformation: Vision 2047. *Lecture Notes in Networks and Systems, 1075 LNNS*, 637 – 655. https://doi.org/DOI: 10.1007/978-981-97-6106-7_38

Khorbaladze, E. (2025). BRICS: A comprehensive mechanism bringing changes into the world financial system and international climate diplomacy. In *The Multilateralism of the New Development Bank on the Sustainable Economic Growth in BRICS Nations and Beyond*. Emerald Publishing., DOI: 10.1108/978-1-83662-292-520251006

Kukreti, M., Sehajpal, S., Tiwari, R., & Sood, K. (2024a). Driving global health and sustainable development goals with smart technology. In *Driving Global Health and Sustainable Development Goals With Smart Technology*. IGI Global., DOI: 10.4018/979-8-3373-0240-9

Kukreti, M., Sehajpal, S., Tiwari, R., & Sood, K. (2024b). Introduction. In *Driving Global Health and Sustainable Development Goals With Smart Technology*. IGI Global.

Kukreti, M., Sehajpal, S., Tiwari, R., & Sood, K. (2024c). Preface. In *Driving Global Health and Sustainable Development Goals With Smart Technology*. IGI Global.

Lavrik, M., & Saurombe, A. (2025). BRICS and Climate Change Law: An Opportunity to Save the Planet. *Potchefstroom Electronic Law Journal*, *28*, 1–51. DOI: 10.17159/1727-3781/2025/v28i0a17634

Leal-Arcas, R., Al Zarkani, M., Jbara, L., Mubwana, R. M., Margaritidou, M., & van der Berg, A. (2022). Climate change mitigation law and policy in the BRICS. In *Research Handbook on Climate Change Mitigation Law*. Edward Elgar Publishing Ltd.

Liu, X., Yuan, X., Ge, X., & Jin, Z. (2024). Adaptation and innovation: How does climate vulnerability shapes corporate green innovation in BRICS. *International Review of Financial Analysis*, *94*. Advance online publication. DOI: 10.1016/j.irfa.2024.103272

Malhotra, R. K., Sharma, C. A. P., Upreti, H., Sharma, S., Prakash, N., & Mittal, S. (2024). Role of Psychological Contract in Employee Engagement in Higher Education: Empirical Study on Uttarakhand. In P. P.K. (Ed.), *E3S Web of Conferences* (Vol. 556). EDP Sciences. https://doi.org/DOI: 10.1051/e3sconf/202455601030

Martins, M. A. B. (2019). The BRICS commitment on climate change: process towards an effective approach in the path of sustainable development. *Contributions to Economics*, 175 – 187. https://doi.org/DOI: 10.1007/978-3-030-02662-2_9

Nawaz, M. A., Seshadri, U., Kumar, P., Aqdas, R., Patwary, A. K., & Riaz, M. (2021). Nexus between green finance and climate change mitigation in N-11 and BRICS countries: Empirical estimation through difference in differences (DID) approach. *Environmental Science and Pollution Research International*, *28*(6), 6504–6519. DOI: 10.1007/s11356-020-10920-y PMID: 32997248

Özen, E., Singh, A., Taneja, S., Rajaram, R., & Davim, J. P. (2024). *Innovating Sustainability Through Digital Circular Economy*. IGI Global., DOI: 10.4018/979-8-3373-0578-3

Sæther, S. R. (2021). Climate policy choices: An empirical study of the effects on the OECD and BRICS power sector emission intensity. *Economic Analysis and Policy*, *71*, 499–515. DOI: 10.1016/j.eap.2021.06.011

Saqib, A., Hussain, I., & Mefteh-Wali, S. (2024). Do stock returns respond to physical and transition climate risks? Evidence from emerging BRICS economies. *Journal of Environmental Management*, *372*. Advance online publication. DOI: 10.1016/j.jenvman.2024.123303 PMID: 39561456

Sehajpal, S., & Sidhu, N. K. (2024). Neuroscience, nursing, and financial innovation: A synergistic approach to optimized patient care and economic well-being. In *Neuromarketing's Role in Sustainable Finance*. IGI Global., DOI: 10.4018/979-8-3693-9117-4.ch008

Shah, M. I., Khan, Z., Moise, M. L., & Abbas, S. (2022). Tourism adaptability amid the climate change and air pollution in BRICS: A method of moments quantile regression approach. *Environmental Science and Pollution Research International*, *29*(57), 86744–86758. DOI: 10.1007/s11356-022-21725-6 PMID: 35794334

Sharma, S., Gupta, C., Khanna Malhotra, R., & Upreti, H. (2024). Sustainable Banking Practices: Impact, challenges and opportunities. In P. P.K. (Ed.), *E3S Web of Conferences* (Vol. 556). EDP Sciences. https://doi.org/DOI: 10.1051/e3sconf/202455601031

Singh, A., Kumar, P., & Taneja, S. (2024). Artificial Intelligence and Machine Learning-Powered Smart Finance. In Taneja, P. K. S., & Singh, A. (Eds.), *IGI Global*. IGI Global., DOI: 10.4018/979-8-3693-3264-1

TANEJA. S., KUMAR, P., REEPU, KUKRETI, M., & OZEN, E. ̈. (2024). *Data Alchemy in the Insurance Industry*. Emerald Publishing Limited. https://doi.org/https://doi.org/10.1108/978-1-83608-582-920241033

Taneja, S., Gupta, S., Singh, A., Bhatnagar, M., & Gaur, A. (2023). Green management - A new paradigm in the world of business. *Green Management - A New Paradigm in the World of Business*, 1–230. https://doi.org/DOI: 10.52305/DLIH6622

Tshikovhi, N., Santos, A., Zou, X., Netswera, F., Yarygina, I. Z., & Divi, S. (2023). COVID-19 and Climate Change in BRICS Nations: Beyond the Paris Agreement and Agenda 2030. In *COVID-19 and Climate Change in BRICS Nations: Beyond the Paris Agreement and Agenda 2030*. Taylor and Francis., DOI: 10.4324/9781032643182

Udeagha, M. C., & Ngepah, N. (2023). Towards climate action and UN sustainable development goals in BRICS economies: Do export diversification, fiscal decentralisation and environmental innovation matter? *International Journal of Urban Sustainable Development*, *15*(1), 172–200. DOI: 10.1080/19463138.2023.2222264

ENDNOTES

1 According to World Bank Database, "GDP per capita is gross domestic product divided by midyear population. GDP is the sum of gross value added by all resident producers in the economy plus any product taxes and minus any subsidies not included in the value of the products. It is calculated without making deductions for depreciation of fabricated assets or for depletion and degradation of natural resources. Data are in constant 2015 U.S. dollars."

2 According to World Bank Database, "General government expenditure on education (current, capital, and transfers) is expressed as a percentage of GDP. It includes expenditure funded by transfers from international sources to government. General government usually refers to local, regional and central governments."

3 According to World Bank Database, "Renewable energy consumption is the share of renewable energy in total final energy consumption."

4 Carbon dioxide emissions are those stemming from the burning of fossil fuels and the manufacture of cement. They include carbon dioxide produced during consumption of solid, liquid, and gas fuels and gas flaring.

Chapter 4
Foundations of Green Human Resource Management (GHRM)

Abdul Ghafoor
Institute of Business Management Sciences, Pakistan

Asia Rafiq
https://orcid.org/0000-0002-9751-3296
Institute of Business Management Sciences, Pakistan

ABSTRACT

While organizations struggle with the intensifying effects of climate change, incorporating sustainability into fundamental business functions has become essential. GHRM then becomes a key framework for aligning human capital with sustainable objectives, developing sustainable culture, and promoting organizational resilience. This chapter discusses the premises of GHRM, exploring its theoretical bases, essential elements, and its central role in driving climate change mitigation and resilience initiatives. Based on theoretical lenses (different theories) GHRM covers the conventional practices such as recruitment, training, performance management, employee compensation and occupational health and safety from a climate resilience approach perspective. The chapter uses Patagonia and Unilever among other good businesses to show how GHRM may drive environmental performance, employee engagement, and corporate resilience. The chapter concludes by describing the road map for GHRM in terms of cooperation, employee empowerment, and link to other CSR activities.

DOI: 10.4018/979-8-3373-2479-1.ch004

INTRODUCTION

Defining Green Human Resource Management (GHRM)

Sustainable HR policies and practices designed to create a culture of sustainability inside a business is what GHRM means. GHRM aims to help an organization's environmental goals by means of several HR operations including recruitment, training, performance management, payment, and staff engagement. By means of GHRM, companies can therefore enable their staff to become proactive members in corporate sustainability initiatives, therefore raising environmental and company performance.

GHRM is the human capital development process that underpins environmental sustainability, according to (Renwick, Redman, and Maguire 2013). This process turns sustainability into a rooted practice that influences company culture, employee behavior, and business operations, not just a policy. Companies using GHRM will find improved resource efficiency, reduced waste, higher staff morale, and better compliance of environmental laws.

How GHRM Differs From Traditional HRM

Green Human Resource Management (GHRM) differs from typical HRM by integrating environmental goals into the primary operations of human resources. Whereas conventional HRM mostly aims to increase employee satisfaction, align human capital with organizational goals, and maximize workforce efficiency, GHRM expands this focus to include environmental sustainability as a main objective. Stated differently, GHRM is about managing individuals in manners that support and promote environmental sustainability, not only people management.

Conventional HRM emphasizes performance indicators, compliance, and organizational efficiency without always considering the environmental impact of policies on human resources. GHRM differs in that it marries environmental values with every HR process. For example, green recruitment looks for people who value sustainability and can assist the business in accomplishing its environmental goals. Green training and development provide employees the resources to follow sustainable practices and knowledge on environmental issues. Green performance management likewise considers staff members' contributions to sustainability projects as well as their job appropriate skills.

Green employee engagement, which promotes a culture of environmental awareness and mutual accountability, and green job design, where roles are organized to minimize environmental impact, are other GHRM techniques. Together, these methods help to create a workforce that is environmentally aware and dedicated to sustainable growth, therefore GHRM is a key instrument in pushing corporate environmental accountability.

Historical Evolution of GHRM

1. Pre1980: First Roots of Environmental Awareness in HRM

Before the 1980s, Human Resource Management (HRM) was mostly concerned with maximizing organizational efficiency and had almost no environmental worries integration.

Particularly with the emergence of the environmental movement and the creation of major environmental legislation (e.g., the Environmental Protection Agency in 1970 in the United States), environmental awareness started to show in the 1970s.

During this time, businesses began to solve environmental problems mostly by means of operational rather than HRM techniques.

2. 1980s1990s: Corporate Environmental Management Integration.

Businesses started to see in the 1980s that corporate strategies could include environmental concerns. It was prompted by sustainability's conceptual development.

HRM began by the late 1980s and early 1990s to move more towards a proactive approach to environmental damage management. Still, though, the emphasis remained mostly on adherence to environmental regulations as opposed to advancing sustainability at a strategic level.

Finally, the idea of "Corporate Social Responsibility" (CSR) began to take off during this time, further affecting HR policies to take ethical and environmental concerns into account in corporate operations.

3. Emergence of Green Human Resource Management: A Modern Development.

Green HRM (GHRM) was established as a rising discipline inside HRM in the early 2000s. Practices of environmental sustainability started to be woven into businesses' human resources policies including in recruitment, training, and performance management (Renwick et al., 2013).

One major advance was the release of papers connecting sustainability with HR policies, underlining how human resources might contribute to the development of a corporate culture of environmental responsibility.

The publication of the Triple Bottom Line idea (Elkington, 1997)—which considers people, planet, and profit—also reinforced the need for HRM to sponsor sustainability initiatives.

Regarding 2008 research, Jabbour & Santos (2008): discussed how HRM policies could aid in sustainable development and organizational climate adaptation.

4. 2010s: GHRM in the Environment of Resilience and Climate Change

Resilience in the face of climate related challenges grew in significance as worldwide awareness of climate change rose. Especially in context of how human resources might help to improve a company's flexibility to climate risks, GHRM developed to cover not only environmental protection but also climate resilience.

Green HRM approaches started to clearly feature initiatives including guide for employees to embrace sustainable acts (e.g., energy conservation), training for staff on environmental dangers, and inclusion of sustainability objectives into the organizational culture (Jabbour &Santos, 2008).

Research on green HRM gave us the idea of aligning HR policies with corporate social responsibility goals and environmental sustainability objectives, which laid the groundwork for its more general application in climate resilience (Renwick et al. 2013).

5. Growing Role of GHRM in Climate Resilience: the 2020s and Thereafter.

Other international agreements including the Paris Agreement emphasized the need on organization adjustment to climate threats, while in the 2020s the urgency of response to climate change became clearer.

In reaction to impacts of climate change, GHRM approaches now include policies for building employee resilience. This includes developing employees' skills for working in a sustainable economy, fostering an attitude of flexibility, and providing industry-specific climate change courses.

Amongst environmental disruptions, Ekechukwu and Simpa (2024) noted that the incorporation of climate resilience into HRM policies and the development of climate aware leadership were essential components of organizational success.

Rising environmental consciousness and climate resiliency have shaped GHRM's growth together changing HR policies meant to promote sustainability and flexibility. Originally oriented toward compliance with environmental regulations, HR has grown to become a significant influence in building organizational resilience against climate change effects so that workers are ready to achieve long-term sustainability goals.

Theoretical Foundations of GHRM

The embedding of sustainability in HRM depends on several theoretical models that define how and why companies develop eco-friendly HR policies.

Resource-Based View (RBV)

The Resource-Based View (RBV) posits that a firm's sustainable competitive advantage is derived from its unique resources and capabilities, which are valuable, rare, inimitable, and non-substitutable (Barney, 1991). One of the key among these resources is human capital, which is most significant in fostering competitive advantage, particularly when it comes to sustainability. Applying RBV to GHRM shows how strategic investment in human capital enables firms to achieve environmental sustainability without affecting business competitiveness.

Human capital, being one of the key organizational assets, encompasses employees' know-how, competence, capabilities, and motivation, all in harmony with each other in advancing sustainability initiatives. Properly equipped staff with competency in sustainability—like ecological awareness, energy efficiency expertise, and reducing wasteful practices—make sure that organizations reap long-term benefits. With the adoption of GHRM management approaches companies can then use their human capital to graft sustainability onto the company rather than just using it as an extra within the core business model.

One of the best examples of GHRM - in practice -- is the Sustainable Living Plan at Unilever, which incorporates sustainability training throughout the organization. Unilever has invested a lot of money in green training programs aimed at making employees aware of the skills and knowledge needed to adopt environmentally sustainable approaches in their work. Through environmental responsibility leadership courses, environmental e-learning modules, and sustainability workshops, the company has been able to reduce its environmental impact while optimizing productivity as well as innovation, a clear, strategic investment in human capital and which has enabled Unilever to be competitive in the global market.

Not only will implementing GHRM save money on operating costs; companies that incorporate sustainability strategies into their HRM strategy also enjoy higher human resource retention rates, more highly-motivated and engaged employees. As per the World Economic Forum report (2021) an expected 70% of employees would prefer to work for an organization that exemplifies a high degree of sustainability commitment to them. If an organization embeds sustainability into its overall HRM strategy, it may be rewarded with higher employee reputation, lower employee turnover, and higher employee loyalty. Workers who consider themselves a part of the company's sustainability vision are more productive, motivated, and motivated, contributing more to the company's long-term competitive advantage.

In conclusion, the Resource-Based View focuses on the key position of human capital in obtaining and sustaining competitive advantage via Green Human Resource Management. By strategic investment in employees' sustainability competencies, organizations are able to enhance innovation, enhance operational efficiency, and solidify their market standing. Organizations integrating sustainability into HRM frameworks contribute not only to environmental protection but also to earning a long-run advantage that their competitors find hard to replicate. As businesses walk through an increasingly sustainability-conscious international economy, utilization of human resources as a major asset through the instrument of GHRM will be key in providing economic as well as ecological success.

Stakeholder Theory

Freeman's (1999) Stakeholder Theory highlights the importance of altering the interests of various stakeholders such as employees, customers, investors, regulators, and communities. Businesses that incorporate environmental sustainability into their strategic mainstreams can build encourage commitment, stakeholder confidence, and eventually boost their status and financial performance. GHRM is vital in ensuring that HR policies allowing employees to be actively involved in, support sustainability goals and contribute to green activities. By integrating sustainability into HR practices, companies can drive long-term business success as well as increase stakeholder relationships as well as. Employees, being the most significant internal stakeholders, are dominant to executing an organization's sustainability agenda.

By integrating sustainability into training, performance management, and employee engagement, HRM endorses a workforce that is not only devoted to corporate sustainability objectives but also environmentally aware. Workers are more engaged and motivated if their job is aligned with substantial environmental goals. For example, one Harvard Business Review study publicized that firms with vigorous stakeholder engagement in sustainability practices had 30% more satisfied employees than their competitors lacking conclusive green strategies(Paine, 2023). This proposes the real advantages of incorporating stakeholder issues into HRM practice, such as enhanced productivity, heightened employee loyalty, and better organizational performance. A well-known example of stakeholder-led sustainability is Patagonia, an organization well-known for its extreme passion for environmental protection.

Patagonia vigorously engages its employees in sustainability initiatives by proposing paid leave for volunteering in environmental projects, thereby aligning workforce activities with its corporate sustainability mission. The firm also donates 1% of its total sales to environmental organizations, strengthening its commitment to green business practices. This attitude not only strengthens Patagonia's brand image but also improves employee engagement, motivation, and loyalty. The synergy with sustainability Patagonia effectively embeds stakeholders' expectations into its business model and ultimately has a competitive advantage in the marketplace. Other stakeholder groups, apart from employees, also place greater emphasis on sustainability, such as investors, customers, and regulators.

In addition, consumers are increasingly buying products from companies that they trust and believe in their respect for social responsibility as well as environmental responsibility. Another factor affecting the financial investment decisions of investors is the consideration of ESG (Environmental, Social and Governance) topics in the financial decision process as they believe organizations that focus on sustainability will tend to be more profitable and sustainable over the long term. Companies that invest carefully to incorporate sustainability into key HRM activities – such as green hiring, sustainability-focused leadership development, and sustainability practices that align with business practices – will be well-positioned to attract ethical investments and become financially sound over the long run..

Ignoring stakeholders' interests in environmental sustainability may incur major threats to firms in terms of legal fines, reputation, and lost customer loyalty. International regulatory authorities are setting increasingly tight environmental principles, and offending businesses risk facing economic and legal penalties. Also, firms that ignore including stakeholders in sustainability may drop their competitive edge as consumers and investors transmit their loyalty towards more responsible companies.

In conclusion, Stakeholder Theory recaps us of the need to regularize and serve the different interests of internal as well as external stakeholders to vintage long-term business success.

Institutional Theory

Institutional Theory presumes that firms are driven by outside pressures like government policy, industry norms, and societal pressures (DiMaggio & Powell, 1983). These outside pressures condition business strategies and stimulate the embracement of green business practices, such as Green Human Resource Management (GHRM). Firms that implement GHRM practices do not just aim to promote internal effectiveness but also to obey environmental legislation, fulfill stakeholder expectations, and enhance corporate legitimacy. With the increase in sustainability issues, firms are obligated to reconcile their HR policies with institutional pressures in order to remain competitive and socially responsible.

GHRM practices have been adopted by several multinational companies, including IKEA and Nestlé, to help them ensure that they comply with international environmental and industry standards as well. IKEA's sustainability policy has included environmental concerns into its supply chain management (see Supply Chain Environmental Control), requiring suppliers to meet rigorous environmental norms. Because of that, IKEA's HRM department trains employees and suppliers on sustainability, in order to ensure that they are in compliance with international environmental norms. Nestlé has developed an expansive sustainability framework, which includes employee training, recycling campaigns, and accountable sourcing standards. These activities are not only in order to get such companies into regulatory compliance but also help to improve the image of these companies as sustainable businesses.

Institutional pressures also create demand for change in the industry larger than the individual company as those companies that excel in sustainability create new paradigms by which other companies can follow. Companies that lead the charge for making the environment a priority can set industry standards, shape public policy and emerge as pioneers in green business.

Institutional Theory stresses the importance of external 'pressure' on corporate sustainability strategies. Through implementation of GHRM strategies in line with government legislation and industry standards and customer/employee expectations, organizations can fulfill regulatory compliance, maintain superior corporate reputation and benefit from being competitive. Organizations that actively integrate sustainability into HRM not only meet institutional demand but are also in a position to secure long-term success, backed by institutional concerns that will serve them well in an increasingly environmentally aware business environment.

One example of investing human capital in sustainability is Google's extensive employee sustainability training programs. This company conducts a range of employee sustainability education and workshops, as well as host different workplace sustainability competitions like the Green Employee Engagement competition. These programs help to foster green thinking in the workplace and also help Google to maintain its corporate sustainability leadership status. Through the inclusion of sustainability in employees development plans Google can embed sustainability into the business and encourage innovation and accountability in the sustainability effort.

Not leveraging the green skills available to them can hurt their ability to compete effectively, especially as business models are shifting toward a green economy. From a global labor supply point of view, the reality is that workers with sustainability education will find themselves in high demand. As the International Labour Organization (ILO) recently concluded, this shift towards a sustainable economy will generate 24 million new jobs by 2030, which highlights the importance of sustainability competencies at the level of the labor market. Undeveloped human capital will simply not have enough the capacity to ride out the next wave in a race towards economically sustainable growth.

Human Capital Theory

Human Capital Theory assumes that employees are assets that bring value through their knowledge, skills, and competencies that help an organization succeed in the long term (Becker, 1964). In the case of GHRM, investing in green training and development helps improve employees' capacity to implement sustainability initiatives, preparing organizations for economic and environmental success. By empowering workers with the skills in sustainability practices, organizations are able to enhance operational effectiveness, minimize waste, and develop a culture of environmental stewardship.

Green training initiatives are essential in cultivating human capital that supports corporate sustainability objectives. Studies show that companies with strong sustainability training programs have higher levels of productivity, higher levels of employee satisfaction, and better financial performance. Well-trained employees who understand environmental conservation practices are able to support sustainability, resulting in cost savings, innovation, and a better reputation for the corporation. In addition, employee training in green practices assists businesses in maintaining compliance with changing environmental regulations, reducing the legal and monetary risks of non-compliance.

Behavioral Theories of Sustainability

The Theory of Planned Behavior (Ajzen, 1991) posits that people's attitude, subjective norm, and perception of behavior control their action. In Green Human Resource Management (GHRM), integration of green values within HRM policy can define behaviors of the workers, support ecological responsibility, and encourage sustainable organization practices. Organizations can nurture more environmentally active and responsible human resource by actually communicating green drives under leadership vision, employee commitment, and incentive pay.

Organizations that embed sustainability into their corporate culture can shape the behavior of employees through specific HR practices like sustainability training, green rewards tied to performance, and environmentally friendly workplace policies. Workers are more likely to internalize and enforce sustainable behaviors if sustainable practices are valued and encouraged. They have shown that companies that adopt sustainability-based HRM policies experience higher levels of employee involvement in green initiatives, higher job satisfaction and greater organization commitment.

One of the most known examples of behavioral sustainability is Tesla, a company that makes it a goal of its employees to work on innovative new solutions to reduce carbon emissions and invent technologies in renewable energy. Tesla's HRM practices explicitly prioritize employees that will make a difference for sustainability such as making the batteries more efficient, providing more charging facilities, and creating alternative energy solutions. Such strategic HRM integration helped boost employee motivation not only at Tesla, but also it helped build Tesla's reputation as a leader in electric vehicles, an example of how behavioral theories can make large-scale environmental change happen.

Furthermore, by employing behavioral sustainability strategies, companies can create an adaptive and resilient workforce in today's increasingly environmental conscious business environment. As stricter environmental regulations regulate operations, and consumer preference shifts toward sustainable products and services, businesses that actively affect their employees' behavior in favor of sustainability gain a competitive edge. Businesses that adopt behavioral sustainability standards effectively will benefit from improved company image and more effective contributing towards sustainability of both environment and bottom line.

Finally, Behavioral Theories of Sustainability emphasize the importance of HRM for influential behavior change and attitudes towards environmental accountability. Through behavioral knowledge and sustainability integration in HRM policies, organizations can engage, motivate, and corporately sustainability-focused staff; through commitment to sustainability through leadership, employee rewards, and cultural reinforcement, organizations can influence sustainable behavioral change to achieve long-term sustainability.

GHRM: PRACTICES, IMPLEMENTATION, AND SOLUTIONS

Key Dimensions of GHRM

In its implementation, a number of dimensions are dealt with that address different dimensions of workforce sustainability:

1. **Green Recruitment and Selection**: Recruiting environmentally-conscious and green-ready employees who possess the necessary skills to contribute to green initiatives.

2. **Green Training and Development (GRD)**: (Related – green manager skills) Develop and impart competencies for employees to limit their environmental impact.
3. **Green Performance Management**: Monitoring and reward employees for their performance towards sustainable initiatives.
4. **Green Award/Compensation**: For recognizing/promotion of employers who make a proactive contribution to the environmental cause.
5. **Green Employee Engagement and Participation**: Promoting employees' active involvement in sustainability initiatives through environmentally friendly management policies and corporate practices.
6. **Green Workplace Policies**: Setting up work policies that will reduce employee carbon footprints, minimize waste, and increase energy efficiency.

All of these areas will be discussed in depth in this chapter to demonstrate how organizations can incorporate sustainability into HRM to produce a positive environmental impact.

The Growing Importance of Environmental Sustainability in Business

The current global business landscape features environmental sustainability as a missing puzzle piece which requires urgent attention. Climate change, depletion of resources, pollution, and loss of biodiversity are prominent environmental issues that have subjected organizations to a lot of pressure to transform their unsustainable practices into sustainable ones. Governments, consumers, investors, and workers are seeking more corporate responsibility for safeguarding the environment. As a result, companies need to incorporate sustainability into their businesses, not just for regulatory compliance but also to address stakeholders' demands and ensure competitiveness within a rapidly changing economic environment.

Transformation towards sustainability is not only a regulatory requirement but also a strategic requirement. Organizations which act early and adopt sustainable practices are often found to have greater brand loyalty, profitability, and stakeholder confidence. Over 90% of CEOs feel that sustainability is the key to their business's success (United Nations Global Compact, 2022). Firms which neglect sustainability will become market-irrelevant and face legal and reputation issue

Key Components of GHRM

Green Recruitment and Selection

Green recruitment-practicing organizations entice environmentally aware talent by including sustainability skills in job adverts and hiring criteria. By guaranteeing that new employees are in line with company sustainability objectives, businesses construct an employee base that actively participates in environmental programs.

For instance, Unilever has collaborated with universities to hire graduates specializing in sustainable business practices. The firm incorporates questions related to sustainability during interviews, evaluating candidates' knowledge of climate change, reducing carbon footprints, and green business operations. Furthermore, recruitment marketing efforts emphasize Unilever's dedication to sustainability, appealing to job hunters who value environmental responsibility in their professional choices.

Another example is IKEA, which obviously has a social conscience - candidates for management positions are asked to show they have an understanding of the company's vision of sustainability and how they can contribute to reducing the company's carbon footprint.

Moreover, companies like Tesla and Siemens have already started using AI-powered recruitment software to assess a candidate's sustainability culture. Advertisement tracking systems using AI-based technology will check whether a candidate has a history of professional projects and projects that relate to environmental activities. Businesses are also actively using social media and online websites to market their sustainability initiatives. Companies on platforms like LinkedIn can showcase their corporate social responsibility (CSR) projects.

Companies that utilize green hiring strategies not only attract environmentally-conscious employees, but also improve their company image. A study by the Society for Human Resource Management (SHRM) found that 74% of candidates consider whether a company's sustainability practices match their personal and corporate values when making their first hire. By becoming green employers, companies are better able to secure top talent and stay on track with their corporate sustainability goals.

Green Training and Development

Integrating environmental responsibility into business operations requires training that is important to sustainability. Improving employee engagement, operational effectiveness and compliance with relevant legal requirements all benefit from investing in training staff on environmentally sustainable practices.

(IKEA and Siemens are two examples of companies that provide employees with on-going sustainability training, which incorporates sustainability topics like sustainable sourcing, minimizing waste and saving energy.) Google, on the other hand, has included hands-on sustainable training in its on-boarding process, which makes sure that every employee knows the environmental policies of the company from day one.

Furthermore, Nestlé also offers an online sustainability learning platform, where workers are able to take courses on climate change, circular economy and sustainable agriculture, in order to become more actively involved in the company's environmental goals by considering sustainability in their decision-making.

In order to involve staff in environmental training companies like Microsoft use gamification. In order to achieve some sustainability goals like reducing paper consumption or using less fossil fuels for transportation staff will get points and incentives.

In addition, when businesses partner with organizations such as the World Wildlife Fund (WWF) or Greenpeace, employees can gain real-world experience that helps to foster a deeper passion for sustainability. Research indicates that businesses with strong green learning programs see higher levels of worker satisfaction and higher levels of productivity. Workers who have a sense of sustainability's importance are more likely to be action-oriented in cutting carbon emissions, which means saving money while improving environmental performance.

Green Performance Management

Competitive performance management systems now rely heavily on Green Key Performance Indicators (GPIs), which measure and reward performance against sustainability targets to maintain environmental alignment between business and individual initiatives.

A good example is the way Tesla awards staff for their efforts to reduce carbon emissions, improve energy efficiency, and create sustainable ideas. Some of the credit given for research and development staff involves creating environmentally friendly battery technologies and recycling manufacturing waste.

Patagonia is another illustration of how employee performance reviews are related to sustainability projects. The organization judges staff members according on how much they get into conservation projects, volunteer work, and adoption of environmentally friendly office rules. Performance based rewards and promotions go to staff members actively supporting sustainability initiatives.

Companies using green performance management systems sometimes use environmental impact scorecards to monitor data including waste minimization, energy savings, and sustainable sourcing compliance. Annual performance assessments include these metrics to help corporate culture emphasize sustainability.

With programs giving immediate feedback and performance analysis, corporations are as using real time monitoring tools to follow staff members' contributions to environmental objectives. A International Journal of Human Resource Management analysis finds that companies using performance reviews with sustainability ratings see a 20% rise in staff participation in green projects. Greater adherence to company sustainability aims results from employees feeling more responsible for their ecological impact.

Green Compensation and Benefits

Organizations offer financial and nonmonetary rewards to inspire staff members to participate in ecological activities. At the office as well as in their own life, green compensation packages help employees to embrace sustainable habits.

A classic illustration of a business including environmental advantages into their compensation scheme is Patagonia. Using public transportation, bicycling to work, or buying electric vehicles earns employees monetary rewards. The company also provides paid time off for staff members who volunteer in environmental preservation projects.

Similarly, Apple offers assistance for staff using solar panels at home or shifting to electric cars. In the company's green initiative, sustainable home improvement reimbursements help to make sustainability a daily part of staff members' lives.

For cutting their carbon impact, some businesses offer carbon credit awards whereby staff members earn credits. One can trade these credits for more vacation time, health benefits, or financial compensation. This strategy not only encourages workers but also underlines corporate sustainability promises.

Corporate are also rolling out flexible benefits packages that let workers apportion some money for environmentally friendly living choices including sustainable travel alternatives or organic food subscriptions. Deloitte's sustainability report says that businesses offering green benefits retain more personnel and have better employee job content. Since their environmental efforts are appreciated, workers are more productive and loyal.

Green Employee Engagement and Participation

Initiatives like Green Teams under employee engagement inspire energetic involvement in environmental programs. Businesses whose staff participates in environmental projects find benefits for greater creativity, more effective collaboration, and a more meaningful life.

Google has, for example, set up an inside sustainable task force working with management to enact environmentally friendly office policies. Staff members present ideas to enhance energy efficiency and lower waste during environmental brainstorming sessions.

General Electric (GE) has launched a staff driven sustainability initiative in which teams vie to create creative green solutions. Senior management helps provide money and help with execution for winning ideas.

At sustainability hackathons, some groups let staff collaborate to create environmentally friendly solutions for company problems. Such events motivate staff to provide innovative ideas for sustainability enhancements through cross functional cooperation.

Furthermore, company volunteering projects let workers engage in wildlife preservation efforts, neighborhood cleanups, and tree planting campaigns. The Harvard Business Review says that businesses with robust employee engagement in sustainability activities claim thirty five percent higher productivity levels than those without environmentally friendly initiatives. Working on sustainability initiatives increases engagement with the mission of one's employer, which in turn improves job happiness and general office morale (Paine, 2023).

Green Workplace Policies and Practices

Green workplace policies can help businesses to effectively lessen their environmental impact. Waste segregation, energy efficient illumination, and paperless offices—all modest but effective actions—help to meet corporate sustainability targets at the same time they lower operating expenses.

Microsoft has, for instance, adopted strict environmentally friendly office rules that have cut energy use by 20%. To reduce commuting emissions and office energy usage, the business promotes flexible hours and working from home.

Many companies, including HP and IBM, have embraced zero waste strategies in support of recycling and the removal of single use plastics in the office. Giving staff members reusable office materials, prohibiting trash coffee cups, and running composting initiatives are among these rules.

Green IT policies developed by businesses like Adobe and SAP maximize data center energy consumption and encourage environmentally friendly digital solutions. Virtualization and cloud computing have improved performance.

According to research, firms with clearly defined green workplace policies achieve great cost savings while improving their business reputation. Companies using energy efficient policies save about 15% every year on operating costs finds a Carbon Trust research.

Why GHRM Matters: The Strategic Role of HR in Sustainability

Human Resource Management (HRM) has a crucial function in creating organizational culture and the conduct of employees. HRM has conventionally been aimed at maximizing workforce productivity, talent management, and labor relations. With companies moving toward sustainability, HR departments need to go beyond their roles to encompass environmental responsibility.

HR practitioners are best suited to drive sustainability across all levels of an organization. They can:

- Hire and train employees who share the company's sustainability values.
- Put policies in place that encourage green behavior within the workplace.
- Monitor and reward employees for their contribution to sustainability efforts.
- Promote organization culture based on long-term ecological planning rather than short-term profit.
- Maintain compliance with environmental laws and sustainability practices.

74 per cent of HR heads say they are 'involved' in decision-making on sustainability in their organizations, according to a survey by the Chartered Institute of Personnel and Development (CIPD, 2021). The incorporation of green values into the processes of HRM is one way organizations can create an environment-friendly workforce.

Correlation between GDPR, ethics management systems and corporate social responsibility. Player as an important player in the alignment of a company's human resource policies with its sustainability and environmental objectives, Green Human Resource Management involves Incorporation of ethics management systems and Corporate Social Responsibility (CSR) becomes very important as businesses increasingly attempt to be sustainable. This part looks at how GHRM, ethics management systems, and CSR relate and how they contribute to shaping a company's greater environment and social influence.

GHRM and Ethics Managing Systems

With the help of ethics management systems (EMS) policies, organizations adhere to ethical norms and principles on a day-to-day basis and create an environment for dealing with the ethical aspects of their business activities. Environmental ethics through the incorporation of environmental ethics into human resource policies ensures the management's accountability towards ethics and in the process ensures that the moral issues surrounding sustainability and environmental impact are included in the corporate culture and behavior of the employees.

GHRM helps corporate ethicalness by thinking about ethical behaviour in the workplace but:

The responsibility of GHRM to have staff on the side who are environment friendly is included in selection of candidates who meet the companies' sustainability criteria.

Knowledge and development Training and improvement of a culture of ethical behaviour with respect to the environment is necessary, as is training courses on sustainability and ethical decision making. Education in social responsibility, resource management and climate change will help staff to make moral decisions that are in line with sustainability goals.

Performance management: Ethical and sustainability performance can be integrated into the employee evaluation process, to reward actions that contribute to environmental conservation, ethical business practices and corporate social responsibility.

By embedding GHRM as part of ethics management systems, organizations can develop a coherent approach to promoting ethical behaviour in a way that encourages environmental sustainability as well–a synergy that assists organizations to achieve sustainability objectives and ethical performance simultaneously.

Limitations of Traditional Corporate Social Responsibility (CSR) Initiatives

Classic corporate social responsibility practices that mostly focus on philanthropy, environmental conservation, and good business practices no longer have the mandate to solve today's sustainability problems. CSR programs can improve corporate reputations by raising public awareness and promoting better practices, but most companies consider CSR only an add-on to their overall corporate strategy. This has led to the growing complaint of "greenwashing," in which companies take greater pride in their advocacy for sustainability and spend less on core business operations.

To go beyond the superficial sustainability efforts, organizations need to create a culture of sustainability across the organization. This includes embedding sustainability as part of corporate strategies, processes, operations, supply chain management and most importantly human resource management. As a result, GHRM has emerged as a highly integrated approach to aligning HRM practices with sustainability efforts.

Human resource management and corporate social responsibility (CSR)

Corporate social responsibility refers to an organization's commitment to a greater good, encompassing the social, environmental and financial consequences of its actions. (Cone, 2017) CSR is a broad scale approach for companies, which requires them to accept accountability for the environmental, social and community consequences of their activities.

As it links human resource policies with organization sustainability activities, GHRM is an important driver of CSR. Companies with connection of GHRM with CSR have the opportunity to:

Encourage a sustainable mindset: GHRM assures that CSR is part of the company's core culture rather than just something that's done at company level. With a focus on environmental stewardship within the HR environment, a business can foster the feeling that sustainability is something they can and should practice on a day to day basis both in the office and out.

Better employee engagement and loyalty: Employees who work with CSR-related activities (i. e. those related to environmental sustainability) may feel closer to their company's social and ethical goals. This could increase job satisfaction, staff retention and corporate loyalty, all of which are important factors to successful CSR implementations in the long term.

Finally GHRM is an important part of defining a company's CSR strategy as it allows a company to transform a theoretical commitment to sustainability into measurable (and effective) action. By aligning GHRM with CSR, employees at all levels are helping ensure sustainable business practices are being implemented, and thus have major social and environmental impact.

How GHRM contributes to the achievement of Sustainable Development Goals (SDGs).

The 17 interrelated objectives of the United Nations' SDGs seek to solve worldwide problems including poverty, inequality, climate change, and environmental destruction. Some of these aspirations are directly or indirectly dependent on the way companies handle their human capital and environmental obligations.

Linkages between GHRM and SDGs

SDG3: Good Health and Wellbeing

Physical and mental wellbeing of employees benefit from GHRM methods that include promotion of a safe and healthy workplace, work life balance encouragement, and reduction of workplace stress. Green businesses also give attention to lowering exposure to toxic materials and enhancing inside air quality.

SDG 4: Education of Excellence.

Such training and development in GHRM would cover awareness campaigns, green skill building, and environmental education. Organizations support lifetime learning and sustainability related capacity development by improving workers' environmental skills.

SD G 5: Equality of Genders

Green HRM encourages fair and inclusive policies that enable women to actively engage in sustainability projects. Environmental leadership positions and green project teams aid in the advancement of diversity and inclusion.

SDG 8: Economic Growth and Decent Employment

GHRM supports ecofriendly employment, fair working conditions, and sustainable economic policies. It promotes innovation, improved skills, and environmentally friendly techniques, therefore supporting green growth and responsible consumption.

SDG 9: Industry, Innovation, and Infrastructure.

Employee driven invention and environmentally friendly creativity help GHRM to promote sustainable industrialization. Adoption of green technologies and resource efficient infrastructure is encouraged by HR policies.

SDG 12: Responsible Production and Consumption

Through employee participation in sustainable actions, green procurement policies, and waste reduction programs, businesses can lessen their ecological footprints and improve resource use.

SDG 13: Climate Action.

Most directly related, GHRM prompts staff to reduce carbon emissions, engage in environmental initiatives, and embrace environmentally conscious habits. HR policies help to create an ambiance of accountability and proactive environmental involvement.

Strategic Role of HR in Help to SDGs

By their actions, HR specialists are indispensable in integrating sustainability into corporate strategy.

- Attracting green talent with strong environmental ethics.
- Design of performance evaluation systems that give incentives for results related to sustainability.
- Creating incentives systems that inspire environmentally concerned actions.
- Establish green organizational cultures that correspond with employee values and SDG objectives.
- Supporting sustainability minded transformation and enabling green change management.

Advantages of Adopting GHRM

The use of GHRM gives companies many key advantages. Businesses that put green HRM methods into practice often see:

Better Employer Image – Companies with strong green programs draw top talent. Studies show that 64% of millennials think about a company's social and environmental promises when picking a job (Cone Communications, 2017).

More Loyal and Involved Workers – People who work in companies committed to being eco-friendly say they like their jobs more and feel more connected to their work.

Following the Rules – Companies that mix being green into how they manage people are ready to follow changing environmental laws, which cuts down on legal troubles and fines.

Saving Money – Using less energy at work, making less trash, and letting people work from home can save a lot of cash.

Better Company Name – Doing things to help the planet makes a company look good, which makes customers more loyal and investors more sure about the business.

New Ideas and Staying Ahead – Companies that push for being green tend to come up with more new ideas using eco-friendly ways to make new products and offer new services.

The Need for Integrating GHRM Into Business Strategy

Companies today face growing pressure to help the environment making HR's job in promoting green efforts more crucial than ever. Green Human Resource Management (GHRM) offers a way to build sustainability into a company's structure making sure workers help reach environmental targets.

This chapter will explore GHRM's basic ideas key HR practices that back sustainability real-life examples, and the ups and downs of putting GHRM to work. By grasping how GHRM can change things, companies can come up with better plans to be sustainable in the long run, which helps both their business and the planet.

Challenges and Barriers in GHRM Implementation

Despite its benefits, GHRM challenges are apparent to businesses. Reluctance to change, limited budgets, inability to measure green HR effectiveness, ineffective implementation of compliance standards – all these factors contribute to the challenges that business facing to effectively apply GHRM techniques and ensure their long-term viability.

1. Resistance to Change

When companies try to implement GHRM, they face an increasing challenge from resistance from both employees and executives. A person might argue that they are unwilling to change because they do not know the best ways to do it. They may be afraid of working more, feeling it will take more time or unsure if sustainability projects will work. Some people do not see sustainability initiatives as a part of the job, not as a necessity. A person will not seem motivated to go on new tasks like recycling, decreasing power usage or taking sustainability workshops.

On the management side, leaders might hesitate to blend sustainability into HRM. They worry about the costs how practical it is, and whether they'll see a return on their investment. Some higher-ups think sustainability programs don't boost the bottom line, so they might not give them much attention. Also, if top brass don't back environmental goals, it can weaken staff buy-in leading to low uptake of green HR policies.

To get around this resistance, companies need to conduct awareness campaigns, train their managers and use change management methods. Since sustainability contributes to long term benefits such as cost savings and branding advantages, companies can incentivize people to get behind GHRM by creating a positive outlook on the future.

2. Financial Constraints

Often, sustainability initiatives require large amounts of investment in terms of training, new technologies and incentives.

Many companies, especially small and medium-sized enterprises (SMEs), find it hard to spend enough money on GHRM in the face of other business concerns: courses for employees, sustainable office facilities and renewable energy sources among other environmentally friendly measures can be expensive.

, green recruiting might cost a lot of money if specialized hiring platforms are needed to evaluate candidates based on their sustainability skills. Also, green training programs will require a higher cost; learning materials must be purchased, professional trainers must be hired, and training tools must be purchased online. In addition, if businesses are giving out money as compensation for green behavior (bonus points for staff that ride public transportation, bike to work, or carpool), they'll have to dedicate a lot of money to keep them going.

But the long-term effects of GHRM are so powerful that any initial cost can be deducted. Studies have shown that businesses embarking on sustainability initiatives experience lower operational costs, higher employee productivity, and improved customer satisfaction. For example, green supply chain management improves resource efficiency, while energy efficient operations result in lower utility costs. So companies can offset initial costs through state grants, corporate sustainability funding efforts, and environmental partnerships.

3. Measuring Green HR Effectiveness

Establishing appropriate Key Performance Indicators (KPIs) for environmental projects is one of the challenges companies face in implementing GHRM. Measuring the effectiveness of green HRM is much harder, and in some cases it lacks regular benchmarks (as in standard HR metrics like employee retention, engagement and productivity). In part, difficulties in measuring the effectiveness of environmental projects result from businesses' difficulty in determining the direct impact of GHRM on sustainability results.

For example, at the organizational level it becomes rather straightforward to track energy savings, waste management, and carbon footprint reduction, but it is difficult, at best, to connect these accomplishments to human resource policies. One would first need to address how sustainability training affected staff behaviour; then organizations could then look at how green pay incentives had a positive impact on environmental impact.

We suggest organizations create SMART (specific, measurable, achievable, relevant, time bound) sustainability metrics that align with the corporate goals to address this challenge. Organizations can conduct green employee engagement surveys, sustainability scans, and participation tracking software to capture the impact of GHRM projects. Organizations can also implement AI-powered analysis and digital HR dashboards to help them more closely follow sustainability KPIs, and provide data-driven reporting on the effectiveness of green HR policies.

4. Regulatory Compliance Complexity

But it also gets harder when you consider that there are very many environmental regulations from multiple jurisdictions that are involved in multi-national companies. Many companies operate in many geographical regions. They may have very different sustainability reporting standards, labor laws and environmental codes, etc. For a global company where the mandate is being abided by in local as well as global jurisdictions, managing these differences can get very complex.

The European Union's Corporate Sustainability Reporting Directive (CSRD) mandates firms disclose comprehensive environmental and social impact metrics; here, the US oversees: publicly traded companies are now required by the Securities and Exchange Commission (SEC) to disclose risks relating to environmental issues. At the same time, nations like China and India do implement stringent environmental compliance regulations (requiring firms to regularly report on their sustainability performance).

Companies should establish a team of sustainability compliance advisers to reduce the chances of noncompliance; such teams should ensure that they are up to date with new legislation and maintain compliance with environmental regulations. Businesses also have a powerful tool to negotiate regulatory complexity: working on regulatory issues with industry groups, environmental consultants and law firms.

Furthermore, companies can use HRM systems to automatically track conformity and simplify sustainability disclosures by incorporating sustainability reporting mechanisms.

To overcome these barriers:

- Strengthen environmental regulations and enforcement while providing incentives such as tax breaks and subsidies for firms adopting green practices (Sulistiawan, et al., 2024).

- Conduct widespread awareness and capacity-building campaigns targeting both employees and organizational leaders to foster a culture of environmental responsibility (Khalid et al., 2025).
- Promote employee involvement in sustainability decision-making processes to build commitment and a strong green workplace culture (Khan, 2022).
- Encourage collaboration between government, industry, academia, and NGOs to create supportive frameworks and knowledge-sharing platforms (Khalid et al., 2025).
- Stimulate research and sector-specific studies to fill knowledge gaps and tailor GHRM strategies to Pakistan's unique context (Khan et al., 2025).
- Such measures can help Pakistani firms overcome financial, cultural, and regulatory challenges to unlock the benefits of GHRM, both for organizational performance and environmental sustainability.

Local GHRM Policies and Their Impact on Climate Resilience in Developing Nations

Climate change provides organizations with major challenges particularly in developing nations, where vulnerabilities are more pronounced given resource constraints and fragile institutions. By means of human capital, GHRM has become a major driver behind organizational climate resilience and environmental sustainability. Especially in varied socioeconomic and cultural settings like Pakistan, the translation of worldwide GHRM policies into local ones is fraught. Particularly in the quest of climate resilience, GHRM practices in Pakistan are shaped by economic, institutional, and cultural influences, therefore critically analyzing the global-local dynamics of GHRM in developing nations.

Global vs. Local GHRM Policies in Developing Nations

Worldwide GHRM Models

GHRM frameworks internationally emphasizes eco-friendly HR practices like:

- Green hiring and talent evaluation
- Environmental education and growth
- Natural performance assessment
- Consistent rewards systems
- Employee participation in climate change adoption

These practices are based on environmental laws, sustainability standards (such as ISO 14001) and CSR expectations in advanced countries. These practices are supported by Strong institutions, legal enforcement, and technological infrastructure.

Local realities in developing countries

Developing countries have the following kind of constraints sometimes:

- Sparse legislative structures or little enforcement of environmental laws
- Insufficient funds for green infrastructure and training.
- Employers and employees alike have little aware of the risks of climate.
- short term business priorities outweigh long term sustainability goals

Therefore, although global GHRM models offer a guide, their use in domestic settings typically needs sensitivity and adjustment to local conditions.

Contextualizing Requirements

For GHRM to effectively increase climate resilience in developing countries, it has to be:

Culturally pertinent: honoring regional values, work ethics, and beliefs
Economically sustainable: guarantee cost effective ecological initiatives
Institutionally supported: adhering to local regulations and administrative capabilities

GHRM and Climate Resilience in Pakistan

Pakistan is one of the most climate vulnerable country and it needs urgently a comprehensive integrated climate resilience adaptation strategies.

The basis of GRHM in the Pakistani environment is integrating environmental sustainability into HR procedures to promote organizational and ecological advantages. Green recruiting, green training and development, green performance appraisal, and employee participation in sustainability initiatives make up GRHRM in Pakistan. This method connects workforce behavior with environmental targets, encourages innovation, lowers operating expenses, and brings businesses into line with both national sustainability objectives and international standards (Hameed & Akram, 2024).

Particularly, research in Pakistan's leather industry, banking sector, small and medium businesses (SMEs) and hospitality reveals that GHRM methods favorably affect employee green behavior, green culture, and organizational performance.

Challenges include a hierarchical organizational culture, low awareness among HR professionals and employees about environmental concerns, and reactive instead of proactive implementation. Strong environmental regulations, subsidies and tax breaks, awareness campaigns, and corporation among government, business, and academia (Hameed & Akram, 2024; Khan et al., 2024).

GHRM in Pakistan is seen as crucial for addressing serious environmental challenges like pollution and resource depletion while enhancing corporate reputation, competitive advantage, and compliance with sustainability standards. The practice holds potential for sustainable growth by embedding environmental values into everyday HR functions, driving both socio-economic and ecological benefits in Pakistani industries (Khan et al., 2025).

For developing resilience at community and organizational level, GHRM provides a vital resource. Its use is nevertheless determined by several environmental aspects:

a) Economic factors

- Financial Constraints: Many companies, especially small and medium sized ones, do not have the funds necessary to support environmental training, green technologies, or green HR programs.
- Focus on Survival: In riotous financial circumstances, businesses focusing on profit and business continuity rather than sustainability impede long term GHRM implementation.
- Dominance of the informal sector: A large part of the labor force works where there are no organized HR rules, let alone green ones.

b) Institutional Contributors

- Weak Environmental Governance: Enforcement is still difficult even though legislation such the Pakistan Climate Change Act 2017 prevail. Institutional assistance in incorporating GHRM into governmental or business plans is small.
- In HR sectors lacking of awareness: Human resources experts are usually not educated in environmental management or climate risk, therefore a knowledge gap and a skill gap occur.
- Lacking Harmony among HR Policies, Labor Laws, and Environmental Plans: There is a disjointed policy landscape.

c) Cultural factors.

- **Collectivist Culture:** Pakistan's collectivist attitude can benefit GHRM since it encourages community values and group accountability. With the right incentive, staff may be deployed on sustainable projects that advance climate resilience.
- **Religious and ethical values:** Islamic standards of stewardship (Khilafah) over the earth can be matched with GHRM methods to inspire sustainable conduct, therefore helping to support moral and spiritual values.
- **Change Resistance:** New, ecology targeted HR policies could be resisted by traditional work methods and top down structures.

Opportunities and Recommendations for Climate Resilient GHRM in Pakistan

Although difficult, there are emerging opportunities to improve GHRM in helping of climate resilience:

- **Green Skill Development:** National and provincial human resource training agencies (e.g., NAVTTC, TEVTA) can include environmental modules into professional and managerial training.
- **Public-private collaboration:** Government agencies together with private organizations can partner with businesses to test green HR programs.
- **Encourage Green Leadership:** By identifying and honoring green champions in businesses, we can create role models for sustainable HR policies.
- **Develop regional GHRM models**: Create local GHRM systems reflecting the social and economic circumstances and cultural values of Pakistan.

The difference in developing countries between local customs and worldwide GHRM practices emphasizes the need of flexible, context dependent techniques. Pakistan's GHRM reach and efficacy are influenced by economic constraints, institutional weakness, and cultural complexity. But by using local assets, increasing awareness, and creating institutional support roles, GHRM can be a critical instrument for improving climate resilience. As the climate crisis worsens, sustainable development and corporate readiness will much depend on green oriented HR professionals and policies.

Emerging Trends and Future Directions in GHRM

The development of Green Human Resource Management (GHRM) is seen to be very important in defining the next generation of labour as businesses keep on incorporating sustainability into their strategies. GHRM will be effective depending on how companies use technology, develop leadership skills, promote business partnerships, and creatively engage staff. Here are some important tendencies and suggestions that will shape the future of GHRM.

Enhancing Employee Engagement through Gamification

Using gamification strategies is among the most successful means of increasing employee engagement in sustainability projects. Including game like features such as rewards, difficulties, leaderboards, and interactive competitions can significantly raise employee incentive to engage in green initiatives.

Organizations should introduce environmental challenges where employee earn points or get acknowledgement for engaging in eco-friendly activities. For instance, organizations can introduce challenges like:

Competitions for "green commutes" encourage employees to share rides, bicycle, or use public transport.

Incentive programs within departments to lower paper usage and enhance recycling efforts are the focus of the "Zero Waste Office" initiatives.

Employee energy efficiency campaigns where they monitor and cut office power use.

Some businesses are starting online platforms where employees can log their environmentally friendly activities and vie with other workers on green leaderboards. Siemens provides an illustration of a business sustainability application that rewards employee involvement in environmental projects based on accomplishments.

Virtual Reality (VR) and Augmented Reality (AR) simulations in Sustainability Training can be applied. Companies can improve employees' sustainability knowledge by means of simulations that educate employees on greenhouse effect, carbon footprints, and environmental friendly workplace practices. Deloitte and PwC have rolled out VR based environmental training modules to engage employees in realistic sustainability scenarios.

Companies should create gamification policies including sustainability challenges, interactive learning modules, and digital engagement tools in order to boost employee engagement with ecofriendly HRM initiatives. A rewards system for employee sustainability would help to promote a culture of environmental responsibility and keep engagement up over time.

Technological integration, leadership training, sector partnership, and staff engagement all define the way forward for Green Human Resource Management (GHRM). Using artificial intelligence driven analytics, leadership training focused on sustainability, cross sector alliances, and gamification strategies will help companies to adjust to changing patterns and drive GHRM success.

For long run sustainability, companies should to focus on:

- Investing digital technology and artificial intelligence to improve sustainability monitoring and decision making.
- Growing green leadership qualities to include business models in environmental policies.
- Working with governmental officials and industry colleagues to define common sustainability targets.
- Gamified sustainability projects engaging staff help to support a long term dedication to ecofriendly behavior.

Embracing these trends and ideas helps businesses to develop a future ready workforce that is environmentally friendly as well as skilled and efficient. By ensuring that companies are aligned with international environmental goals, persistent bodily integrity will help keep businesses resilient, competitive, and forward.

Integration of AI and Digitalization

AI and digital tools are reshaping HRM by means of improved decision making, predictive employee engagement, and dynamic workforce management. By following sustainability Key Performance Indicators (KPIs), finding green skill gaps, and encouraging environmentally friendly office behavior, the use of AI driven HR analytics can improve GHRM.

Regarding Green Recruitment and Performance Management: Companies can use AI driven recruitment software to evaluate a candidate's sustainability related skills and environmental knowledge. By using AI driven screening techniques to examine past projects, qualifications, and behavioral reactions to environmental circumstances, a better fit between employee values and corporate sustainability objectives can result. AI powered performance management systems can also monitor staff involvement in green projects, such energy saving initiatives, attendance in environmental training courses, and waste reduction operations. Some companies have incorporated artificial intelligence chat bots to give instant feedback on staff members' sustainability actions and provide individualized recommendations for enhancing their eco consciousness at work.

Companies should use digital platforms for employee engagement based on sustainability on which staff can track their environmental effect, contribute ideas, and participate in virtual green contests. Enterprises like Unilever and Google have rolled out internal sustainability applications whereby staff can track their carbon footprints, engage in environmentally friendly activities, and earn incentives for sustainable behavior. Furthermore, remote work is possible with cloud based cooperation technology lowering total office energy use and commute emissions. Companies including Microsoft and Salesforce have introduced AI powered sustainability dashboards where workers may monitor their real time carbon footprint reduction efforts.

Advanced HR analytics allow companies to gauge the effectiveness of GHRM initiatives. Real time information on employee sustainability engagement levels, environmental training efficacy, and organizational carbon emission reduction can be analyzed by AI powered systems. Using these ideas, corporations are able to make educated decisions to advance their GHRM policies. By predicting the effects of particular green projects as well as helping HR leaders to distribute resources wisely, predictive analytics enable them to develop more effective sustainability programs.

Using machine driven learning management systems (LMS), AI Powered Learning and Development for Sustainability can offer employees custom sustainability courses depending on their past learning experiences and roles. Through adaptive learning methods, these systems guarantee that workers interact with the most pertinent environmental information including energy saving ideas, circular economy concepts, and sustainable purchasing policies. IBM and Nestlé are among the businesses using AI driven LMS systems that suggest personalized career path based courses on sustainability—thus encouraging a culture of constant environmental learning.

Organizations need sustainable monitoring, data forecasting, and digital engagement from artificial intelligence driven HR solutions they purchase. Creating AI powered staff sustainability dashboards will improve accountability and openness in company environmental projects. AI powered behavioral analytics can also assist HR departments in spotting and cultivating green leaders inside the company, therefore bolstering long term dedication to sustainability objectives.

Green Leadership Development Programs

Driving organizational sustainability depends significantly on leadership. Future GHRM plans should concentrate on creating leaders able to successfully integrate environmental policies into corporate models. GLDPs—Green Leadership Development Programs—organizations provide opportunities for managers and executives with the expertise and tools needed to promote sustainability. *[Comment: This is a strong emerging trend. Consider adding a brief example from a Pakistani or regional company if available.]*

Many organizations have included sustainability modules in leadership training programs. Executive education initiatives from firms such as Apple, Tesla and Microsoft emphasize sustainable innovation, carbon neutrality policies, and awareness of climate change.

Many organizations are now appointing Chief Sustainability Officers (CSOs) to lead sustainability initiatives from an executive level. These leaders drive cultural change toward sustainability, guarantee legal compliance, and line corporate policies with environmental objectives.

Future GHRM policies should include sustainability KPIs in leadership evaluations. Managers should be assessed on their help with green initiatives such as lowering energy use in their departments, encouraging sustainable supply chain methods, and involving staff in environmental programs.

Companies should integrate sustainability into their leadership development plans by means of custom green leadership courses, mentoring sessions, and interdisciplinary sustainable initiatives. Further improving the effectiveness of green leadership is to offer managers who meet sustainability goals incentives.

Green Human Resource Management Adoption in Pakistani SMEs

Small and Medium Enterprises (SMEs) in Pakistan represent a critical segment, employing over 80% of the non-agricultural workforce and contributing around 40% to the nation's GDP. However, sustainable GHRM adoption within SMEs is still at an emerging stage with significant challenges related to resource limitations and environmental awareness (Larik et al., 2024).

A recent empirical study on SMEs in Sindh highlights promising adoption trends where integration of green recruitment, green training, and green performance management fosters a green organizational culture. This green culture mediates and enhances organizational environmental performance. For instance, SMEs that incorporated environmental considerations into their performance evaluation systems observed improved employee engagement and resource-conserving behaviors. Nonetheless, only around 25% of these SMEs formally documented environmental policies, signaling room for improvement (Larik et al., 2024).

Actionable Steps for Pakistani SMEs to Adopt GHRM:

Green Recruitment and Selection:

Integrate environmental responsibility criteria into job descriptions and selection processes.

Highlight the organization's environmental commitments to attract eco-conscious candidates.

Green Training and Development:

Conduct regular workshops and training sessions focusing on sustainability practices such as waste reduction, energy conservation, and responsible resource use.

Encourage knowledge sharing of green practices among employees to enhance environmental awareness.

Green Performance Management:

Incorporate sustainability goals and green behaviors as key performance indicators in employee appraisals.

Reward and recognize employees who contribute to environmental initiatives, building motivation.

Cultivating a Green Organizational Culture:

Leadership should model eco-friendly behaviors and communicate the importance of sustainability consistently.

Establish forums or committees for employees to suggest and participate in environmental improvements.

Seeking External Support:

Leverage government or NGO programs offering subsidies, training grants, or consultancy for green initiatives.

Build networks with other SMEs to share best practices and collectively address challenges (Yousfani, & Shaikh, 2023).

Table 1. Metrics to Track GHRM Effectiveness in SMEs:

GHRM Area	Actionable Steps for SMEs	Key Metrics / KPIs	Source(s)
Green Recruitment & Selection	Add sustainability criteria in job ads and interviews; onboard new staff with eco-values.	% of job postings with green criteria; % of recruits with eco-awareness.	Yousfani & Shaikh, 2023; Javed et al., 2024
Green Training & Development	Provide low-cost sustainability workshops (waste reduction, energy saving); refresher trainings quarterly.	% of employees completing training; employee knowledge improvement (pre-/post-surveys).	Larik et al., 2024; Javed et al., 2024
Green Performance Management	Integrate eco-criteria into performance appraisals; reward employees adopting green behaviors.	% of appraisals with green KPIs; # of employees rewarded for eco-behavior.	Javed et al., 2024
Employee Engagement & Innovation	Establish green taskforces/ committees; encourage staff-led eco-projects and idea sharing.	Participation rate in green programs; # of employee-driven innovations.	Larik et al., 2024;
Leadership & Role Modeling	SME leaders act as sustainability role models (reduce waste, promote eco-behavior).	Employee feedback on leadership eco-behavior; observed leader practices.	Khan et al., 2025
Green Organizational Culture	Promote eco-values via internal campaigns; provide resources (e.g., recycling bins).	Employee survey scores on sustainability culture; # of green tools/resources provided.	Yousfani & Shaikh, 2023
Policy & Awareness Building	Draft simple eco-policies (paper use, energy saving); conduct awareness drives.	# of policies adopted; employee awareness levels (survey results).	Khan et al., 2025; Mannan, et al., 2023

Industry Collaboration and Policy Advocacy

Companies need to engage in cross industry partnerships and advocacy of policies for GHRM to realize its full potential. By means of cooperation with legislators, industry associations, and universities, defining typical sustainability standards and best practices will ensure a consistent approach to the deployment of green HRM.

Public-private sustainability initiatives should see organizations engaged in multi stakeholder sustainability conferences and industry consortia. The United Nations Global Compact, for instance, offers a forum for companies to work together on environmental projects, share sustainable research, and adopt green HRM standards globally known.

Companies working in multiple regions struggle to follow several environmental standards or norms. Increased cooperation with nonprofits and government bodies will help companies in negotiating compliance issues and arguing for legislation that advances corporate sustainability objectives.

Many colleges nowadays provide special programs on green HRM and sustainability. To create tailored sustainability training programs, research initiatives, and talent pipelines that support company green objectives, businesses should engage academic institutions.

Companies should start cross-sector sustainability networks, engage in policy advocacy campaigns, and work with colleges to craft evidence based GHRM policies. Organizations that fund think tanks focusing on sustainability will be better able to stay ahead of legislative shifts and business dynamics.

CONCLUSION

Summary

Green Human Resource Management (GHRM) was discussed in this chapter together with its part in matching employee management with environmental sustainability. Alongside case studies of companies that have effectively incorporated sustainability into their HR strategies, GHRM initiatives, obstacles, and future directions are also considered. Summarize the major findings and review their management effects.

- Integrating climate resilience and sustainability factors with HR policies has become a need of time. Organizations have to incorporate environmental responsibility into their HR policies so that sustainability becomes a fundamental business value rather than a one off project. Fostering a green workplace culture depends much on well communication and strong guidance.
- Sustainable workforce development depends on green recruitment and training. Firms should change job descriptions to draw ecological intelligent employees and introduce green training courses meant to help staff members with sustainability capabilities. Good training guarantees staff have the tools to help meet the objectives of company sustainability.
- Performance management linked to sustainability heightens accountability. Implementing KPIs related to environmental impact helps measure employees' contributions to sustainability efforts. Rewarding performance evaluations and incentives for green projects helps to spur constant improvement.
- Green workplace policies help to raise employee engagement and functional efficiency. Reducing operational costs, supporting electric vehicle usage, flexible work schedule, and waste reduction projects, among other green incentives, helps to create a staff driven by sustainability.

- GHRM rollout could be optimized by technology. Gamification, AI driven HR analytics, and digital engagement tools can help to boost involvement in sustainability projects and monitor the efficiency of GHRM techniques. Using technology to support sustainable office practices will help managers to advance their career.
- Dealing with obstacles sets long term success. Adoption of GHRM may be slowed by resistance to change, lack of funds, and legal entanglements. To properly overcome these obstacles, managers have to actively involve staff, find outside relationships, and secure financing.
- Implementing GHRM best policies will help managers to raise sustainability performance, encourage creativity, and promote employee wellbeing. Companies giving GHRM priority will be better placed strictly to meet regulations, appeal to top talent, and sustain long term competitiveness in an environmentally conscious environment.

Review Questions

1. What is GHRM, and why is it important for modern organizations?
2. GHRM departs from conventional HRM techniques in what way?
3. How do the main elements of GHRM help corporate sustainability by themselves?
4. Regarding recruitment and selection procedures, how might companies incorporate sustainability? Give instances.
5. How can performance management systems include sustainability measures?
6. By what way does an organization's environmental goals benefit from green training and development initiatives?
7. What challenges companies face in implementing GHRM and how these challenges can be addressed?
8. Is rewarding sustainable work behaviors with financial incentives a workable plan? Justify your opinion?
9. Can SMEs effectively use GHRM or is it better suited for big firms? Analyze.
10. How might developments in technology—like big data and artificial intelligence—improve GHRM deployment?
11. How do GHRM policies take shape according to CSR projects and government laws?
12. Forecast the possible influence on companies of GHRM transformation in the next ten years.

REFERENCES

Barney, J. (1991). Firm resources and sustained competitive advantage. *Journal of Management, 17*(1), 99–120.

Becker, G. S. (1964). Human capital. The University of Chicago Press Chicago and London Cone Communications. (2017). 2017 Cone Communications CSR Study. http://www.conecomm.com/research-blog/2017-csrstudy

DiMaggio, P. J., & Powell, W. W. (1983). The iron cage revisited: Institutional isomorphism and collective rationality in organizational fields. *American Sociological Review*, *48*(2), 147–160.

Ekechukwu, D. E., & Simpa, P. (2024). A comprehensive review of renewable energy integration for climate resilience. *Engineering Science & Technology Journal*, *5*(6), 1884–1908.

Elkington, J. (1997). The triple bottom line. *Environmental management: Readings and cases, 2*, 49-66.

Freeman, R. E. (1999). Divergent stakeholder theory. *Academy of Management Review*, *24*(2), 233–236.

Hameed, H., & Akram, K. (2024). Adoption of Green Human Resource Management Practices in Pakistan: Case of Leather Industry. *Pakistan Social Sciences Review*, *8*(4), 314–329.

Jabbour, C. J. C., & Santos, F. C. A. (2008). The central role of human resource management in the search for sustainable organizations. *International Journal of Human Resource Management*, *19*(12), 2133–2154.

Javed, S., Yingjun, Z., Kumar, N., & Meghwar, S. (2024). Green human resource management practices and organizational sustainability performance in Pakistan's healthcare sector: A descriptive analysis. *International Journal of Research in Business & Social Science*, *13*(1).

Khalid, U., Rizvi, S. A. A., Abideen, Z. U., Imran, M., & Raza, H. (2025). Exploring the Obstacles to the Implementation of Green HRM Practices for Sustainable Small and Medium Enterprises in Pakistan. *Journal of Management & Social Science*, *2*(1), 122–137.

Khan, A. J. (2022). Green human resource practices: A sustainable approach to increase employee performance. *iRASD. Journal of Management*.

Khan, K., Gogia, E. H., Shao, Z., Rehman, M. Z., & Ullah, A. (2025). The impact of green HRM practices on green innovative work behaviour: Empirical evidence from the hospitality sector of China and Pakistan. *BMC Psychology*, *13*(1), 96. PMID: 39901289

Khan, K., Khan, Q., Jalaly, S., Shams, M. S., & Shah, T. A. (2024). Green human resource management and green service behaviour in organisations: Examining underlying mechanism. *SA Journal of Human Resource Management*, *11*(1), 113–129.

Khan, Z. N., Iqbal, I., & Saba, S. (2025). The Impact of Green Human Resources Management to Enhance Pro-Environmental Behaviour among Organizational Employees. *Pakistan Social Sciences Review*, *9*(2), 306–321.

Larik, K. A., Mangi, M., & Ahmed, S. I. (2024). Green Initiatives in Pakistani Universities: Nexus among Human Resource Practices, Employee Environment Behavior, and Innovative Work with Mediating Role of Work Engagement in Pakistani Universities. *Business Review*, *19*(1), 68–86.

Mannan, M. M. J., Aiman, A. N., Saba, S. I., & Kiani, M. B. K. (2023). In-depth review of green human resource management along with latest initiatives. *NUST Business Review, 5*(1).

Paine, L. S. (2023). What Does "Stakeholder Capitalism" Mean to You. *Harvard Business Review*, (September-October), 108–119.

Patagonia, Inc. (2023). *Our Responsibility Programs*. Retrieved from https://www.patagonia.com/our-responsibility-programs.html

Renwick, D. W., Redman, T., & Maguire, S. (2013). Green human resource management: A review and research agenda. *International Journal of Management Reviews*, *15*(1), 1–14.

Sulistiawan, J., Herachwati, N., & Khansa, E. J. R. (2024). Barriers in adopting green human resource management under uncertainty: The case of Indonesia banking industry. *Journal of Work-Applied Management*.

Tesla, Inc. (2023). *2023 Tesla Impact Report*. Retrieved from https://www.tesla.com/impact and https://www.tesla.com/ns_videos/2023-tesla-impact-report.pdf

Unilever. (2010). *Sustainable Living Plan*. Retrieved from https://www.unilever.com/planet-and-society/

Yousfani, K., & Shaikh, E. (2023). Identifying Factors Affecting Green HRM Practices And Its Impact On Organizations' Sustainability Of Multinational Corporations (Mncs). *NUST Business Review, 5*(2).

ADDITIONAL READING

Daily, B. F., & Huang, S. (2001). Achieving Sustainability through Attention to Human Resource Factors in Environmental Management. *International Journal of Operations & Production Management, 21*(12), 1539–1552.

Dumont, J., Shen, J., & Deng, X. (2017). Effects of Green HRM Practices on Employee Workplace Green Behavior: The Role of Psychological Green Climate and Employee Engagement. *Journal of Organizational Behavior, 38*(7), 1–18.

Jabbour, C. J. C., & de Sousa Jabbour, A. B. L. (2016). *Green Human Resource* Management and Green Supply Chain Management: Linking Two Emerging Agendas. *Journal of Cleaner Production, 112*, 1824–1833.

Jackson, S. E., Renwick, D. W. S., Jabbour, C. J. C., & Muller-Camen, M. (2011). State-of-the-Art and Future Directions for Green Human Resource Management: Introduction to the Special Issue. *German Journal of Research in Human Resource Management, 25*(2), 99–116.

Kamboj, J. (2024). Mapping the green human resource management practices: A systematic scoping review and its implications for employees' well-being. *Human Systems Management, 43*(6), 1062–1081.

Li, M., Martínez-del-Río, J., Li, P., & Guthrie, J. P. (2024). Exploring the antecedents of green human resource management: A path dependence perspective. *Journal of Business Research, 181*, 114743.

Patagonia, Inc. (2023). *Our Responsibility Programs.* Retrieved from https://www.patagonia.com/our-responsibility-programs.html

Renwick, D. W. S., Redman, T., & Maguire, S. (2013). *Green Human Resource* Management: A Review and Research Agenda. *International Journal of Management Reviews, 15*(1), 1–14.

Tang, G., Chen, Y., Jiang, Y., Paille, P., & Jia, J. (2018). Green Human Resource Management Practices: Scale Development and Validity. *Asia Pacific Journal of Human Resources, 56*(1), 31–55.

Tesla, Inc. (2023). *2023 Tesla Impact Report.* Retrieved from https://www.tesla.com/impact

Unilever. (2010). *Sustainable Living Plan.* Retrieved from https://www.unilever.com/planet-and-society/

Zibarras, L. D., & Coan, P. (2015). HRM Practices Used to Promote Pro-Environmental Behavior: A UK Survey. *International Journal of Human Resource Management*, *26*(16), 2121–2142.

KEY TERMS AND DEFINITIONS

GHRM (Green Human Resource Management): Human resources policies underpin workplace environmental sustainability.

CSR (Corporate Social Responsibility): CSR helps companies to meet their ethical obligations and to have a beneficial societal influence. A business's attempts to operate in an ecologically and socially responsible manner.

Climate Resilience: The ability to cope with and recover from the effects of climate change.

Environmental Sustainability: Preserving the environment for the future by prudent use of natural resources.

Gamification: Using games like incentives where points are scored personally to increase self-motivation and work engagement. Making activities more fun with game like elements (e.g., points, incentives).

Green Behaviour: Environment friendly actions entail recycling and energy conservation.

Green Culture: A workplace culture that promotes and validates environmental friendly policies and values.

Chapter 5
Strategic Human Resource Management for the Advancement of Climate Change Resilience and Mitigation

Shivansu Sachan
https://orcid.org/0000-0002-9893-8018
CSJM University, India

Aishwarya Arya
https://orcid.org/0009-0000-4236-3983
CSJM University, India

Ankit Kumar
https://orcid.org/0009-0009-5002-4522
CSJM University, India

Abhishek Khandelwal
Food Craft Institute, Khajuraho, India

Mohammad Badruddoza Talukder
https://orcid.org/0009-0008-1662-9221
International University of Business Agriculture and Technology, Bangladesh

ABSTRACT

This chapter explores the integration of climate change mitigation and resilience strategies within organizational frameworks through strategic human resource man-

DOI: 10.4018/979-8-3373-2479-1.ch005

agement (HRM). By aligning HR practices with sustainability goals, organizations can foster a workforce equipped to address environmental challenges. Key strategies include developing green competencies, promoting eco-conscious behaviors, and enhancing employee engagement in sustainability initiatives. The chapter examines how HRM can drive organizational resilience by embedding climate-focused policies in recruitment, training, performance management, and leadership development. It highlights the role of HR in cultivating a culture of environmental stewardship, enabling organizations to adapt to climate-related risks while contributing to global sustainability efforts. Practical implications and future research directions are discussed to advance the intersection of HRM and climate action.

1. INTRODUCTION

The escalating impacts of climate change—ranging from extreme weather events to rising sea levels—pose serious challenges for societies and economies worldwide. Governments, civil society, and businesses must act collectively to mitigate environmental damage and build adaptive resilience. However, the role of human capital in this global response remains underemphasized. Strategic Human Resource Management (SHRM), traditionally focused on aligning HR practices with organizational objectives, can evolve into a critical driver of environmental sustainability.

While environmental management has often been confined to operations, facilities, or supply chain departments, the HR function can play a central role in fostering a culture of sustainability and climate awareness. SHRM can facilitate the recruitment, development, and retention of employees who are aligned with environmental goals, thereby embedding sustainable thinking into the core of business practice. Moreover, HR strategies can support organizational resilience, preparing firms to adapt to climate-induced disruptions. Climate change represents one of the most urgent and complex global challenges of the 21st century. The increasingly frequent and severe impacts—ranging from rising sea levels, extreme weather events, water shortages, biodiversity loss, and environmental degradation—threaten not only ecosystems and human health but also the continuity of business and economic systems. According to the Intergovernmental Panel on Climate Change (IPCC, 2023), without significant and sustained global efforts to mitigate greenhouse gas emissions and enhance adaptation strategies, the risks to both the natural environment and socioeconomic structures will escalate substantially. This existential threat compels organizations across all sectors to re-examine their operational models and contribute meaningfully to both climate change mitigation and resilience building. In this context, the role of human capital and, more specifically, Strategic Human Resource Management (SHRM), has gained renewed importance.Traditionally, efforts to address climate change within organizations have been largely technical and operational in nature, focusing on sustainable infrastructure, energy efficiency, waste reduction, and green supply chain management. These are undoubtedly critical components of climate action; however, they often overlook the pivotal role that human resources (HR) can play in enabling and sustaining such transformations. Employees, after all, are the ones who implement sustainability practices, develop innovations, and embody the organization's values in their daily work. Their beliefs, behaviors, and level of engagement ultimately determine the success or failure of environmental strategies. Despite this, human resource management has remained peripheral in many corporate sustainability agendas.Strategic Human Resource Management, with its emphasis on aligning HR practices with long-term organizational goals, offers a powerful mechanism to incorporate climate objectives directly into the employee lifecycle—from recruitment and training to performance management and leadership development. SHRM transcends routine administrative tasks by strategically shaping employee capabilities, motivations, and values in ways that support the broader vision of environmental sustainability. The inclusion of climate goals within SHRM frameworks not only enhances the environmental performance of organizations but also contributes to building resilient workforces capable of adapting to the ongoing disruptions posed by climate change.In parallel with the evolution of SHRM, a related field—Green Human Resource Management (GHRM)—has emerged to specifically address the integration of environmental concerns into HR policies and practices. GHRM focuses on creating environmentally responsible organizational cultures and promoting employee behaviors that support ecological sustainability. It includes initiatives such as green hiring practices, sustainability-focused training, eco-friendly workplace policies, and incentives tied to environmental performance. The synergy between SHRM and GHRM provides a holistic and long-term approach to embedding climate resilience and mitigation into the core fabric of organizational functioning.While GHRM has gained momentum in recent years, much of the literature remains conceptual or fragmented, with limited empirical integration into broader climate strategies. Moreover, many organizations, particularly in developing countries, have yet to systematically incorporate HR-driven sustainability practices. This is due to a combination of factors, including lack of awareness, limited institutional support, fragmented policy environments, and insufficient research linking HR interventions to climate outcomes. Consequently, there is a pressing need to articulate a comprehensive framework that positions SHRM not as a peripheral player but as a central agent in the global response to climate change.

This paper aims to fill this gap by critically exploring the intersections between SHRM, climate change mitigation, and resilience. It builds upon existing theoretical perspectives, such as the resource-based view (RBV) and institutional theory, to argue that human capital is not only a key resource but also a strategic asset in climate action. From this vantage point, SHRM can be leveraged to develop environmental competencies, foster climate-conscious behaviors, and support organizational agility in the face of climate disruptions. Furthermore, this paper advocates for the mainstreaming of SHRM into climate governance at both the organizational and policy levels, recognizing that people-centered strategies are vital for achieving the United Nations Sustainable Development Goals (SDGs), particularly Goal 13 on climate action and Goal 8 on decent work and economic growth. Remote work, digital transformation, and climate-induced migration are reshaping labor markets in unprecedented ways. SHRM must therefore evolve to address not just the environmental footprint of organizations but also the social and psychological impacts of a changing climate on their workforce. This calls for HR professionals to develop competencies in climate literacy, risk management, occupational health, and social equity.The transition to a low-carbon economy is creating new green jobs while rendering certain roles obsolete. As industries such as fossil fuels decline and renewable energy, green construction, and sustainable agriculture rise, HR departments are at the forefront of managing this workforce transformation. Strategic reskilling, succession planning, and leadership development become crucial for ensuring that employees can thrive in this new context. SHRM thus becomes an enabler of the just transition—ensuring that climate action does not come at the cost of social and economic inequality.

2. THEORETICAL FOUNDATIONS OF SHRM AND ENVIRONMENTAL SUSTAINABILITY

As climate change intensifies, organizations are increasingly required to address their environmental impact alongside their traditional financial goals. Strategic Human Resource Management (SHRM), originally focused on achieving competitive advantage through people, is now being repositioned to integrate sustainability objectives—particularly those related to climate resilience and environmental stewardship. This section explores the theoretical underpinnings of SHRM, its linkage to environmental objectives, and the emergence of Green Human Resource Management (GHRM) as a specialized approach within the broader SHRM paradigm.

2.1 Strategic Human Resource Management (SHRM)

Strategic Human Resource Management (SHRM) represents a paradigm shift from operational HRM, emphasizing long-term alignment between human resources and organizational strategy. Rooted in the foundational work of Wright and McMahan (1992), SHRM is defined as "the pattern of planned human resource deployments and activities intended to enable an organization to achieve its goals." This strategic approach underscores the proactive configuration of HR systems to support an organization's mission, values, and long-term plans.In traditional settings, SHRM has focused primarily on productivity, innovation, and workforce engagement as mechanisms for achieving competitive advantage. These goals were typically embedded within business-centric priorities such as profitability, efficiency, and market share. However, in response to evolving global risks—including ecological degradation and climate change—SHRM is increasingly being called upon to integrate sustainability dimensions. Scholars now argue that people management should not only serve economic goals but also contribute to broader societal objectives, such as reducing carbon emissions and conserving natural resources (Jabbour & Santos, 2008).This transition is reflected in an expanded view of human capital that includes environmental competencies such as sustainability literacy, pro-environmental behavior, and climate-related problem-solving skills. In this context, SHRM becomes a lever for achieving environmental objectives by embedding sustainability values into the organizational fabric. This includes creating HR systems that attract environmentally conscious employees, train them in eco-friendly practices, and reward behaviors aligned with climate change mitigation and resilience.

Moreover, SHRM is instrumental in creating organizational cultures that support environmental ethics. The notion of culture-building through HR practices—such as performance appraisal, reward systems, and leadership development—extends beyond conventional business goals to embrace values of environmental responsibility. For example, performance metrics may now include indicators such as reduction in energy use or contribution to sustainability initiatives. In this way, SHRM serves not only to optimize labor productivity but also to cultivate a workforce committed to sustainable development.

2.2 Linking SHRM with Environmental Objectives

The conceptual linkage between SHRM and environmental sustainability can be understood through several theoretical lenses. One of the most influential is the Resource-Based View (RBV) of the firm, which argues that organizations achieve sustained competitive advantage by leveraging resources that are valuable, rare, inimitable, and non-substitutable (Barney, 1991). Human resources, when strategically managed, fulfill these criteria. Employees' knowledge, attitudes, and skills become critical assets that shape an organization's capability to engage in environmental innovation and responsiveness.

From the RBV perspective, SHRM practices that cultivate environmental competencies—such as green training, sustainability-focused recruitment, and eco-leadership development—can generate a workforce that provides a strategic edge in addressing environmental challenges. For instance, organizations with employees who can devise energy-efficient processes or design low-emission products are better positioned to respond to regulatory demands and consumer preferences for sustainability. These capabilities are difficult for competitors to replicate, thereby securing long-term advantage.Complementing the RBV is Institutional Theory, which posits that organizations are embedded within broader societal and regulatory frameworks that influence their behavior. Institutional pressures—categorized as coercive (legal regulations), mimetic (industry benchmarking), and normative (social expectations)—shape organizational responses to environmental issues. SHRM plays a critical role in enabling organizations to adapt to these pressures by institutionalizing environmentally responsible practices at the workforce level.For example, organizations facing increasing stakeholder demands for environmental accountability may revise their HR policies to include sustainability performance reviews, promote eco-conscious leadership, or create green job roles. In doing so, HR departments act as institutional agents that translate external pressures into internal behavioral norms. Through this process, organizations do not merely comply with environmental mandates but embed sustainability into their organizational identity and culture.Another relevant perspective is Stakeholder Theory, which emphasizes the importance of addressing the interests of all parties affected by organizational activities—not just shareholders. Employees, customers, suppliers, governments, and communities are increasingly concerned with environmental performance. SHRM enables organizations to meet these expectations by aligning workforce capabilities and values with stakeholder demands for sustainable operations. For instance, involving employees in green initiatives through participatory mechanisms not only enhances environmental outcomes but also builds trust and goodwill among stakeholders.In practice, the linkage between SHRM and environmental goals manifests through cross-functional collaboration between HR departments and environmental management units. Joint efforts can produce integrated strategies where HR serves as a catalyst for sustainability, facilitating initiatives such as energy audits, green innovation labs, and carbon footprint reduction programs. Such integration requires rethinking the role of HR as merely administrative to that of a strategic partner in climate governance.

2.3 Emergence of Green Human Resource Management (GHRM)

The growing intersection of HRM and environmental management has led to the emergence of Green Human Resource Management (GHRM)—a subset of SHRM explicitly designed to foster environmental sustainability. According to Renwick et al. (2013), GHRM refers to "the use of HRM policies to promote the sustainable use of resources within business organizations and, more generally, to promote environmental sustainability." GHRM extends conventional HR functions into green domains by integrating environmental objectives into the full spectrum of HR activities.

Green Recruitment and Selection

GHRM begins with the recruitment process, wherein organizations prioritize candidates who exhibit environmental values, sustainability knowledge, and a willingness to engage in eco-friendly practices. Job descriptions are rewritten to emphasize green roles and responsibilities, and selection criteria include assessments of environmental awareness. This ensures that sustainability is embedded from the earliest point of entry into the organization.Organizations also use employer branding to position themselves as green employers, thereby attracting applicants who are motivated by environmental purpose. This alignment enhances person–organization fit, increasing the likelihood of long-term employee engagement in sustainability efforts.

Green Training and Development

Once recruited, employees undergo green training programs that build their environmental competencies. These may include workshops on climate change, waste reduction, sustainable procurement, and energy efficiency. In more advanced settings, organizations partner with academic or non-governmental organizations to offer certifications in environmental management or corporate sustainability.Training also plays a vital role in fostering a culture of continuous improvement in environmental performance. By institutionalizing learning pathways around sustainability, GHRM enables employees to become change agents who innovate solutions to environmental challenges. Moreover, training promotes systems thinking—encouraging employees to understand how their actions impact broader ecological systems.

Green Performance Management and Rewards

Performance management systems under GHRM are adapted to include green KPIs (Key Performance Indicators). Employees are assessed not only on conventional productivity metrics but also on their contribution to environmental objectives. For example, an operations manager may be evaluated on the successful reduction of energy use, while a procurement officer might be assessed on sustainable sourcing practices.Reward systems are similarly aligned to incentivize green behavior. These may include bonuses for achieving sustainability targets, public recognition for eco-initiatives, or non-monetary rewards such as additional time off for participation in environmental volunteering. This alignment ensures that sustainability is not a peripheral concern but a core aspect of organizational success.

Green Employee Involvement and Engagement

GHRM fosters a participatory approach to environmental management by encouraging employee involvement in green decision-making and initiatives. Organizations may establish "green teams" or sustainability task forces comprising cross-functional members who collaborate on eco-projects such as waste reduction campaigns or green innovation challenges.Employee suggestion systems, sustainability hackathons, and open forums also enable employees to contribute ideas and feedback on environmental issues. This participatory culture not only enhances the quality of green initiatives but also strengthens employee commitment to organizational goals. Research suggests that employees who are actively engaged in environmental initiatives experience higher levels of job satisfaction and organizational citizenship behavior (Boiral & Paillé, 2012).

Building Resilient Organizational Cultures

In addition to promoting environmental sustainability, GHRM contributes to organizational resilience in the face of climate-related disruptions. Through proactive talent management and capacity-building, organizations are better equipped to anticipate, absorb, and adapt to environmental shocks. This may include developing contingency plans, reskilling employees for green transitions, or embedding climate risk assessment into strategic HR planning.Organizations that embrace GHRM also signal to external stakeholders—including investors, regulators, and customers—that they are serious about sustainability. This enhances reputational capital and can open access to green finance or public procurement opportunities. Furthermore, by aligning with international frameworks such as the UN Sustainable Development Goals (SDGs), GHRM reinforces an organization's social license to operate.

3. CLIMATE CHANGE MITIGATION: SHRM'S ROLE

The global imperative to combat climate change necessitates the active participation of all organizational functions, with Human Resource Management (HRM) playing a critical and often underutilized role. Strategic Human Resource Management (SHRM) can act as a fulcrum in aligning human capital with organizational environmental objectives, especially in terms of climate change mitigation.

SHRM strategies support the cultivation of a sustainability-oriented organizational culture, facilitate the integration of green values into the workforce, and reinforce climate-responsible behaviors at all levels of operation. This section explores how key HRM functions—recruitment, training, performance management, employee engagement, and workplace design—can be strategically directed toward climate change mitigation.

3.1 Green Recruitment and Selection

The foundation of any successful sustainability strategy lies in acquiring the right talent. Green recruitment and selection involve not only attracting candidates who possess technical competencies but also those who are committed to sustainability values. Through intentional recruitment processes, organizations can embed environmental priorities into their workforce DNA from the outset.Green recruitment practices involve multiple layers. Firstly, job descriptions are tailored to highlight environmental responsibilities, such as reducing carbon emissions, participating in waste management programs, or adhering to sustainable supply chain protocols. Secondly, organizational branding during recruitment communicates environmental commitments—for instance, through sustainability certifications, ecological impact reports, or climate pledges. This branding not only appeals to eco-conscious candidates but also strengthens the employer brand in an increasingly sustainability-driven job market (Renwick et al., 2013).Multinational corporations such as Unilever and IKEA exemplify best practices in green recruitment. These companies prioritize candidates with demonstrated interests in sustainability, often assessed through behavioral interviews, prior volunteering, or participation in green initiatives. For example, Unilever's "Sustainable Living Plan" influences its hiring criteria, ensuring new recruits align with its environmental mission. The strategic implication is that such alignment reduces the onboarding time needed for green acclimatization and enhances long-term commitment to environmental performance (Jabbour & de Sousa Jabbour, 2016).

3.2 Training and Capacity Building for Environmental Performance

While recruitment introduces sustainability values, training reinforces and deepens them. Green training programs aim to raise awareness about climate change, promote understanding of sustainable practices, and enhance the competencies required for environmentally responsible behavior. Effective training supports both individual behavioral change and organizational transformation.

3.3 Performance Management and Incentives

An organization's performance management system serves as a powerful mechanism for signaling what behaviors are valued. By integrating environmental metrics into performance appraisals, SHRM sends a clear message that sustainability is not peripheral but central to organizational success.Green performance indicators can be both quantitative and qualitative. Quantitative metrics might include energy savings achieved, reduction in paper usage, or decrease in travel-related emissions. Qualitative assessments might evaluate an employee's participation in green initiatives, contributions to sustainability committees, or adherence to eco-friendly practices. These metrics can be cascaded from organizational climate goals into individual KPIs, ensuring alignment and accountability at all levels (Jackson et al., 2011).The effectiveness of performance-based green HRM is enhanced when linked to tangible incentives. Organizations such as Adobe Systems have set precedents by embedding sustainability into their performance reviews. Employees are encouraged to track and report their carbon footprint, and departments are rewarded for achieving emissions reduction targets. This results in not only improved environmental outcomes but also in higher employee motivation and organizational commitment (Tang et al., 2018). Importantly, incentives do not have to be financial. Recognition programs, awards, or development opportunities tied to environmental contributions can be equally effective. The strategic goal is to integrate environmental accountability into the fabric of organizational performance, rather than treating it as an ancillary concern.

3.4 Encouraging Employee Engagement in Sustainability

Employee engagement in sustainability initiatives is a vital component of successful climate change mitigation. Engaged employees tend to exhibit higher levels of discretionary behavior, champion green practices, and promote a shared sense of purpose within the organization.SHRM can foster environmental engagement through a variety of participatory mechanisms. These include establishing green teams, where cross-functional groups identify and implement sustainability projects; supporting environmental volunteering days; and providing digital platforms for sustainability idea sharing and peer-to-peer learning. The participatory nature of such platforms enhances ownership and reinforces sustainability as a collective endeavor.Organizational case studies reveal that bottom-up engagement complements top-down mandates. For instance, Google encourages employees to propose green innovations, ranging from energy-saving algorithms to workplace waste reduction strategies. Such engagement not only improves environmental outcomes but also enhances employee satisfaction and retention (Bansal, 2005).Fostering a sustainability identity within the organization reinforces consistent pro-environmental behavior.

When employees see themselves as stewards of climate responsibility, they are more likely to internalize green norms and act accordingly—even outside formal incentive structures. This internalization is key to long-term cultural transformation (Ones & Dilchert, 2012).

3.5 Green Workplace Design and Flexibility

The physical and structural design of the workplace has a substantial impact on an organization's carbon footprint. SHRM has a unique role to play in advocating for and managing green workplace innovations that simultaneously reduce emissions and improve employee well-being.

One of the most significant developments in this space is the promotion of remote and hybrid work models. By reducing daily commuting, organizations contribute to substantial cuts in transportation-related greenhouse gas emissions. Flexible work policies, when designed strategically, also support work-life balance, thereby enhancing employee morale and productivity (Miller & Monaghan, 2017).Green commuting incentives such as subsidies for electric vehicle use, bike-to-work programs, and public transit allowances can shape employee travel behavior in favor of low-carbon options. For instance, Salesforce offers green travel reimbursements as part of its climate action plan, effectively aligning personal and organizational climate goals.Paperless systems and digital HR processes also represent significant opportunities. Transitioning from paper-based to cloud-based HR systems not only improves efficiency but drastically cuts down resource use. Digital tools for onboarding, performance reviews, and payroll reduce material consumption and align with broader digital sustainability strategies (Meuer et al., 2020).HR professionals can advocate for green building certifications (e.g., LEED or BREEAM) and contribute to the creation of eco-conscious workspaces with energy-efficient lighting, HVAC systems, and recyclable materials. These infrastructural changes are essential for aligning organizational operations with climate mitigation goals and demonstrate a visible commitment to sustainability.

4. CLIMATE RESILIENCE: SHRM'S STRATEGIC CONTRIBUTION

Climate resilience—the capacity of systems, organizations, and individuals to anticipate, absorb, adapt to, and recover from climate-induced shocks—has become a defining concern for contemporary organizations. The strategic human resource management (SHRM) function plays a pivotal role in strengthening this resilience by embedding adaptive capacity within the workforce. Rather than treating environmental uncertainty as an operational obstacle, SHRM can transform it into a domain for strategic development and innovation.

4.1 Building Adaptive Capacity

Adaptive capacity refers to the organizational and individual ability to adjust to potential damage, take advantage of opportunities, or respond to consequences of climate variability and change. SHRM builds adaptive capacity by fostering a culture of continuous learning, flexibility, and preparedness. This can be achieved through tailored training programs focused on environmental risk awareness, change management, and systems thinking.

4.2 Workforce Planning and Climate Risk

Effective workforce planning that integrates climate risk is central to building resilience. This is particularly important in industries heavily reliant on manual labor or outdoor activity. HR professionals must incorporate climate data—such as temperature trends, storm projections, and water availability—into workforce scheduling, safety planning, and talent deployment.For instance, in agriculture, unpredictable rainfall patterns and heatwaves can affect both crop yield and worker productivity. HR strategies must therefore include rotational workforce models, remote monitoring systems, and responsive health protocols. By aligning HR planning cycles with environmental forecasting, organizations ensure they are not caught off guard during critical periods.Logistics and supply chain organizations are also vulnerable to climate-induced delays and labor shortages. HR must develop contingency staffing plans that include cross-training, temporary workforce pools, and geographical mobility support. These actions reduce exposure to single points of failure and maintain business continuity.

4.3 Mental Health and Employee Wellbeing

Climate change is not only a physical threat but also a psychological one. Terms like "eco-anxiety" and "climate grief" are now widely recognized, particularly among younger generations who feel the weight of a deteriorating planet. Moreover, employees directly exposed to natural disasters, displacement, or resource scarcity often suffer from post-traumatic stress disorder (PTSD) or chronic stress conditions. SHRM can strategically address this by embedding mental health into the core of resilience strategies. Initiatives include offering counseling services, facilitating access to trauma support specialists, and incorporating stress-reduction practices like mindfulness and wellness coaching into the workplace. For example, organizations like the International Federation of Red Cross and Red Crescent Societies (IFRC) have integrated psychological first aid training as a core component of employee preparedness in disaster-prone zones.In leadership development, emotionally intelligent managers must be trained to recognize signs of distress, encourage open dialogue, and cultivate a psychologically safe environment. Mental wellbeing must not be treated as a reactive service but rather as a proactive competency that contributes to both individual and organizational resilience.

4.4 Diversity, Equity, and Inclusion (DEI) in Resilience Planning

Climate impacts are not evenly distributed. Marginalized communities—including low-income groups, racial minorities, people with disabilities, and women in vulnerable economies—are disproportionately affected. Therefore, resilience planning must be intersectional and equity-focused, ensuring that all demographic groups have equal access to safety, opportunity, and resources.SHRM plays a transformative role by embedding DEI principles into climate adaptation policies. HR can ensure inclusive decision-making by involving diverse employee voices in resilience strategy formulation. For example, inclusive town halls, employee resource groups (ERGs), and participatory workshops allow for localized understanding of risks and foster community-driven solutions.Moreover, access to climate resilience training, remote work technology, and relocation support must be equitably distributed. Organizations that overlook DEI in resilience strategies risk not only reputational damage but also increased internal vulnerability. A DEI-centered SHRM ensures that adaptation resources reach those who need them most, while also leveraging the cognitive and experiential diversity of the workforce to enhance collective problem-solving.

5. ORGANIZATIONAL CASE STUDIES: BEST PRACTICES IN GREEN SHRM

While theoretical and conceptual frameworks offer valuable insights into the potential of Strategic Human Resource Management (SHRM) to drive environmental sustainability, the translation of these concepts into practice is best illustrated through case studies. These organizational narratives highlight the tangible ways in which SHRM can operationalize green policies, foster environmental stewardship, and promote climate resilience and mitigation across diverse sectors and geographic contexts. The following case studies illuminate exemplary practices from both private and public sectors that align human capital management with ecological goals.

5.1 Interface Inc.: Pioneering Carbon-Neutral HRM in Manufacturing

Organizational Overview and Sustainability Vision

Interface Inc., a global leader in modular carpet manufacturing, has established itself as a pioneer in aligning sustainability with core business operations. The company's "Climate Take Back" initiative, launched in 2016, marked a transformative shift toward becoming a carbon-negative enterprise. Central to this mission is the deliberate integration of sustainability objectives within human resource functions—a strategy that exemplifies Green SHRM in action.

HRM-Embedded Climate Strategies

Interface's HRM architecture incorporates environmental key performance indicators (KPIs) into employee evaluation systems, embedding sustainability into individual performance metrics. This approach ensures that ecological goals are not relegated to a separate sustainability department but become part of each employee's professional responsibilities. New employees participate in structured green onboarding sessions, where climate-related organizational values are communicated early and emphatically. These sessions include workshops on circular economy principles, sustainable supply chain logistics, and waste minimization techniques.

Learning and Development: Empowering Employees

The HR function at Interface collaborates with operations and technical divisions to design and deliver bespoke training modules focused on circular manufacturing and carbon footprint reduction. These learning interventions aim not only to enhance technical proficiency but also to cultivate a sense of environmental responsibility and ownership among employees. A recurring module involves scenario-based simulations on sustainable product end-of-life strategies, encouraging employees to innovate around reuse, recycling, and biodegradability.

Outcomes and Impact

The outcomes of Interface's Green SHRM approach are measurable and impactful. Since the mid-1990s, the company has achieved a 96% reduction in greenhouse gas (GHG) emissions from its operations. Employee-driven innovations have also led to a significant reduction in landfill waste and water usage. By positioning HRM as a central actor in climate leadership, Interface has fostered a culture where sustainability is a shared norm and not a top-down directive. The organization's experience affirms that HRM, when aligned with ecological imperatives, can serve as a transformative agent for climate action in industrial sectors.

5.2 Infosys: Training for Energy Efficiency in the IT Sector

Contextual Background

Infosys, a leading Indian multinational corporation in the information technology (IT) services and consulting space, presents a compelling example of how SHRM can advance climate goals in a high-tech, knowledge-intensive industry. While IT operations are often perceived as less carbon-intensive than manufacturing, the large energy demands of data centers, computing infrastructure, and sprawling campuses contribute significantly to emissions. Recognizing this, Infosys implemented the "Green Infosys" initiative, spearheaded jointly by its HR and sustainability departments.

Strategic HR Interventions for Energy Conservation

The HR function took a leadership role in embedding sustainability into training and leadership development programs. All new managers undergo mandatory training on energy-efficient IT practices, while senior leaders are trained in carbon accounting and responsible procurement. A key innovation was the introduction of a "Green Champion" program, wherein employees volunteer to monitor and promote sustainable practices in their respective departments.HR has also played a crucial role in developing gamified digital dashboards that track individual and team-level energy consumption metrics. These dashboards promote behavioral change by offering real-time feedback and providing incentives such as digital badges, recognition in internal newsletters, and performance-based rewards for energy-saving achievements.

Employee Development and Cultural Shifts

Environmental literacy is integrated into broader upskilling programs. Courses in green computing, digital waste minimization, and sustainable procurement are available through the company's internal learning platform. Participation in these modules is tracked and included in performance appraisals, further solidifying the link between environmental goals and career development.

Cultural transformation has been a key focus of HR interventions. Through storytelling and internal communication campaigns, Infosys positions sustainability as a shared value and strategic priority. Annual green week celebrations, eco-friendly hackathons, and department-level competitions promote employee engagement and reinforce the organization's climate ethos.

Tangible Benefits and Recognition

Infosys's Green SHRM initiatives have yielded concrete outcomes. All company buildings have achieved LEED Platinum certification, and the carbon intensity per employee has reduced significantly over the past decade. These achievements have been recognized globally, positioning Infosys as a leader in corporate sustainability. Crucially, these results were driven not just by technological innovation but by deliberate HR strategies that fostered sustainable behaviors, mindsets, and competencies among the workforce.

5.3 Patagonia: Embedding Climate Ethics through Human-Centric HRM

Organizational Values and HRM Alignment

Patagonia, a U.S.-based outdoor clothing and gear company, is often cited as an exemplar of corporate environmentalism. The firm's HR strategy is unique in that it aligns ethical climate commitments with organizational design, employee engagement, and governance processes. Rather than treating sustainability as a separate function, Patagonia integrates ecological consciousness into every dimension of its human capital strategy.

Purpose-Driven Recruitment and Onboarding

Patagonia's recruitment philosophy is rooted in values-based hiring. Job advertisements explicitly state the organization's environmental activism, and candidates are assessed not only on technical competencies but also on their alignment with Patagonia's climate mission. The onboarding process includes not only orientation to the company's history and operations but also training on its environmental campaigns and policy advocacy.

Employee Engagement and Advocacy

Employees are encouraged and financially supported to engage in environmental activism. The company provides paid leave for staff who wish to participate in protests, attend environmental justice conferences, or engage in community conservation efforts. This practice positions HRM as a facilitator of civic engagement and sustainability leadership. Patagonia has established employee-led advisory councils that influence sustainability decisions and environmental reporting. These councils receive administrative support from the HR department, ensuring that climate governance includes diverse voices and reflects internal consensus. HR is also instrumental in implementing the "1% for the Planet" program, which allows employees to nominate and vote on environmental NGOs that receive company donations.

Organizational Culture and Leadership Development

Leadership development at Patagonia emphasizes climate literacy, systems thinking, and ethical leadership. Middle and senior managers undergo scenario planning workshops that assess supply chain disruptions under different climate models. This training enables them to make proactive, resilient decisions in their strategic functions.

The company's culture—rooted in transparency, activism, and stewardship—is reinforced through HR-led storytelling, cross-functional mentorship programs, and symbolic rituals such as "Zero-Waste Days." These practices foster a sense of community around sustainability and encourage employees to view climate action as an integral part of their professional identity.

Outcomes and Reputation

Patagonia's HR-driven climate initiatives have helped it build a loyal workforce, reduce turnover, and attract top talent aligned with its environmental values. In addition to operational efficiencies, the company has earned a reputation as a socially and environmentally responsible brand. Its success demonstrates how HRM can transcend administrative functions to become a strategic partner in advancing planetary health.

5.4 City of Melbourne: Climate-Resilient Workforce Planning in the Public Sector

Institutional Context and Climate Risks

Municipal governments face growing responsibilities in climate adaptation, particularly as urban areas become increasingly vulnerable to extreme weather events. The City of Melbourne has emerged as a global leader in climate-responsive public governance by integrating SHRM principles into municipal workforce planning. This is particularly notable given the complex and risk-laden contexts of city administration, where employee safety and service continuity are critical.

Integrating Climate Data into HR Forecasting

The HR department at the City of Melbourne collaborates with climate scientists and urban planners to incorporate meteorological data into workforce planning models. This includes forecasting workforce needs in anticipation of rising urban heat, flooding events, and air quality deterioration. The city has established a dedicated climate workforce unit within its HR structure, which focuses on long-term talent planning, skill gap analysis, and emergency staffing.

Preparedness, Flexibility, and Employee Wellbeing

The City of Melbourne has developed emergency response protocols that are embedded in HR policies. Frontline staff such as sanitation workers, utility crews, and traffic managers undergo regular climate risk training, which includes emergency response simulations and heat-stress management. Flexible scheduling policies—such as staggered shifts and telecommuting options—are activated during extreme weather events to protect worker health while maintaining essential services.HR has also invested in health infrastructure to mitigate climate impacts on vulnerable employees. These include hydration stations at worksites, mobile first-aid clinics, and climate shelters during heatwaves. Mental health services are also provided, with an emphasis on climate anxiety and burnout, particularly for employees involved in disaster response.

Learning and Capacity Building

Learning modules for city employees cover climate science basics, the implications of climate risks for urban systems, and scenario-based planning. HR works with educational institutions to provide continuous learning opportunities, including short courses, certifications, and workshops on climate resilience. Importantly, climate competencies are now included in the job descriptions and promotion criteria for many public service roles.

Public Accountability and Organizational Resilience

The City of Melbourne regularly publishes workforce climate adaptation reports as part of its broader sustainability disclosures. These reports include KPIs related to HR-led initiatives such as emergency training coverage, employee engagement in climate programs, and service continuity during crises. As a result, the city has positioned itself not only as a steward of the public good but also as a resilient employer that protects and empowers its workforce in the face of escalating environmental challenges.

6. CHALLENGES IN INTEGRATING SHRM AND CLIMATE STRATEGIES

Despite the proven benefits, integrating SHRM with climate strategies remains complex and fraught with structural and behavioral obstacles. These challenges, unless addressed systemically, may hinder progress toward sustainability and resilience goals.

6.1 Organizational Silos

A key barrier lies in the siloed structure of most organizations. Sustainability is often viewed as the domain of the environmental, health, and safety (EHS) or corporate social responsibility (CSR) departments, while HR is confined to operational functions. This fragmentation leads to missed opportunities for synergy and innovation.

Breaking down these silos requires structural reform. Cross-departmental task forces, shared performance indicators, and co-owned sustainability goals can foster collaboration. For instance, co-designing training programs between HR and sustainability officers ensures both technical rigor and behavioral alignment.Leadership commitment is crucial. When top management signals the strategic importance of green HRM, it legitimizes interdepartmental cooperation and resource allocation.

6.2 Skill Gaps in HR Departments

Most HR professionals are trained in traditional disciplines such as labor law, compensation, and recruitment, with little exposure to environmental science or sustainability frameworks. This skills deficit undermines the ability of HR departments to drive climate-responsive strategies.To close this gap, HR curricula—both academic and professional—must evolve. Certification programs on sustainable HRM, partnerships with environmental organizations, and internal learning academies can empower HR staff with the necessary competencies. The rise of roles such as "Sustainability HR Partner" reflects this trend.Organizations like the Society for Human Resource Management (SHRM) and Chartered Institute of Personnel and Development (CIPD) are beginning to offer green HR modules, but adoption remains uneven.

6.3 Resistance to Change

Behavioral resistance is another obstacle. Employees may perceive sustainability initiatives as added burdens or misaligned with their core duties. Managers may fear disruptions to performance metrics or loss of control over traditional workflows. Overcoming resistance requires strategic communication, psychological safety, and structured incentives. Narratives that link sustainability to organizational success, personal wellbeing, and job security are more effective than compliance-based mandates. Participatory approaches—such as crowd sourcing green ideas or involving staff in policy design—build ownership and acceptance.Pilot projects and early successes can be used as change catalysts. By highlighting quick wins, HR can build momentum for broader transformation.

6.4 Measuring Impact

Finally, the absence of standardized metrics to assess SHRM's climate impact impedes strategic integration. While carbon accounting is common in operational contexts, few organizations have HR-focused KPIs that capture resilience, adaptation readiness, or behavioral change.Developing such metrics is essential. Possible indicators include the percentage of employees trained in sustainability, absenteeism during climate events, employee engagement in green initiatives, or even climate literacy scores. Impact dashboards that integrate HR data with environmental outcomes can support evidence-based decision-making.Organizations like the Global Reporting Initiative (GRI) and Task Force on Climate-related Financial Disclosures (TCFD) are moving toward more integrated disclosures, but HR contributions remain underrepresented. Academic researchers and practitioners must collaborate to develop robust, scalable, and sector-specific metrics.

7. FUTURE DIRECTIONS FOR RESEARCH AND PRACTICE

As the nexus between climate change and strategic human resource management (SHRM) continues to evolve, new research avenues and practical innovations are emerging. The complexity and urgency of environmental sustainability necessitate a more profound and nuanced understanding of how SHRM strategies can be optimized to support mitigation and adaptation goals. This section outlines several promising directions for future research and practice, rooted in empirical inquiry, sector-specific models, cross-cultural dynamics, and integration with corporate sustainability frameworks.

7.1 Empirical Studies on SHRM–Climate Outcomes Linkages

One of the most critical needs in this domain is the generation of robust empirical evidence demonstrating the direct and indirect impacts of SHRM on climate-related outcomes. While theoretical models and conceptual frameworks abound, there is a paucity of quantitative research that rigorously tests the efficacy of green HR practices on tangible environmental performance indicators. Future research should focus on longitudinal studies and experimental designs to establish causality between HR interventions—such as green recruitment, eco-centric training programs, and environmentally linked performance incentives—and metrics like carbon footprint reduction, energy conservation, and sustainable resource use.Researchers should seek to disentangle the mediating and moderating mechanisms that influence these relationships. For example, organizational culture, leadership support, and employee environmental values may shape the effectiveness of SHRM interventions. Advanced statistical techniques such as structural equation modeling (SEM) and multilevel modeling (MLM) can be employed to explore these dynamics in depth.

7.2 Development of Sector-Specific SHRM Models

Climate-related risks and opportunities differ significantly across industries, necessitating customized SHRM frameworks tailored to sectoral contexts. High-emission sectors such as manufacturing, construction, energy, transportation, and agriculture face unique environmental challenges that require context-specific HR strategies. For instance, in the construction sector, SHRM may focus on promoting safety, energy-efficient practices, and sustainable materials training. In contrast, agriculture may benefit from SHRM approaches that enhance local ecological knowledge, seasonal workforce resilience, and sustainable land use practices.

Future studies and policy development should prioritize the creation of industry-specific SHRM toolkits. These models should incorporate relevant regulatory frameworks, stakeholder expectations, and technological innovations unique to each sector. Additionally, partnerships between academia, industry associations, and government bodies can facilitate the development and testing of these tailored frameworks in real-world environments.

7.3 Cross-Cultural and Regional Perspectives

Another promising research frontier lies in understanding how cultural and regional contexts influence the design and implementation of green SHRM practices. The effectiveness of climate-oriented HR initiatives is often mediated by cultural values, employee beliefs, and societal norms. For example, collectivist cultures may respond more positively to community-focused sustainability initiatives, whereas individualistic cultures may emphasize personal accountability and rewards.Cross-cultural comparative studies are essential to uncover these nuances and inform globally adaptable SHRM strategies. Additionally, research should explore the role of national institutional environments—such as legal frameworks, labor market dynamics, and education systems—in shaping green HRM adoption. A regional analysis of green SHRM across the Global North and South can also highlight disparities in resource availability, regulatory maturity, and organizational readiness for sustainability transitions.

7.4 Integration with Environmental, Social, and Governance (ESG) Reporting

In the era of integrated corporate reporting and responsible investment, there is growing recognition of the role that HR plays in achieving Environmental, Social, and Governance (ESG) objectives. However, the HR function is often underrepresented in ESG disclosures, leading to a lack of visibility and accountability regarding human capital's contribution to sustainability.Research and practice should focus on developing standardized metrics and indicators that link HR outcomes to environmental performance. This includes tracking training hours devoted to sustainability, green leadership development initiatives, employee participation in climate programs, and retention of sustainability-skilled talent. Embedding these metrics into ESG reporting frameworks will not only enhance transparency but also position HR as a strategic driver of corporate sustainability. Collaborative efforts with organizations such as the Global Reporting Initiative (GRI) and the Sustainability Accounting Standards Board (SASB) can support this integration.

8. POLICY AND STRATEGIC RECOMMENDATIONS

To operationalize the potential of SHRM in addressing climate change, a robust policy ecosystem is required—one that supports organizations across sectors and sizes in adopting and institutionalizing green HR practices. The following strategic recommendations are designed to inform governments, regulatory bodies, international agencies, and corporate stakeholders on key actions that can accelerate the transition toward sustainable human resource management.

8.1 Mandate Sustainability Training Across Sectors

One of the foundational steps to climate-resilient HRM is widespread sustainability training. Regulatory authorities should make it mandatory for organizations—across public and private sectors—to integrate climate literacy into employee development programs. This training should encompass both general awareness (e.g., climate science, sustainability principles) and job-specific competencies (e.g., sustainable procurement, green innovation, eco-efficiency practices).National qualification bodies and industry certification agencies can play a pivotal role in setting curriculum standards and accreditation mechanisms. Embedding sustainability into training not only enhances employee knowledge but also fosters a culture of shared responsibility, empowering the workforce to contribute actively to mitigation and adaptation goals.

8.2 Develop National Green HRM Frameworks

Governments should spearhead the creation of national-level green HRM frameworks that provide standardized guidelines, tools, and resources to support organizations in integrating environmental considerations into HR functions. These frameworks should be designed collaboratively with input from academia, industry, labor unions, and civil society.Such frameworks can include sample job descriptions for green roles, templates for sustainability-focused performance appraisals, recruitment protocols that prioritize eco-conscious candidates, and training modules on green competencies. Tailoring the framework for different sectors, organizational sizes, and regional contexts will enhance applicability and adoption.

8.3 Encourage Interdepartmental Collaboration

Sustainability is inherently cross-functional, and HR cannot operate in isolation. Policymakers and organizational leaders should institutionalize mechanisms for interdepartmental collaboration by establishing formal green committees or task forces. These committees should include representatives from HR, sustainability, operations, finance, and innovation departments.The goal is to ensure alignment between environmental goals and people management strategies. For example, a sustainability goal to reduce waste should be supported by HR initiatives such as staff incentives for eco-efficiency, training on waste management, and inclusion of green KPIs in appraisals. Cross-functional collaboration ensures consistency, efficiency, and shared ownership of climate objectives.

8.4 Offer Incentives for Green HR Initiatives

To catalyze the adoption of green SHRM, governments and development agencies should provide financial and non-financial incentives for organizations that implement sustainable HR practices. This could include tax breaks, grants, or preferential treatment in public procurement for companies that demonstrate excellence in green HRM.Recognition programs such as national sustainability awards or green employer certifications can also motivate firms to innovate in their HR strategies. These incentives should be tied to clearly defined criteria and measurable outcomes, encouraging accountability and continuous improvement.

8.5 Promote Green Leadership Development

Leadership is a critical enabler of organizational change, and building a cadre of sustainability-oriented leaders is essential for advancing SHRM goals. Policies should encourage or mandate leadership development programs that integrate environmental stewardship, systems thinking, and change management.Universities, business schools, and training institutions should embed sustainability modules into executive education and management courses. In parallel, organizations should identify and nurture internal sustainability champions who can inspire and guide their teams toward greener behaviors and strategic alignment.

8.6 Establish Metrics and Benchmarks for HR Sustainability Performance

One of the challenges in scaling green SHRM practices is the lack of standardized performance metrics. Policymakers and professional HR bodies should collaborate to develop universally applicable benchmarks for assessing the environmental dimension of HRM.Suggested metrics include the proportion of staff trained in sustainability, employee engagement in green programs, HR's contribution to ESG outcomes, and workforce resilience indicators. These metrics can be integrated into national sustainability indices and used for benchmarking, reporting, and policy evaluation.

8.7 Foster International Cooperation on Green SHRM

Climate change is a global issue, and fostering international collaboration is vital for scaling effective SHRM practices. Governments, international organizations, and research institutions should facilitate knowledge exchange platforms such as global forums, regional summits, and online communities of practice focused on green HRM.Bilateral and multilateral partnerships can also support the co-development of training curricula, research projects, and technical assistance programs. For example, joint research initiatives between developed and developing countries can help localize SHRM strategies and bridge capacity gaps.

8.8 Integrate Climate Considerations in Labor Laws and Employment Standards

To institutionalize the climate-HR nexus, labor legislation and employment policies must explicitly recognize and support green roles and sustainability responsibilities. Labor ministries and legal bodies should update employment standards to include climate resilience provisions—such as rights to a safe environment, environmental responsibility clauses, and recognition of green job classifications.This legal recognition enhances the legitimacy of SHRM interventions and protects workers engaged in climate-critical roles. It also obligates employers to invest in sustainability training, workplace adaptation, and equitable transitions for vulnerable workers.

8.9 Fund Academic and Applied Research in Green SHRM

To expand the evidence base and foster innovation, public and private funding agencies should invest in interdisciplinary research on green SHRM. Priority should be given to action research that explores the implementation, outcomes, and scalability of SHRM practices in real organizational settings.Research funding should also support the development of digital tools, decision-support systems, and HR analytics platforms that facilitate the integration of climate data into human resource decision-making. Collaborative research centers, university–industry partnerships, and global research networks can serve as incubators for high-impact knowledge creation.

8.10 Support SMEs in the Green Transition

Small and medium enterprises (SMEs) often lack the capacity and resources to implement sophisticated SHRM practices. Policymakers should design targeted programs to support SMEs in adopting sustainable HR strategies. This includes subsidies for green training programs, toolkits tailored to small business contexts, and access to sustainability consultants or HR experts.Public–private partnerships can create shared platforms or networks that allow SMEs to pool resources, share best practices, and access affordable training solutions. Supporting SMEs is particularly important in developing economies where they form the backbone of employment and are critical to national climate objectives.

9. CONCLUSION

The climate crisis demands not only technological and policy solutions but also a transformation of how organizations manage their people. Strategic Human Resource Management offers a powerful, yet underutilized, lever to align human capital with environmental goals. By embedding sustainability into recruitment, training, performance, and workplace culture, SHRM can contribute to both mitigation and resilience. The integration of SHRM with environmental sustainability is not merely an operational adjustment—it represents a fundamental rethinking of the role of people in addressing climate change. Drawing from theories such as RBV, institutional theory, and stakeholder theory, SHRM provides a powerful framework for aligning human resource strategies with ecological goals. The emergence of GHRM further operationalizes this alignment through targeted practices that infuse sustainability across the employee lifecycle.By embedding environmental values into HR systems, organizations can leverage their human capital to drive innovation, enhance adaptability, and contribute meaningfully to global climate action. This shift also redefines competitive advantage in the 21st century—not solely as a function of financial performance, but as a measure of resilience, responsibility, and sustainability. The convergence of climate imperatives and human resource management opens a transformative pathway for organizations to become agents of sustainability. Realizing this potential requires a concerted effort across academia, industry, and policy spheres. By advancing empirical research, promoting cross-sectoral innovation, and implementing forward-looking policy instruments, the integration of SHRM into climate change strategies can be both deepened and accelerated. These future directions and recommendations provide a roadmap for embedding environmental consciousness into the very fabric of human capital management—paving the way for a more resilient, sustainable, and inclusive organizational future.This paper has shown that HR is not just an administrative function—it is a strategic enabler of climate action. With appropriate frameworks, incentives, and collaboration, SHRM can help build organizations that are both competitive and climate-conscious. In a warming world, the future of work must also be the future of sustainability.

REFERENCES

Bansal, P. (2005). Evolving sustainably: A longitudinal study of corporate sustainable development. *Strategic Management Journal, 26*(3), 197–218.

Barney, J. (1991). Firm resources and sustained competitive advantage. *Journal of Management, 17*(1), 99–120.

Boiral, O., & Paillé, P. (2012). Organizational citizenship behaviour for the environment: Measurement and validation. *Journal of Business Ethics, 109*(4), 431–445.

Intergovernmental Panel on Climate Change (IPCC). (2023). [*Synthesis Report of the IPCC Sixth Assessment Report.* Geneva: IPCC.]. *Climatic Change*, •••, 2023.

Jabbour, C. J. C., & de Sousa Jabbour, A. B. L. (2016). Green human resource management and green supply chain management: Linking two emerging agendas. *Journal of Cleaner Production, 112*, 1824–1833.

Jabbour, C. J. C., & Santos, F. C. A. (2008). The central role of human resource management in the search for sustainable organizations. *International Journal of Human Resource Management, 19*(12), 2133–2154.

Jackson, S. E., Renwick, D. W. S., Jabbour, C. J. C., & Muller-Camen, M. (2011). State-of-the-art and future directions for green human resource management: Introduction to the special issue. *German Journal of Human Resource Management, 25*(2), 99–116.

Meuer, J., Koelbel, J., & Hoffmann, V. H. (2020). On the nature of corporate sustainability. *Organization & Environment, 33*(4), 319–341.

Miller, K., & Monaghan, P. (2017). Flexible working arrangements: Green benefits and challenges. *Journal of Environmental Management, 193*, 69–78.

Ones, D. S., & Dilchert, S. (2012). Environmental sustainability at work: A call to action. *Industrial and Organizational Psychology: Perspectives on Science and Practice, 5*(4), 444–466.

Renwick, D. W. S., Redman, T., & Maguire, S. (2013). Green Human Resource Management: A review and research agenda. *International Journal of Management Reviews, 15*(1), 1–14.

Tang, G., Chen, Y., Jiang, Y., Paillé, P., & Jia, J. (2018). Green human resource management practices: Scale development and validity. *Asia Pacific Journal of Human Resources, 56*(1), 31–55.

Wright, P. M., & McMahan, G. C. (1992). Theoretical perspectives for strategic human resource management. *Journal of Management*, *18*(2), 295–320.

Chapter 6
Promoting Climate Change Mitigation and Resilience Strategies Through Human Resource Management:
Performance Management and Rewards for Sustainability

Cansu Tunahan
https://orcid.org/0000-0001-8953-9739
Istanbul Rumeli Üniversitesi, Turkey

ABSTRACT

In line with sustainable development goals principles, the long-term value creation potential of businesses has started to be measured not only by economic outputs but also by environmental and social impacts (Elkington, 1997). Accordingly, it is necessary to restructure performance management and reward systems in order for organizations to effectively implement their sustainability strategies. Aligning performance management processes with sustainability goals is reported to increase employee motivation and organizational commitment, (Warhurst, 2002). In this context, rewarding sustainability-oriented performance is not limited to financial incentives, but is also supported through intangible rewards. In order for these systems to be implemented effectively, it is necessary to integrate strategic sustainability goals with corporate vision and actions, measure sustainable performance

DOI: 10.4018/979-8-3373-2479-1.ch006

indicators, create an organizational culture conducive to stakeholder engagement, base leadership activities on sustainability principles, and establish a fair reward system.

INTRODUCTION

In today's world, where unlimited wants and needs are tried to be met with limited resources, the rapid depletion of resources is an extremely important problem. In order to solve this problem, states have come together and tried to produce solutions to protect the resources of future generations. Since the 1960s, the activities of people and governments have been shaped in line with the aim of protecting nature and future generations (Elkington, 2006). These new behavioral patterns based on sustainability have closely affected the activities of businesses.

Especially with the awareness of the consumer group and the prominence of sustainability as a competitive factor, this issue has become of vital importance for businesses. Within the scope of sustainability, businesses first focused on using resources effectively in production and management processes. Later, they started to integrate sustainability into all their processes due to factors such as meeting customer expectations, creating an effective image, high profitability and strategic reasons. However, for the successful implementation of sustainability, employee engagement and motivation play as important and determining role as management decisions (Merriman et al., 2016). In order for employees to be motivated about sustainability, it is necessary to increase employee awareness (Das & Singh, 2016). Human Resource Management plays an important role in raising employee awareness and ensuring real environmental sustainability (Singh et al., 2020).

The 21st century sustainability approach goes beyond traditional corporate performance evaluation paradigms. In this context, redesigning performance management systems in line with sustainability principles increases the long-term value creation potential of organizations and plays a critical role in meeting stakeholder expectations. Sustainability-focused restructuring of performance management is seen as a fundamental tool for achieving behavioral and cultural transformation within the organization.

In order to increase the contribution of employees to sustainability goals and to ensure that corporate strategy is effectively reflected on the ground, performance-based reward systems should also be considered from a sustainability perspective. As a matter of fact, recent studies have shown that rewarding sustainability performance with tangible and intangible incentives strengthens employee motivation and organizational commitment (Eccles et al., 2014).

This book chapter discusses the theoretical foundations of sustainability-oriented performance management and reward systems. First, the theoretical relationship between sustainability and performance management is evaluated. Then, sustainable performance indicators and reward systems based on sustainability goals are discussed.

Throughout the chapter, it is emphasized that performance management and reward systems developed from a sustainability perspective not only provide environmental and social benefits to businesses, but also offer strategic value in terms of long-term competitive advantage and reputation management. Finally, the chapter discusses the main challenges faced in implementing sustainability-oriented performance and reward systems and suggests solutions.

SUSTAINABILITY

The concept of sustainability was first used in 1712. However, the concept gained popularity about 250 years later (Scoones, 2007). The concept was defined by WCED (World Commission on Environment and Development, 1987) as "meeting the needs of the present without endangering the ability of future generations to meet their needs". The first UNCED (United Nations Conference on Environment and Development, 1992) held in Rio in 1992 underlined the importance of the issue and the concept of sustainability became a global issue (Scoones, 2016).

The concept of sustainability focuses on the human connection with nature to achieve justice between nature and humans (Baumgärtner & Quaas, 2010). In this sense, sustainability represents the ability to make something work, to repair it, and thus to create a balance between the present and the future (Wilkinson et al., 2001).

SUSTAINABILITY IN THE BUSINESS WORLD

The fundamental nature of sustainability is to evaluate the success of the business not only in terms of material resources but also in social and environmental terms (Amjad et al., 2021). In this perspective, sustainability brings advantages not only to society but also to businesses, such as reduced transaction costs, higher brand reputation and competitive advantage (Dyllick & Muff, 2016). The widespread adoption of sustainability in the business world is due to factors such as compliance with increasing environmental regulations, cost, risk of resource depletion, and the level of public awareness (Jones et al., 2008). Businesses that invest in environmental sustainability gain competitive advantage and more benefits by involving stakeholders in sustainability activities (Saunila et al., 2024). This has led managers to pay more attention to the issue and to think that sustainability will have a significant impact on their future survival (Danciu, 2013).

The effective realization of environmental sustainability at the corporate level is not limited to technology and process-oriented innovations, but also requires the strategic role of human resources management. Human resources management is positioned as a key transformative element that shapes employee behavior, strengthens organizational values and is fundamental to achieving environmental sustainability goals. In this context, human resource functions (recruitment, training and development, performance management and rewarding) need to be structured in line with environmental sustainability principles. In particular, through green HRM practices, it becomes possible to raise environmental awareness among employees, encourage environmentally friendly business behaviors and ensure the adoption of a sustainable business culture. In addition to training and development activities, it is important to develop performance management and reward systems based on sustainability criteria in order to strengthen employees' motivation towards environmental goals and increase their awareness of environmental responsibility. Thus, it will be possible for sustainability strategies to go beyond rhetoric and become firmly rooted in business culture. GHRM plays an extremely important role in enabling businesses to achieve this goal by incorporating environmentally conscious policies into their human resources processes.

GREEN HUMAN RESOURCE MANAGEMENT

GHRM (Green Human Resource Management), which covers the strategies of organizations that promote environmental sustainability, addresses the relationship between HRM (Human Resource Management) activities and the environment (Renwick et al., 2013; Teixeira et al., 2016). In this sense, it is aimed to carry out business objectives by supporting employees' environmentally friendly behaviors (Rouholamin et al., 2023). GHRM includes activities such as environmentally sensitive employee selection processes, programs to increase employee environmental awareness, sustainable workplace practices, green performance and environmental management (Rajabpur et al., 2022).

GHRM is shaped with reference to sustainable development theory, corporate sustainability approach and stakeholder theory perspectives. In this context, the principles of conducting economic activities without harming the environment, adding social and environmental sustainability to business goals in addition to profitability, and being responsible to stakeholders in addition to shareholders appear in an integrated structure (Molina-Azorin et al., 2021; Kim et al., 2019). Thanks to these practices, businesses increase their environmental performance through energy saving, waste management and low carbon footprint (Miah et al., 2024). In addition, corporate reputation increases in the eyes of the public, and the loyalty and motivation of both internal and external stakeholders increase. An organizational structure that is environmentally sensitive and adopts ethical values increases employee job satisfaction (Naseer et al., 2023; Zoogah, 2011).

Effective execution of the GHRM process is extremely important as it encourages green behaviors through green reward systems and increases employee motivation to achieve environmental goals (Renwick et al., 2013). In order to be able to talk about an effective GHRM process, it is extremely important to shape organizational goals by evaluating the environment, organize activities to instill sustainability values in employees, encourage such activities by including sustainable activities in performance evaluation and reward processes, and ensure that employees take part in decision-making processes (Tran, 2023; Acquah et al., 2021).

PERFORMANCE MANAGEMENT

Due to the expansion of the fields of activity in today's enterprises, the increase in the number of employees and the dynamic structure of the sector, the evaluation of employee performance has become extremely important (Saratun, 2016). Especially after the realization that labor costs constitute almost half of the total cost with globalization, the importance of employee management has increased even more (Datta, 2012). Basically, performance management, which aims to improve human performance by developing people, determines performance criteria, monitors the relationship between these criteria and implementation, and reports the relationship between these two (Rowland & Hall, 2014; Li et al. 2024).

In the most basic sense, performance management is a system that measures the efficiency and effectiveness of employees and directs employee performance in line with strategic decisions (George et al., 2016). It covers the processes of determining, developing and directing the performance of employees in line with business goals (Auginis et al., 2012). Through performance management, it is aimed for employees to reveal their capacities effectively. In this way, a favorable environment is created in businesses that can continuously improve performance. Thanks to this environment, customer satisfaction can be achieved, interaction with the public can increase and thus competitive advantage can be achieved (Muhammad et al., 2021; Jakobsen et al., 2019; Glendinning, 2002).

THE BENEFITS OF PERFORMANCE MANAGEMENT FOR BUSINESSES

Performance management, which has been widely used since the 1990s from the management of businesses to the management of governments, is based on the monitoring of goals set by top managers (Burns & Zgiren, 2010). Management should evaluate and guide employee performance from a sustainability perspective (Amjad et al., 2021). In this sense, human resources should manage key performance indicators in line with sustainable goals while organizing performance management processes. In order for this process to be carried out successfully, awareness should be raised, employees should be encouraged and informed (Prasad, 2013). In addition, knowing the vision and strategies of the organization, its structure and business processes is extremely important for the efficiency of the process (Hendry et al., 2000). According to Simpson & Simpson (2022), effective performance management provides organizations with employee motivation, productivity, teamwork, effective feedback system and the opportunity to achieve desired results.

Considering performance management from a sustainability perspective is based on the view that environmental performance measures should be integrated into the overall performance indicators of the business and the culture of sustainability should be spread throughout the organization (Mutingi et al., 2014; Chen et al., 2022). Sustainable performance management has a structure that supports long-term strategic goals, enables employee personal development, and identifies strategies for the relationship between corporate social responsibility and environmental sustainability (Gadenne et al., 2012). The focus of sustainable performance management is to maintain employee welfare, ethical values, environmental impact and corporate growth. In this context; increasing energy efficiency, reducing greenhouse gas emissions, managing waste systems and optimizing resource utilization are aimed (De Mendonca et al., 2019). In order to achieve these goals, it is necessary to continuously monitor the employee, provide continuous feedback, and determine measurable criteria for performance (Rahman et al., 2020). In this respect, sustainable performance management, unlike traditional performance management processes, involves high employee involvement. It increases employee commitment to the process through feedback-oriented evaluation systems and self-assessment processes.

Sustainable performance management, which supports stakeholder engagement, corporate sustainability and human capital, has a holistic approach that includes environmental, social and ethical dimensions (Longoni & Cagliano, 2016). In this sense, the criteria that can be considered important in performance management are based on economic, environmental and social performance criteria (Marucci et al., 2024).

PERFORMANCE MANAGEMENT PROCESS

Through sustainable performance management, businesses can gain advantages such as long-term resilience to competitive conditions, employee satisfaction and motivation, increased environmental and social impact, and reputation. The first thing to do in this context is to set specific, measurable, achievable, realistic and timely (SMART) goals in line with the United Nations Global Compact and sectoral regulatory standards (Epstein & Buhovac, 2014). Next, key performance indicators (KPIs) with economic, social, and environmental coverage appropriate to the objectives should be determined. These indicators include conditions such as energy consumption, carbon footprint, waste management, resource utilization, as well as employee satisfaction, occupational health and safety, diversity and inclusion (Renwick et al., 2013). However, according to the results of the study conducted by Marucci et al. (2024), determining KPIs is not sufficient. It is also extremely important to explain KPIs in detail. Data on the identified indicators need to be collected regularly from reliable sources. In this process, measurement processes can be made more effective by utilizing digital systems, automation and sustainability software. The collected data is analyzed by comparing it with past performance, sectoral norms and regulations (Searcy, 2012). The data obtained are reported in accordance with international standards. Thanks to these reports, the company has the opportunity to provide transparent and accountable information about its sustainability performance to internal and external stakeholders (Schaltegger & Wagner, 2006). At the end of the process, sustainability strategies are reviewed in line with the performance results, performance deviations are analyzed and corrective and preventive actions are planned. At this stage, continuous improvement in sustainability performance should be ensured by receiving feedback from employees and other stakeholders (Sebhatu, 2009).

Effective measurement of sustainable performance management not only reduces environmental and social risks, but also provides businesses with opportunities to create long-term strategic value (Ding, 2005). Well-defined indicators, regular data analysis and reporting practices based on international standards are essential tools for businesses to achieve their sustainability goals. While this process strengthens internal decision-making mechanisms, it also provides important contributions in terms of gaining the trust of external stakeholders and creating competitive advantage (Dubey et al., 2015). In this context, environmental sustainability-oriented performance management represents a holistic management approach that strengthens the economic value creation processes of businesses while fulfilling their responsibilities to society and the environment. In summary, performance management in GHRM processes monitors the extent to which employees carry out their activities in an environmentally conscious manner and focuses on encouraging employees to adopt such behaviors. However, performance management alone may not be sufficient to achieve this goal. In particular, reward systems are extremely important for green practices to go beyond a managerial strategy and become a culture embraced by employees.

REWARDING SYSTEM

Environmental sustainability is becoming increasingly important among the strategic priorities of businesses. Global climate change, resource depletion and increasing environmental regulations lead businesses to sustainability-oriented practices; in this context, human resource management practices are aligned with environmental goals. The adoption of a sustainable system by employees leads to environmentally friendly performance of employees and efficient implementation of business strategies (Mehak & Batcha, 2024).

Reward systems developed to increase employees' contributions to environmental performance play a critical role in the sustainability strategies of businesses (Das & Dash, 2024; Chiang, 2005). From a theoretical perspective, these systems are shaped on the basis of behavioral approaches such as expectancy theory, reinforcement theory and self-determination theory and encourage employees' environmental behaviors (Gagne et al., 2018; Mansaray, 2019; Fernandez et al., 2003).

According to the results of a study conducted by Kelly et al. (2023); rewarding activities carried out within the organization have a positive effect on the employee both functionally and symbolically. In addition, rewarding activities allow individuals with high organizational commitment to establish a stronger bond with the business. Another study reveals that reward is a motivator for employees in sustainable processes, but if the reward is not motivating enough, the employee is relatively less interested in sustainability (Amoako et al., 2023). However, it should be taken into account that excessive use of extrinsic rewards can reduce intrinsic motivation and reward mechanisms should be designed on the basis of holistic, fair and measurable criteria (Cameron, 2001).

TYPES OF REWARDS

Sustainability-oriented rewards include intrinsic and extrinsic motivation elements; employee engagement is increased through practices such as financial incentives, bonuses, promotion opportunities and public recognition of employees (Mehak & Batcha, 2024). In this case, the employee is expected to benefit from the reward system by avoiding undesirable behaviors that are contrary to sustainability (Jamal et al., 2021). Until the 1950s, rewards were expected to include only financial elements, while other elements negatively affected motivation (Rowland & Hall, 2014). According to the results of a study conducted in the 2000s, a performance-based salary increase increased employee motivation to a certain extent (Sayım, 2010). Today, in addition to material reward options, the importance of using rewards that emotionally nurture employees has been recognized. Rewards used within the scope of environmental sustainability are generally classified as intrinsic and extrinsic (Ye & Li, 2024).

Intrinsic rewards increase voluntary participation in environmental actions because they are aligned with an individual's values and beliefs. Such rewards include elements such as employees' awareness of environmental responsibility, desire to contribute to sustainability policies, and personal satisfaction from environmental achievements (Deci et al., 1981). Extrinsic rewards, on the other hand, are based on increasing an individual's motivation through extrinsic factors. These types of rewards include financial and social incentives when the employee develops the desired level of performance and action (Zhou et al., 2011). Salary increases, bonuses, seniority and performance raises, promotions and title changes, taking part in prestigious projects, job security, recognition, award ceremonies, flexible working hours can be given as examples of extrinsic rewards (Mottaz, 1985).

Sustainability-oriented rewarding activities provide businesses with long-term employee motivation and loyalty. In this sense, it offers the opportunity to work with environmentally conscious and responsible employees. The presence of environmentally responsible employees provides businesses with the opportunity for innovation and continuous development. These activities carried out in the light of ethical values strengthen corporate reputation and support long-term financial performance. This increases resource efficiency and reduces waste and costs. Finally, it prevents legal risks by increasing the likelihood that businesses will comply with any current and future legal requirements.

CONTRIBUTIONS OF PERFORMANCE MANAGEMENT AND REWARD SYSTEM IN SUSTAINABILITY TO BUSINESSES

Sustainability, which has become a fundamental component of strategic management processes, has ceased to be a concept that businesses address only within the scope of social responsibility. In this context, performance management and reward systems play a supporting role for the sustainability of businesses (Ngare et al., 2018). These activities, which are carried out effectively, offer various contributions to businesses and provide competitive advantage (Saratun, 2016; Kusnawan et al., 2025).

Sustainable performance management and reward systems are based on eKPIs such as carbon footprint, energy consumption, waste management, biodiversity (Hristoy & Chirico, 2019). The practices increase employees' commitment to their jobs and reinforce their sense of organizational belonging (Johanson et al., 2022). This increases employee motivation and increases social and environmental contributions (Schiehll & Morissette, 2000).

Businesses that can effectively embed sustainability values within the organization are considered reliable and reputable by consumers and investors. Thus, the corporate reputation and brand value of the business are strengthened. Especially the effective and fair execution of rewarding activities is an important source of motivation in behavioral changes (Meirinhos et al., 2023). Such practices increase sustainability awareness at the individual and organizational level, while supporting innovation and improving performance (Antoni et al., 2017). Tangible and intangible rewards, such as bonus systems, recognition programs and career development opportunities, increase employee engagement in environmental sustainability activities (Grynko et al., 2017).

Effective management of environmental sustainability performance and successful implementation of reward systems provide businesses with reputational advantage in the market. Environmentally friendly practices are decisive in consumer preferences and increase brand value (Baldassare & Campo, 2016). Systematic monitoring of environmental performance makes it easier for businesses to comply with current and future environmental regulations. It also supports the early identification of environmental risks and the development of proactive strategies to address these risks (Ekins & Zenghelis, 2021). Environmental performance improvements through energy efficiency, waste reduction and optimizing resource use reduce costs and strengthen the financial sustainability of businesses (Matzembacher & Meira, 2019).

Performance management and reward systems that focus on environmental sustainability increase employee commitment to the organization and encourage a sustainability-focused transformation in corporate culture (Bolis et al., 2021). Tracking sustainability performance indicators triggers environmental innovation and contributes to continuous improvement of processes. This increases both environmental and economic sustainability (Matzembacher & Meira, 2019; DesJardins, 2005).

The effective establishment of performance management and reward systems focused on environmental sustainability ensures not only the fulfillment of environmental obligations, but also the rebuilding of the long-term success of the business on the axis of sustainability. These systems minimize environmental risks, increase competitiveness and build stakeholder trust. Therefore, integrating an environmental sustainability approach with strategic performance management and human resources practices will be a fundamental element in the sustainable business models of the future.

KEY CHALLENGES AND SOLUTIONS FOR THE IMPLEMENTATION OF SUSTAINABILITY-FOCUSED PERFORMANCE MANAGEMENT AND REWARD SYSTEMS

The effective implementation of sustainability-oriented performance management and reward systems in corporate structures poses various methodological and structural challenges. The primary problem is the lack of standardized and comparable, objective, transparent and cross-sectoral metrics for measuring the quantitative and qualitative dimensions of sustainability performance. As a result of this lack, uncertainties may arise (Searcy, 2012). The nature of sustainability, which is not limited to financial metrics, has an environmental, social and managerial scope. This can make it complex to measure and reward employees' level of contribution to sustainable goals (Linnenluecke & Griffiths, 2010). Moreover, sustainability-based processes require taking into account multidimensional and long-term goals. In this sense, it is extremely difficult to ensure the effective participation of internal and external stakeholders (Flammer, 2015). Short-term performance pressures in organizations can negatively affect environmental orientation in the long term (Costa & Opera, 2025).

The first thing that is expected to be done as a solution to these problems is to examine indicators in line with global standards and integrate them into corporate performance indicators and to associate sustainability-based outputs with more concrete values (Grimm et al., 2016). The main benefit of this process is to ensure that sustainability goals permeate the corporate culture and accelerate organizational adaptation (Garbie, 2015). In order to create organizational adaptation, employees need to be raised awareness through training and development activities (Craig & Allen, 2013). In addition, considering reward criteria not only from a financial perspective but also from a sustainability perspective increases employee commitment to the goals and helps sustainability to become a corporate culture (Saxena & Khanna, 2024). Finally, regular feedback and checks can reinforce the effectiveness of the process and employee motivation (Boeske, 2023).

CASE STUDIES IN SUSTAINABILITY-FOCUSED PERFORMANCE MANAGEMENT AND REWARDS

This section provides a comparative analysis of case studies highlighted in the literature on sustainability-focused performance management and reward systems. Table 1 systematically evaluates four studies conducted in different countries and sectors in terms of their sample, main findings, and sustainability trends. The evaluation findings are presented in Table 1.

Table 1. Trends in Sustainable Performance Management in Case Studies

Study	Sample	Key Findings	Sustainability Trend
Amjad et al. (2021)	Textile Industry Executives in Pakistan	- GHRM activities positively impact organizational sustainability. - Employee performance plays a mediating role in the relationship between GHRM and organizational sustainability. - GHRM is an innovative approach in developing countries.	- Performance management and rewards should include not only employee satisfaction but also environmental and organizational sustainability. - Using environmental indicators in performance management creates sustainable performance management. - In developing countries, viewing GHRM as an emerging paradigm offers a new approach to sustainability strategies.
Marucci et al. (2024)	All employees and managers in all departments at the Italian Food Company	- KPIs are used in performance evaluation and rewards. - Businesses have developed management practices based on KPI trends. - Integrating KPIs into employee rewards improves environmental performance. - Environmental indicators are not just a reporting tool, but a strategic management mechanism.	- Integrating environmental performance measurement and reward systems can make strategies more sustainable. - Dashboards and KPIs are important for tracking sustainability goals and motivation. - Management by objectives is important for green management.

Table 2. Trends in Sustainable Performance Management in Case Studies (Continued)

Study	Sample	Key Findings	Sustainability Trend
Shahedul Quader et al. (2016)	SMEs in the retail and manufacturing sectors	- SMEs have begun to be more sensitive to environmental issues. - However, they need to exert more effort to take environmentally conscious steps. - Environmental performance varies according to the conditions of each SME. - Entrepreneurs can focus not only on profitability but also on environmental concerns.	- The adoption of environmental management systems such as ISO 14001 by SMEs is becoming widespread. - Sustainability provides SMEs with competitive advantage and profitability. - SMEs need flexible sustainability strategies. - Reward systems can improve environmental performance.
Doni & Gasperini (2014)	Etica SGR, Generali Group, and Vancity employees	- Businesses should provide opportunities to create value for internal stakeholders. - Integrated reporting systems increase efficiency by making communication more transparent. - Interdepartmental barriers should be reduced, intellectual capital should be managed, and performance and reward systems should be integrated with KPIs.	- Sustainability reporting is important for internal management and performance management. - A link should be established between KPIs and reward systems. - Measuring intellectual capital and adding it to performance criteria can be effective.

Table 1 compares four different academic studies addressing sustainability-focused performance management and reward systems. The table systematically explains the studies' samples, main findings, and sustainability trends, revealing the diversity in the field. Upon examination of the table, a significant relationship is established between GHRM processes and organizational sustainability. Businesses have begun to use the organization's sustainability performance not only as a reporting tool but also as a reward and management decision tool. This situation has enabled SMEs to take important steps towards sustainability thanks to standards such as ISO 14001. Furthermore, the expansion of reporting processes from external stakeholders to internal stakeholders is shaping performance management and reward processes.

THE ROLE OF DIGITAL TECHNOLOGIES AND ARTIFICIAL INTELLIGENCE IN SUSTAINABILITY-FOCUSED PERFORMANCE MANAGEMENT

The introduction of Artificial Intelligence (AI) into the business world has profoundly impacted organizational structures, making concepts such as innovation, efficiency, and sustainability increasingly important. This has led to the development of Sustainable HRM and increased the use of AI in performance management processes. As a result, businesses have optimized resource utilization and achieved energy savings in their operations (Thangaraju & Palani, 2024). Businesses can now measure their environmental impact more transparently, particularly through big data, machine learning, and digital monitoring tools. This means that digitalization and AI technologies not only increase efficiency in operational processes but also enable compliance with global sustainability standards (Sahlin & Angelis, 2019).

Digital technology-supported performance management can facilitate tracking employees' contributions to sustainability-focused projects and their overall performance, thereby enabling the creation of a fair reward system. Furthermore, natural language processing and machine learning-based tools reveal how sustainability strategies are perceived by employees by tracking internal communication flows (Shrestha & Kumar, 2022).

Therefore, digital tools and AI technologies not only improve sustainability-focused performance management processes, but also transform them. With these technologies, businesses can take a more holistic view of sustainability and evaluate their employees in this regard.

GLOBAL STANDARDS AND REPORTING FRAMEWORKS SUPPORTING SUSTAINABLE PERFORMANCE MANAGEMENT AND REWARDS

AI technologies are crucial for performance management and reward processes, as are global standards and reporting frameworks. In particular, the ISO 26000 Social Responsibility Standard and GRI (Global Reporting Initiative) Standards play an effective role in measuring, reporting, and rewarding companies' sustainability practices.

ISO 26000, which represents the social sustainability of businesses, addresses management processes related to topics such as ethics, employee rights, and environmental relations from a performance management perspective. Thus, businesses carry out performance management processes not only from a financial perspective but also from a social perspective. This allows businesses to integrate their reward systems with sustainability goals (Chakroun, et al., 2020).

The GRI Standards, which address the environmental, social, and governance (ESG) performance of businesses, are important for ensuring that sustainability reports can be evaluated in a transparent and comparable manner (Adams, et al., 2022). GRI indicators can be used to establish sustainable employee performance metrics and reward mechanisms. This allows individual and corporate goals to be integrated with global sustainability standards (Luo & Tang, 2023).

Thanks to these standards, businesses can more effectively meet stakeholder expectations and utilize performance management and reward processes within the organization in a fairer and more effective manner. Effective use of these standards by businesses can increase employee motivation and also provide a competitive advantage (Dissanayake, 2021).

CONCLUSION

Globalizing world conditions have brought environmental problems such as climate change and rapid consumption of natural resources. These environmental problems have brought about changes in many aspects from individual life to government policies. Changes have also affected the business world and the concept of sustainability has become important for businesses. Businesses need to revise their activities in accordance with sustainability in order to survive, gain competitive advantage and achieve their strategic goals. Employee participation is extremely important for the effective use of sustainability.

HRM activities have been revised to enable employees to work in line with sustainable goals and strategies to increase employee motivation are being developed to improve environmental performance. These activities, known as GHRM, aim to raise employee awareness, encourage environmental behavior and instill a sense of environmental responsibility. In these conditions where the organizational structure is rebuilt, performance management and reward systems are used to bring employee performance to the desired level.

Thanks to the sustainability-focused performance evaluation and rewarding system, ethical business practices are reinforced, corporate social responsibility is increased and long-term competitive advantage is achieved. Thanks to processes organized in the light of important indicators such as energy efficiency, waste management, resource saving and carbon footprint reduction, strategic advantages such as increased productivity, cost savings and brand reputation are achieved. In addition, the reward system encourages active employee participation and instills a sense of responsibility at the corporate level. An effective and balanced reward system increases employee loyalty and job satisfaction.

Performance management and reward systems, which provide great benefits to organizations, also have certain challenges. Improperly structured performance criteria and reward mechanisms can be a source of short-term motivation for employees. In such a case, the employee's intrinsic motivation for sustainability, which he/she really needs to develop, may disappear. Failure to ensure a fair structure within the organization may cause sustainable goals to take a back seat. This may cause employees to develop resistance to sustainable policies and lead to a decrease in productivity. In addition to all these possibilities, if the knowledge of the employee and the organization is low, it is extremely difficult to carry out the processes effectively.

In order for performance management and reward systems to be used effectively, businesses must first define their vision clearly and consistently. In this context, clearly defining sustainability goals and performance criteria supports the process. In addition, under the leadership of an effective leader, employees should be integrated into the process and all processes should be carried out transparently. Continuous training and development activities are important to ensure that employees are fully integrated. It can be functional in terms of predicting possible problems that may be encountered in the process, preparing measures and revising processes.

It is very important for businesses to adopt sustainability-oriented performance management and reward systems to fulfill their environmental and social responsibilities and for long-term profitability. Effective execution of the processes provides businesses with brand reputation, employee loyalty and competitive advantage. Thanks to the effective use of GHRM processes, businesses gain a holistic and innovative business model and gain strategic benefits.

REFERENCES

Acquah, I. S. K., Agyabeng-Mensah, Y., & Afum, E. (2021). Examining the link among green human resource management practices, green supply chain management practices and performance. *Benchmarking*, *28*(1), 267–290. DOI: 10.1108/BIJ-05-2020-0205

Adams, C. A., Alhamood, A. M., & He, X. (2022). The development and implementation of GRI standards: practice and policy issues. In *Handbook of accounting and sustainability* (pp. 26–43). Edward Elgar Publishing., DOI: 10.4337/9781800373518.00009

Amjad, F., Abbas, W., Zia-Ur-Rehman, M., Baig, S. A., Hashim, M., Khan, A., & Rehman, H. U. (2021). Effect of green human resource management practices on organizational sustainability: The mediating role of environmental and employee performance. *Environmental Science and Pollution Research International*, *28*, 28191–28206. DOI: 10.1007/s11356-020-11307-9

Amoako, D. K., Zakuan, M. N., Okyere-Kwakye, E., & Tetteh, F. K. (2023). Effect of Training and Reward on Social Sustainability in Ghana's Cocoa Supply Chain: The Role of Green Buyer-Supplier Relationship. *Journal of International Food & Agribusiness Marketing*, *35*(2), 212–243. DOI: 10.1080/08974438.2021.1981511

Antoni, C. H., Baeten, X., Perkins, S. J., Shaw, J. D., & Vartiainen, M. (2017). Reward management. *Journal of Personnel Psychology*. Advance online publication. DOI: 10.1027/1866-5888/a000187

Baumgärtner, S., & Quaas, M. (2010). What is sustainability economics? *Ecological Economics*, *69*(3), 445–450. DOI: 10.1016/j.ecolecon.2009.11.019

Boeske, J. (2023). Leadership towards sustainability: A review of sustainable, sustainability, and environmental leadership. *Sustainability*, *15*(16), 12626. DOI: 10.3390/su151612626

Bolis, I., Morioka, S. N., Leite, W. K. D. S., & Zambroni-de-Souza, P. C. (2021). Sustainability is all about values: The challenges of considering moral and benefit values in business model decisions. *Sustainability*, *13*(2), 664. DOI: 10.3390/su13020664

Burns, J. P., & Zhiren, Z. (2010). Performance management in the government of the People's Republic of China: Accountability and control in the implementation of public policy. *OECD Journal on Budgeting*, *10*(2), 7.

Cameron, J. (2001). Negative effects of reward on intrinsic motivation—A limited phenomenon: Comment on Deci, Koestner, and Ryan (2001). *Review of Educational Research, 71*(1), 29–42. DOI: 10.3102/00346543071001029

Chakroun, S., Salhi, B., Ben Amar, A., & Jarboui, A. (2020). The impact of ISO 26000 social responsibility standard adoption on firm financial performance: Evidence from France. *Management Research Review, 43*(5), 545–571. DOI: 10.1108/MRR-02-2019-0054

Chen, W., Zhang, Q., Wang, C., Li, Z., Geng, Y., Hong, J., & Cheng, Y. (2022). Environmental sustainability challenges of China's steel production: Impact-oriented water, carbon and fossil energy footprints assessment. *Ecological Indicators, 136*, 108660. DOI: 10.1016/j.ecolind.2022.108660

Chiang, F. (2005). A critical examination of Hofstede's thesis and its application to international reward management. *The International Journal of Human Resource Management, 16*(9), 1545-1563. https://doi.org/DOI: 10.1080/09585190500239044

Costa, M. D., & Opare, S. (2025). Impact of corporate culture on environmental performance. *Journal of Business Ethics, 196*(1), 61–92. DOI: 10.1007/s10551-024-05674-3

Craig, , C., & Allen, M. (2013). Sustainability information sources: Employee knowledge, perceptions, and learning. *Journal of Communication Management (London), 17*(4), 292–307. DOI: 10.1108/JCOM-05-2012-0035

Danciu, V. (2013). The sustainable company: New challenges and strategies for more sustainability. *Theoretical and Applied Economics, 20*(9), 7–26.

Das, S., & Dash, M. (2024). Green compensation and reward system: A novel approach towards the growth and sustainability of organization. *Journal of Advanced Zoology, 45*(2), 1213–1219.

Das, S. C., & Singh, R. K. (2016). Green HRM and organizational sustainability: An empirical review. *Kegees Journal of Social Science, 8*(1), 227–236.

Datta, P. (2012). An applied organizational rewards distribution system. *Management Decision, 50*(3), 479–501. DOI: 10.1108/00251741211216241

De Mendonca, T. R., & Zhou, Y. (2019). Environmental performance, customer satisfaction, and profitability: A study among large US companies. *Sustainability, 11*(19), 5418. DOI: 10.3390/su11195418

Deci, E. L., Nezlek, J., & Sheinman, L. (1981). Characteristics of the rewarder and intrinsic motivation of the rewardee. *Journal of Personality and Social Psychology, 40*(1), 1. DOI: 10.1037/0022-3514.40.1.1

DesJardins, J. (2005). Business and environmental sustainability. *Business & Professional Ethics Journal, 24*(1/2), 35–59. https://www.jstor.org/stable/27801371

Ding, G. K. (2005). Developing a multicriteria approach for the measurement of sustainable performance. *Building Research and Information, 33*(1), 3–16. DOI: 10.1080/0961321042000322618

Dissanayake, D. (2021). Sustainability key performance indicators and the global reporting initiative: Usage and challenges in a developing country context. *Meditari Accountancy Research, 29*(3), 543–567. DOI: 10.1108/MEDAR-08-2019-0543

Doni, F., & Gasperini, A. (2014). The internal integrated reporting and the value creation. A case study approach. In *Knowledge and Management Models for Sustainable Growth. Proceedings. 11-13 June Matera Italy* (Vol. 9, pp. 2658-2696). Institute of Knowledge Asset Management.

Dubey, R., Gunasekaran, A., & Chakrabarty, A. (2015). World-class sustainable manufacturing: Framework and a performance measurement system. *International Journal of Production Research, 53*(17), 5207–5223. DOI: 10.1080/00207543.2015.1012603

Dyllick, T., & Muff, K. (2016). Clarifying the meaning of sustainable business: Introducing a typology from business-as-usual to true business sustainability. *Organization & Environment, 29*(2), 156–174. DOI: 10.1177/1086026615575176

Eccles, R. G., Ioannou, I., & Serafeim, G. (2014). The impact of corporate sustainability on organizational processes and performance. *Management Science, 60*(11), 2835–2857. DOI: 10.1287/mnsc.2014.1984

Ekins, P., & Zenghelis, D. (2021). The costs and benefits of environmental sustainability. *Sustainability Science, 16*, 949–965. DOI: 10.1007/s11625-021-00910-5

Elkington, J. (1997). The triple bottom line. *Environmental management: Readings and cases, 2*, 49-66.

Elkington, J. (2006). Governance for sustainability. *Corporate Governance, 14*(6), 522–529. DOI: 10.1111/j.1467-8683.2006.00527.x

Epstein, M. J., & Rejc Buhovac, A. (2014). A New Day for Sustainability. *Strategic finance, 96*(1).

Fernández, E., Junquera, B., & Ordiz, M. (2003). Organizational culture and human resources in the environmental issue: A review of the literature. *International Journal of Human Resource Management, 14*(4), 634–656. DOI: 10.1080/0958519032000057628

Flammer, C. (2015). Does corporate social responsibility lead to superior financial performance? A regression discontinuity approach. *Management Science, 61*(11), 2549–2568. DOI: 10.1287/mnsc.2014.2038

Gadenne, D., Mia, L., Sands, J., Winata, L., & Hooi, G. (2012). The influence of sustainability performance management practices on organisational sustainability performance. *Journal of Accounting & Organizational Change, 8*(2), 210–235. DOI: 10.1108/18325911211230380

Gagné, M., Deci, E. L., & Ryan, R. M. (2018). Self-determination theory applied to work motivation and organizational behavior. *The SAGE handbook of industrial, work & organizational psychology: Organizational psychology, 2*, 97-121.

Garbie, I. H. (2015). Sustainability awareness in industrial organizations. *Procedia CIRP, 26*, 64–69. DOI: 10.1016/j.procir.2015.03.003

George, R. A., Siti-Nabiha, A. K., Jalaludin, D., & Abdalla, Y. A. (2016). Barriers to and enablers of sustainability integration in the performance management systems of an oil and gas company. *Journal of Cleaner Production, 136*, 197–212. DOI: 10.1016/j.jclepro.2016.01.097

Glendinning, P. M. (2002). Performance management: Pariah or messiah. *Public Personnel Management, 31*(2), 161–178. DOI: 10.1177/009102600203100203

Grimm, J. H., Hofstetter, J. S., & Sarkis, J. (2016). Exploring sub-suppliers' compliance with corporate sustainability standards. *Journal of Cleaner Production, 112*, 1971–1984. DOI: 10.1016/j.jclepro.2014.11.036

Grynko, T., Krupskyi, O. P., Koshevyi, M., & Maximchuk, O. (2017). Tangible and intangible rewards in service industries: Problems and prospects.

Hendry, C., Woodward, S., Bradley, P., & Perkins, S. (2000). Performance and rewards: Cleaning out the stables. *Human Resource Management Journal, 10*(3), 46–62.

Jakobsen, M., James, O., Moynihan, D., & Nabatchi, T. (2019). JPART virtual issue on citizen-state interactions in public administration research. *Journal of Public Administration: Research and Theory, 29*(4), e8–e15. DOI: 10.1093/jopart/muw031

Jamal, T., Zahid, M., Martins, J. M., Mata, M. N., Rahman, H. U., & Mata, P. N. (2021). Perceived green human resource management practices and corporate sustainability: Multigroup analysis and major industries perspectives. *Sustainability, 13*(6), 3045. DOI: 10.3390/su13063045

Johanson, U., Aboagye, E., & Yao, J. (2022). Integrating business model for sustainability and performance management to promote occupational health and safety—A discussion of value. *Frontiers in Sustainability, 3*, 950847. DOI: 10.3389/frsus.2022.950847

Jones, P., Clarke-Hill, C., Comfort, D., & Hillier, D. (2008). Marketing and sustainability. *Marketing Intelligence & Planning, 26*(2), 123–130. DOI: 10.1108/02634500810860584

Kelly, K., Liu, W., & Presslee, A. (2023). The interactive effect of organizational identification and reward type on reward valuation. *Contemporary Accounting Research, 40*(3), 1733–1759. DOI: 10.1111/1911-3846.12868

Kim, Y. J., Kim, W. G., Choi, H. M., & Phetvaroon, K. (2019). The effect of green human resource management on hotel employees' eco-friendly behavior and environmental performance. *International Journal of Hospitality Management, 76*, 83–93. DOI: 10.1016/j.ijhm.2018.04.007

Kusnawan, A., Riyadi, S., Lestari, S. D., & Mudjijah, S. (2025). Green Corporate Performance Management: A Strategic Approach to Environmental and Organizational Sustainability. *Primanomics: Jurnal Ekonomi & Bisnis, 23*(1), 256–272. DOI: 10.31253/pe.v23i1.3523

Li, W., Lee, H., Liu, Y., Tang, T., & Zheng, G. (2024). Reward exploitation or exploration? The effects of local government sustainability strategies and performance information on citizen evaluation. *Public Management Review*, •••, 1–22. DOI: 10.1080/14719037.2024.2399147

Linnenluecke, M. K., & Griffiths, A. (2010). Corporate sustainability and organizational culture. *Journal of World Business, 45*(4), 357–366. DOI: 10.1016/j.jwb.2009.08.006

Longoni, A., & Cagliano, R. (2016). Human resource and customer benefits through sustainable operations. *International Journal of Operations & Production Management, 36*(12), 1719–1740. DOI: 10.1108/IJOPM-11-2014-0564

Luo, L., & Tang, Q. (2023). The real effects of ESG reporting and GRI standards on carbon mitigation: International evidence. *Business Strategy and the Environment, 32*(6), 2985–3000. DOI: 10.1002/bse.3281

Mansaray, H. E. (2019). The role of human resource management in employee motivation and performance-An overview. *Budapest International Research and Critics Institute (BIRCI). Journal*, *2*(3), 183–194. DOI: 10.33258/birci.v2i3.405

Matzembacher, D. E., & Meira, F. B. (2019). Sustainability as business strategy in community supported agriculture: Social, environmental and economic benefits for producers and consumers. *British Food Journal*, *121*(2), 616–632. DOI: 10.1108/BFJ-03-2018-0207

Mehak, S. S., & Batcha, H. M. (2024). Analyzing the influence of green human resource practices on organizational sustainability: The role of green attitudes and performance of employees'. *Environment, Development and Sustainability*, •••, 1–34. DOI: 10.1007/s10668-024-05157-5

Meirinhos, G., Cardoso, A., Neves, M., Silva, R., & Rêgo, R. (2023). Leadership styles, motivation, communication and reward systems in business performance. *Journal of Risk and Financial Management*, *16*(2), 70. DOI: 10.3390/jrfm16020070

Merriman, K. K., Sen, S., Felo, A. J., & Litzky, B. E. (2016). Employees and sustainability: The role of incentives. *Journal of Managerial Psychology*, *31*(4), 820–836. DOI: 10.1108/JMP-09-2014-0285

Miah, M., Szabó-Szentgróti, G., & Walter, V. (2024). A systematic literature review on green human resource management (GHRM): An organizational sustainability perspective. *Cogent Business & Management*, *11*(1), 2371983. DOI: 10.1080/23311975.2024.2371983

Molina-Azorin, J. F., López-Gamero, M. D., Tarí, J. J., Pereira-Moliner, J., & Pertusa-Ortega, E. M. (2021). Environmental management, human resource management and green human resource management: A literature review. *Administrative Sciences*, *11*(2), 48. DOI: 10.3390/admsci11020048

Mottaz, C. J. (1985). The relative importance of intrinsic and extrinsic rewards as determinants of work satisfaction. *The Sociological Quarterly*, *26*(3), 365–385. DOI: 10.1111/j.1533-8525.1985.tb00233.x

Muhammad, S., Khan, I., & Hameed, F. (2021). The Impact of Performance Management System on Employees' Performance. *International Journal of Business and Management Sciences*, *2*(3), 38–47.

Mutingi, M., Mapfaira, H., & Monageng, R. (2014). Developing performance management systems for the green supply chain. *Journal of Remanufacturing*, *4*, 1–20. DOI: 10.1186/s13243-014-0006-z

Naseer, S., Song, H., Adu-Gyamfi, G., Abbass, K., & Naseer, S. (2023). Impact of green supply chain management and green human resource management practices on the sustainable performance of manufacturing firms in Pakistan. *Environmental Science and Pollution Research International*, *30*(16), 48021–48035. DOI: 10.1007/s11356-023-25409-7

Ngare, I., Wemali, E., Gichuki, C., Gichuki, N., Kipkemoi, I., & Karanja, J. (2018). Integrating environmental concerns in enterprises and businesses for sustainable development. *International Journal of Environmental Sciences*, •••, 3.

Prasad, R. S. (2013). Green HRM-partner in sustainable competitive growth. *Journal of Management Sciences and Technology*, *1*(1), 15–18.

Rahman, M., Aziz, S., & Hughes, M. (2020). The product-market performance benefits of environmental policy: Why customer awareness and firm innovativeness matter. *Business Strategy and the Environment*, *29*(5), 2001–2018. DOI: 10.1002/bse.2484

Renwick, D. W., Redman, T., & Maguire, S. (2013). Green human resource management: A review and research agenda. *International Journal of Management Reviews*, *15*(1), 1–14. DOI: 10.1111/j.1468-2370.2011.00328.x

Rouholamin, A., Varposhti, S., Alavian, A., & Talebi, S. (2023). Relationship Between Green Intellectual Capital, Green Human Resource Management, Sustainable Supply Chain Management, and Sustainable Performance. *Journal of Organizational Behavior Research*, *8*(2-2023), 91-106. https://doi.org/DOI: 10.51847/teEVWNYvUZ

Rowland, C., & Hall, R. (2014). Management learning, performance and reward: Theory and practice revisited. *Journal of Management Development*, *33*(4), 342–356. DOI: 10.1108/JMD-08-2012-0110

Sahlin, J., & Angelis, J. (2019). Performance management systems: Reviewing the rise of dynamics and digitalization. *Cogent Business & Management*, *6*(1), 1642293. DOI: 10.1080/23311975.2019.1642293

Saratun, M. (2016). Performance management to enhance employee engagement for corporate sustainability. *Asia-Pacific Journal of Business Administration*, *8*(1), 84–102. DOI: 10.1108/APJBA-07-2015-0064

Saunila, M., Ukko, J., & Kinnunen, J. (2024). Sustainability partnership as a moderator in the relationship between business sustainability and firm competitiveness. *Business Strategy and the Environment*, *33*(2), 123–133. DOI: 10.1002/bse.3493

Saxena, M., & Khanna, R. (2024). Green pay and reward systems: Antecedents for sustainable employee environmentism in green human resource management. *Journal of Southwest Jiaotong University*, *59*(2), 270–294.

Sayım, K. Z. (2010). Pushed or pulled? Transfer of reward management policies in MNCs. *International Journal of Human Resource Management*, *21*(14), 2631–2658. DOI: 10.1080/09585192.2010.523580

Schaltegger, S., & Wagner, M. (2006). Integrative management of sustainability performance, measurement and reporting. *International Journal of Accounting. Auditing and Performance Evaluation*, *3*(1), 1–19. DOI: 10.1504/IJAAPE.2006.010098

Schiehll, E., & Morissette, R. (2000). Motivation, measurement and rewards from a performance evaluation perspective. *Revista de Administração Contemporânea*, *4*, 7-24. https://doi.org/DOI: 10.1590/S1415-65552000000300002

Scoones, I. (2007). Sustainability. *Development in Practice*, *17*(4-5), 589–596. DOI: 10.1080/09614520701469609

Scoones, I. (2016). The politics of sustainability and development. *Annual Review of Environment and Resources*, *41*(1), 293–319. DOI: 10.1146/annurev-environ-110615-090039

Searcy, C. (2012). Corporate sustainability performance measurement systems: A review and research agenda. *Journal of Business Ethics*, *107*, 239–253. DOI: 10.1007/s10551-011-1038-z

Sebhatu, S. P. (2009). *Sustainability Performance Measurement for sustainable organizations: beyond compliance and reporting*. Service Research Centre Karlstad University.

Shahedul Quader, M., Kamal, M. M., & Hassan, A. E. (2016). Sustainability of positive relationship between environmental performance and profitability of SMEs: A case study in the UK. *Journal of Enterprising Communities: People and Places in the Global Economy*, *10*(2), 138–163. DOI: 10.1108/JEC-05-2014-0008

Shrestha, M., & Kumar, R. (2022). Performance management in digital transformation: A strategy for sustainability. In *Handbook of Research on Digital Transformation Management and Tools* (pp. 1–23). IGI Global Scientific Publishing., DOI: 10.4018/978-1-7998-9764-4.ch001

Simpson, G., & Simpson, R. (2022). Managing the performance management process: The importance of employees' performance in determining the organisation's viability and efficiency. *Journal of Human Resource & Leadership*, *6*(4), 42–56. DOI: 10.53819/81018102t2100

Singh, S. K., Del Giudice, M., Chierici, R., & Graziano, D. (2020). Green innovation and environmental performance: The role of green transformational leadership and green human resource management. *Technological Forecasting and Social Change*, *150*, 119762. DOI: 10.1016/j.techfore.2019.119762

Teixeira, A. A., Jabbour, C. J. C., de Sousa Jabbour, A. B. L., Latan, H., & De Oliveira, J. H. C. (2016). Green training and green supply chain management: Evidence from Brazilian firms. *Journal of Cleaner Production*, *116*, 170–176. DOI: 10.1016/j.jclepro.2015.12.061

Thangaraju, K., & Palani, P. (2024). Integrating Sustainability into Ai-Driven HRM Practices: Examining Employee Engagement and Organizational Performance in the IT Industry. *Pakistan Journal of Life and Social Sciences*, *22*(2). Advance online publication. DOI: 10.57239/PJLSS-2024-22.2.001008

Tran, N. K. H. (2023). An empirical investigation on the impact of green human resources management and green leadership on green work engagement. *Heliyon*, *9*(11).

Warhurst, A. (2002). Sustainability indicators and sustainability performance management. *Mining, Minerals and Sustainable Development [MMSD] project report, 43*, 129.

Wilkinson, A., Hill, M., & Gollan, P. (2001). The sustainability debate. *International Journal of Operations & Production Management*, *21*(12), 1492–1502. DOI: 10.1108/01443570110410865

World Commission on Environment and Development (WCED). (1987). *Our Common Future*. Oxford University Press.

Ye, W., & Li, Z. (2024). The impact of socially responsible human resource management on employees' environmental citizenship behavior: The role of psychological safety and contingent reward leadership. *SAGE Open*, *14*(2). Advance online publication. DOI: 10.1177/21582440241254272

Zhou, Y., Zhang, Y., & Montoro-Sánchez, Á. (2011). Utilitarianism or romanticism: The effect of rewards on employees' innovative behaviour. *International Journal of Manpower*, *32*(1), 81–98. DOI: 10.1108/01437721111121242

Zoogah, D. B. (2011). The dynamics of Green HRM behaviors: A cognitive social information processing approach. *German Journal of Human Resource Management*, *25*(2), 117–139. DOI: 10.1177/239700221102500204

Chapter 7
Green Leadership and Cultural Transformation

Satya Subrahmanyam
https://orcid.org/0000-0003-0441-2742
Holy Spirit University of Kaslik, Lebanon

ABSTRACT

The urgent challenge of climate change necessitates a shift in leadership and organizational culture to foster sustainability. This chapter explores Green Leadership as a transformative approach that integrates environmental responsibility into organizational goals. It examines the role of Cultural Transformation in embedding sustainability within corporate ethos through leadership-driven strategies. Theoretical frameworks, including Transformational Leadership and Ecocentric Leadership, are analyzed alongside HRM initiatives that promote sustainable practices. Case studies illustrate successful implementations, highlighting challenges such as resistance to change and greenwashing. The chapter concludes with insights on emerging sustainability trends, ESG frameworks, and HRM's role in driving innovation. By aligning leadership, culture, and sustainability, organizations can become proactive agents of environmental change and resilience.

1. INTRODUCTION

Corporates must act swiftly and decisively in response to environmental threats as the climate change catastrophe worsens. According to Alemu (2025) and Khaddage-Soboh et al. (2024), corporates are required to do more than just comply with climate change regulations; they should implement policies that make sustainability their top priority. Environmental consciousness aside, this change is essential for

DOI: 10.4018/979-8-3373-2479-1.ch007

staying competitive and resilient in the face of a dynamic global market (Mushtaq & Akhtar, 2024).

1.1 The Evolving Role of Leadership and HRM in Sustainability

The key to developing sustainability and organizational resilience now lies in leadership and HRM. According to recent studies, leaders play a pivotal role in creating a sustainable culture by establishing a vision and motivating followers to work together to achieve that vision (Nakra & Kashyap, 2024). Human resource management makes this goal a reality by strengthening an organization's resilience to regulatory, environmental, and operational shocks via the incorporation of green principles into hiring, training, and performance evaluation (Mushtaq & Akhtar, 2024). Research has shown that sustainable human resource management techniques improve organizational resilience and have a positive impact on society, the economy, and the environment.

1.2 Defining Green Leadership and Cultural Transformation

The term "green leadership" describes a style of management that encourages creativity, works to reduce waste, and unites workers in a common effort to protect the environment. According to Khaddage-Soboh et al. (2024), this kind of leadership is characterized by a focus on sustainability rather than environmental stewardship as an afterthought. In this setting, cultural transformation refers to the process of altering the beliefs, conventions, and practices inside an organization in order to back and maintain environmental programs. According to Srivastava et al. (2024), a green organizational culture acts as a go-between for leadership and environmental performance. It does this by establishing a common set of values that influences how employees think about and act in regard to sustainability.

1.3 Scope and Significance

To achieve environmental performance and resilience, organizations must strategically align their leadership behaviors, HRM practices, and organizational values. This is what green leadership and cultural change are all about. According to research conducted by Khaddage-Soboh et al. (2024) and Srivastava et al. (2024), a green transformational leadership style and an environment that promotes sustainability are crucial for attaining significant and long-term results. (Nakra & Kashyap, 2024; Mushtaq & Akhtar, 2024) In this age of ecological complexity, organizations that effectively include these factors are in a better position to innovate, react to external shocks, and preserve stakeholder confidence.

1.4 Chapter Objectives and Structure Overview

This chapter aims to:

- Contextualize the urgency of climate change and the necessity for organizational transformation.
- Examine the evolving roles of leadership and HRM in promoting sustainability and resilience.
- Define and explore the interplay between green leadership and cultural transformation.
- Present empirical insights and practical implications for embedding sustainability into organizational DNA.

The chapter is structured to first establish the climate crisis context, then analyze the theoretical and practical dimensions of green leadership and HRM, followed by a discussion of cultural transformation processes, and finally, to synthesize actionable recommendations for organizations seeking to lead in sustainability.

2. CONCEPTUAL FOUNDATIONS OF GREEN LEADERSHIP

Green leadership represents a progressive paradigm that integrates environmental consciousness into leadership philosophy and practice. It emphasizes ethical responsibility, sustainability-driven decision-making, and long-term ecological stewardship. This section explores the theoretical roots, defining attributes, and competencies of green leaders who inspire organizations to align strategic objectives with environmental and social sustainability goals.

2.1 Defining Green Leadership

A paradigm of leadership known as "green leadership" incorporates ecological principles into an organization's goals, strategies, and culture with a focus on long-term environmental sustainability. To tackle the complicated issues of climate change and environmental degradation, green leadership is becoming more and more acknowledged as crucial in modern enterprises (Zhong et al., 2023). Its reach goes beyond mere compliance and includes actively altering organizational practices, engaging stakeholders, and fostering innovation to accomplish social, economic, and ecological objectives in the long run (Fry & Egel, 2021). According to Albadali et al. (2024), green leadership is crucial because it may increase organizational resilience when faced with environmental uncertainty, stimulate green innovation, and encourage employee participation in sustainability efforts.

Green leadership is distinct from traditional and transformational leadership paradigms. Traditional leadership often emphasizes efficiency, hierarchy, and short-term performance, with limited focus on environmental or social outcomes (Boeske & Murray, 2022). Transformational leadership, while more dynamic, centers on inspiring and motivating followers toward organizational change and innovation. Transformative concepts provide the foundation of green leadership, which is distinguished from other styles by a strong emphasis on environmental protection, systemic thinking, and the pervasive incorporation of sustainability into every facet of an organization's operations (Ali et al., 2025). Leadership now has to consider the "triple bottom line" of social, economic, and environmental performance, which is causing a paradigm change (Fry & Egel, 2021).

2.2 Theoretical Underpinnings

The conceptual foundations of green leadership are rooted in several interrelated theories. Sustainability leadership theories emphasize the need for leaders to balance competing demands and paradoxes - such as short-term profitability versus long-term ecological health - by adopting a holistic, systems-oriented perspective (Sajjad et al., 2024). These theories highlight the importance of ethical values, stakeholder engagement, and the co-creation of sustainable futures (Fry & Egel, 2021).

A green leader should first have a firm grasp of the models of transformational and servant leadership. In order to motivate people to accept sustainable objectives, transformational leadership is essential (Begum et al., 2022). This kind of leadership is marked by vision, inspiration, intellectual stimulation, and personalized concern. The ethical imperatives of sustainability are very congruent with servant leadership's emphasis on self-determination, service to others, and ethical accountability (Boeske & Murray, 2022). Green leadership is informed by both paradigms, which highlights the leader's role in creating a collective sense of mission and accountability for environmental results.

Ecocentric leadership models and systems thinking perspectives further enrich the theoretical landscape. Ecocentric leadership places the natural environment at the center of decision-making, advocating for organizational practices that respect ecological limits and promote regeneration (Biswas et al., 2022). Systems thinking encourages leaders to recognize the interconnectedness of organizational, social, and ecological systems, enabling more adaptive and resilient responses to sustainability challenges (Sajjad et al., 2024). These perspectives underscore the need for leaders to move beyond linear, siloed approaches and embrace complexity in pursuit of sustainable transformation.

2.3 Core Competencies of Green Leaders

Effective green leaders demonstrate a set of core competencies that enable them to drive sustainability-oriented change. First, they possess a clear environmental vision and strategic foresight, articulating a compelling narrative that aligns organizational goals with broader ecological imperatives (Fry & Egel, 2021; Begum et al., 2022). This vision-setting is critical for mobilizing resources, inspiring commitment, and guiding innovation toward green outcomes (Alabdali et al., 2024).

Second, green leaders exhibit strong ethical and moral responsibility toward sustainability. They model integrity, transparency, and accountability in decision-making, ensuring that environmental considerations are embedded in organizational values and practices (Boeske & Murray, 2022; Biswas et al., 2022). This ethical orientation fosters trust among stakeholders and legitimize the organization's sustainability efforts.

Finally, green leadership is based on three pillars: inclusion, empowerment, and innovation. Green leaders provide an environment where workers are encouraged to actively engage, learn, and solve sustainability-related problems creatively (Priyadarshini et al., 2023; Begum et al., 2022). They foster an environment that welcomes all people, respects all opinions, and works to bring together different departments and stakeholder groups. Promoting innovation entails integrating new technology and methods that improve environmental performance, encouraging creative involvement, and encouraging green thinking.

Table 1. Core Competencies of Green Leaders

Core Competency	Description
Environmental Vision and Strategic Foresight	Green leaders articulate a compelling environmental vision that aligns organizational goals with global ecological imperatives. They demonstrate foresight by anticipating sustainability challenges and opportunities, mobilizing resources, inspiring commitment, and guiding innovation toward green outcomes.
Ethical and Moral Responsibility	Effective green leaders uphold ethical standards by modeling integrity, transparency, and accountability. They ensure that sustainability principles are embedded in organizational decision-making, thereby fostering stakeholder trust and legitimizing sustainability initiatives.
Empowerment, Inclusivity, and Innovation	Green leaders empower employees through participatory opportunities, learning, and creative problem-solving in sustainability projects. They foster an inclusive culture that values diversity and collaboration. Furthermore, they encourage innovation by promoting green thinking, creative engagement, and the adoption of new technologies that enhance environmental performance.

Green leadership represents a transformative approach that integrates environmental vision, ethical responsibility, and inclusive innovation to drive sustainable organizational change. Its conceptual foundations draw from and extend traditional leadership theories, emphasizing systems thinking, stakeholder engagement, and the pursuit of the triple bottom line.

3. ORGANIZATIONAL CULTURE AND ENVIRONMENTAL SUSTAINABILITY

How sustainability ideals are perceived and implemented inside organizations is greatly influenced by organizational culture. Employees are more likely to become involved, come up with new ideas, and dedicate themselves to green activities in an organization that values environmental responsibility. This section delves into the ways in which HRM practices, together with shared ideas and values, may foster an eco-conscious company identity that promotes sustainability in the long run.

3.1 Understanding Organizational Culture

Organizational culture is often described as values, beliefs, norms and behaviors that are shared by the members of an organization and determine how they interact and collaborate (Roscoe et al., 2019). The model by Schein defines culture as operating in the three levels, namely artifacts (observable structures and processes), espoused values (proclaimed strategies and objectives), and basic underlying assumption (strongly held beliefs and unspoken perceptions). All these layers shape the way things are done in organizations, and how the employees perceive and react to different situations (Piwowar-Sulej, 2020).

Cultural norms are very critical in influencing environmental attitudes and behavior. Once the culture of sustainability is implemented within an organization, employees have a higher chance of embracing environment-friendly behaviors, e.g. using less energy, generating less waste and other green projects. These norms are passed on during daily practice, communication by the leadership and by means of peers, which helps confirm the necessity of environmental responsibility and the coordination of individual activities with the objectives of the sustainability of organizations (Ahmad et al., 2023).

3.2 Green Organizational Culture

Values, concepts, and beliefs that direct all organizational behaviors toward environmental friendliness characterize a green or sustainability-oriented corporate culture (Roscoe et al., 2019). Beyond maximizing profits, this kind of culture prioritizes reducing harm to the environment and increasing good contributions to ecological health (Al-Swidi et al., 2021). Important characteristics include a commitment to environmental protection, a set of shared ideas and values, a way to publicly recognize and honor green efforts, and tangible symbols like green office spaces or eco-certifications (Ahmad et al., 2023).

Shared values, beliefs, and rituals are essential in fostering eco-consciousness among employees. When environmental values are deeply rooted and consistently communicated, they become part of the organizational identity, shaping how employees perceive their roles and responsibilities (Roscoe et al., 2019). Rituals - such as sustainability awards, green team meetings, or Earth Day celebrations - reinforce these values and create a sense of collective purpose. The alignment between corporate identity and a green image is crucial, as organizations with a strong green culture are more likely to be perceived as authentic and trustworthy by stakeholders, enhancing their reputation and competitive advantage (Piwowar-Sulej, 2020).

Table 2. Key Dimensions of a Green Organizational Culture

Dimension	Description
Shared Values and Beliefs	A green culture is built on collective values and beliefs emphasizing environmental stewardship, ethical responsibility, and long-term ecological balance. These values guide decision-making and shape how employees view their roles within the organization.
Rituals and Practices	Organizational rituals such as sustainability awards, Earth Day celebrations, or green team meetings reinforce environmental commitment and create a sense of shared purpose among employees.
Symbols and Artifacts	Visible expressions of environmental commitment—such as eco-certifications, green office spaces, or sustainability logos—serve as tangible representations of organizational values and inspire eco-conscious behavior.
Identity and Reputation Alignment	When an organization's internal culture aligns with its green identity, it enhances authenticity and stakeholder trust, strengthening the firm's reputation and competitive advantage.

3.3 HRM as a Cultural Enabler

Practice in human resource management (HRM) plays a role in instilling values to environment in organizational culture. Green HRM initiatives Green recruitment, training, performance management, and rewards contribute to imbuing the employees with pro-environmental values, beliefs, and behaviors (Roscoe et al., 2019; Ali et al., 2024). Employment of people with good environment values will lead to cultural fit and by employing specific training and development programs, environmental competencies and awareness will be developed (Subrahmanyam, 2025). The performance management systems which praise and appreciate the green behaviors also enforce the culture that is desired (Roscoe et al., 2019).

One of the outcomes of green HRM is employee engagement. Employees should be empowered and incentivized to involve themselves in sustainability initiatives and would internalise green values and pro-environmental behaviour in the course of their daily work (Gazi et al., 2024). The concepts of green teams, feedback, and non-monetary rewards (e.g., recognition, emotional salary) can instill a sense of ownership and intrinsic motivation and thus sustainability becomes a common organisational project. HRM acts as an interface between strategic sustainability objectives and the realities on the ground with employees in which the values of the environment are not just talked about but exercised both on the ground and throughout the organization (Ali et al., 2024).

Table 3. Green HRM Practices as Cultural Enablers

HRM Practice	Description	Cultural Impact
Green Recruitment and Selection	Hiring individuals with strong pro-environmental values and attitudes to ensure alignment with the organization's sustainability goals.	Strengthens cultural fit and embeds environmental values from the outset.
Green Training and Development	Providing education and skills enhancement focused on sustainability practices and environmental management.	Builds employee competence and awareness, reinforcing sustainability as part of daily work behavior.
Green Performance Management	Evaluating employee performance based on environmental objectives and sustainable work practices.	Encourages accountability and consistency in pro-environmental behavior across the organization.
Green Rewards and Recognition	Implementing monetary and non-monetary incentives for employees who contribute to sustainability initiatives.	Reinforces desired behaviors and strengthens emotional and moral engagement with green values.
Employee Empowerment and Participation	Establishing green teams, feedback systems, and participatory mechanisms to involve employees in sustainability decisions.	Promotes ownership, intrinsic motivation, and collective responsibility for environmental goals.

A sustainable organizational culture emerges when environmental values become deeply rooted in collective norms, beliefs, and behaviors. By aligning culture with green principles, organizations cultivate eco-conscious identities that drive long-term sustainability. Human Resource Management plays a pivotal role as a cultural enabler - embedding environmental responsibility through recruitment, training, and reward systems that reinforce commitment to sustainable development.

4. THE NEXUS BETWEEN GREEN LEADERSHIP AND CULTURAL TRANSFORMATION

Green leadership and cultural transformation are deeply intertwined, with leadership serving as a primary catalyst for embedding sustainability into organizational values, norms, and behaviors. The synergy between leadership and human resource management (HRM) is essential for institutionalizing a green culture and overcoming resistance to change, ultimately aligning organizational vision and operations with sustainability goals.

4.1 Leadership as a Catalyst for Cultural Change

Leaders become instrumental in development and change of organizational culture as their values, norms and behaviors are significantly influenced by them either directly or indirectly. According to the social learning theory, employees monitor the behavior of leaders and follow it, which is especially likely when the latter persistently exemplifies pro-environmental values and practices (Al-Swidi et al., 2021). Namely, green transformational leaders serve as role models since they show their commitment to sustainability through their actions, communication, and decision-making (Sanchez-Garcia et al., 2024). Through a well-defined vision of sustainability and turning it into an organizational value system, the leaders will create a common value system that will focus on being environmentally responsible (Alemu, 2025).

Communication is also an effective way of ensuring sustainability by leaders who make certain that the environment goals are integrated in the strategic objectives and daily operations. This is a form of sustainability modeling that does not only influence the attitude of employees, but also encourages voluntary green practices, including organizational citizenship behavior towards the environment (Priyadarshini et al., 2023). Green leadership is further enhanced by the fact that passionate employees empowered with motivation have been demonstrated to intermoderate and intermediatethe relationship between leadership and pro-environmental behaviours (Althnayan et al., 2022).

4.2 The Process of Cultural Transformation

Cultural transformation toward sustainability typically unfolds in stages, beginning with awareness and progressing to commitment and institutionalization. Initially, leaders raise awareness of environmental issues and the need for change, often through education, communication, and symbolic actions (Alemu, 2025). As awareness grows, organizations move toward commitment, where sustainability becomes a core value reflected in policies, rituals, and shared beliefs (Sánchez-García et al., 2024).

Aligning organizational vision, strategy, and operations with sustainability goals requires deliberate change management. Leaders must integrate sustainability into strategic planning, performance metrics, and operational processes, ensuring coherence across all levels of the organization (Fry & Egel, 2021). Overcoming cultural inertia and resistance to green change is a significant challenge, often necessitating comprehensive change management strategies that address employee concerns, provide training, and create incentives for sustainable behaviors (Sánchez-García et al., 2024). Dialogical approaches and collaborative practices can help break down silos, reduce resistance, and foster a sense of ownership among employees.

4.3 Synergizing Leadership and HRM

The collaboration between leadership and HRM is crucial for institutionalizing a green culture. HRM practices such as green recruitment, training, performance management, and rewards embed environmental values into the fabric of the organization (Ahmad et al., 2023). Leadership-driven HRM programs, including sustainability-focused training and competency frameworks, ensure that leaders and employees alike possess the skills and motivation necessary for environmental responsibility (Sánchez-García et al., 2024).

Examples of HR-driven leadership programs include integrating sustainability criteria into leadership development, performance appraisals, and promotion decisions, as well as offering training in social responsibility and workplace ethics (Zhang et al., 2021). These initiatives reinforce the alignment between leadership behaviors and organizational culture, creating a green work climate that inspires voluntary green behaviors among employees (Zhang et al., 2021).

Figure 1. The Nexus Between Green Leadership and Cultural Transformation

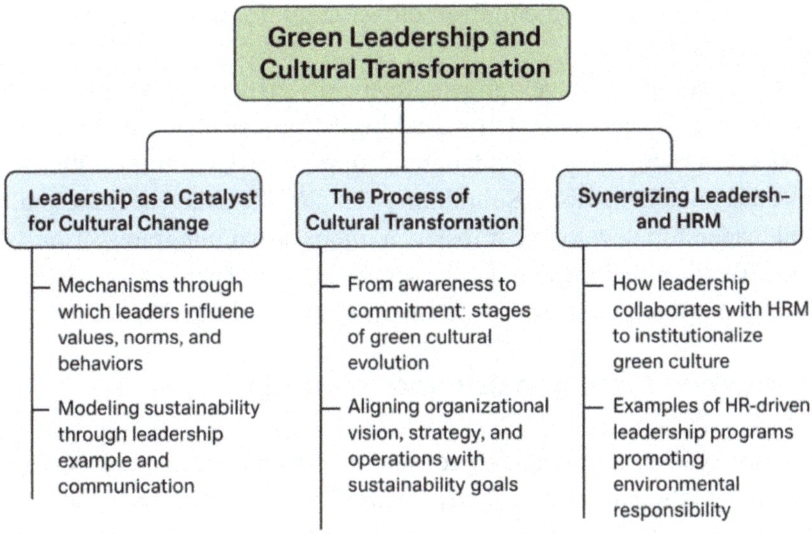

Author, (2025)

The nexus between green leadership and cultural transformation is characterized by a dynamic interplay in which leaders act as catalysts for change, HRM institutionalizes green values, and both work together to overcome resistance and embed sustainability into the organizational DNA.

5. STRATEGIES FOR CULTIVATING GREEN LEADERSHIP AND CULTURE

Cultivating green leadership and a sustainability-oriented culture requires a comprehensive approach that integrates talent management, performance systems, employee engagement, and organizational learning. Recent research highlights the importance of aligning leadership, HRM practices, and organizational values to drive environmental performance and innovation across sectors.

5.1 Talent Acquisition and Development

Recruitment strategies that prioritize sustainability values are foundational for building a green workforce. Organizations increasingly seek candidates who demonstrate environmental awareness and a commitment to sustainability, using green job descriptions and selection criteria to attract talent aligned with ecological goals. Green recruitment not only ensures cultural fit but also signals the organization's environmental priorities to potential employees, reinforcing its green employer brand (Jamal et al., 2021).

Leadership training and development are equally critical. Programs focused on green competencies - such as systems thinking, ethical decision-making, and environmental stewardship - equip leaders to champion sustainability initiatives and model eco-conscious behaviors (Subrahmanyam, 2025). Green servant and transformational leadership development fosters a mindset that integrates sustainability into strategic decision-making and daily operations, amplifying the impact of green culture throughout the organization (Hasanein & Metwally, 2025).

5.2 Performance Management and Rewards

The inclusion of environmental objectives in performance appraisals is a strong tool of matching goals of individuals and organizations. Green performance management systems consider workers and executives according to how they contribute to sustainability, like resource use, waste disposal, and involvement in green activities (Amjad et al., 2021; Paille et al., 2020). Such systems ensure that environmental responsibility is an important part of the job performance that promotes accountability and continuous improvement.

Green behaviors are further strengthened with incentive systems that encourage innovation and collaboration that is sustainability-oriented. Corporates use monetary and non-monetary incentives - bonuses, recognition systems, and promotion possibilities - to reward employees that share in environmental agendas (Jamal et al., 2021). This can be encouraged by incentives that foster innovation culture such that teams are encouraged to come up with and implement green solutions together (Paille et al., 2020).

5.3 Employee Involvement and Engagement

Employee involvement is central to embedding sustainability in organizational culture. Green teams, participatory decision-making, and suggestion schemes empower employees to contribute ideas and take ownership of environmental initiatives (Paillé et al., 2020). This participative approach not only enhances problem-solving and innovation but also increases employee engagement and commitment to sustainability.

Internal communication and storytelling play a vital role in cultural transformation. Leaders and HR professionals use storytelling to share success stories, lessons learned, and the organization's sustainability journey, making abstract values tangible and relatable (Hasanein & Metwally, 2025). Regular communication through newsletters, meetings, and digital platforms keeps sustainability top-of-mind and reinforces shared values across the workforce (Jamal et al., 2021).

5.4 Organizational Learning and Continuous Improvement

The process of green leadership and culture requires the organizational learning systems to stay sustainable. The skills of environmental management can be learned and shared by employees on knowledge management platforms, sustainability training, and cross-functional learning opportunities (Subrahmanyam et al., 2024). These systems facilitate creating green competencies and nurturing a culture of constant improvement.

Environmental innovation requires creating experimentation and the loop of feedback. Organizations promote trying out new concepts, failure, and trial, which enables them to develop an adaptable environment that will react to the changing challenges of sustainability (Alabdali et al., 2024). Feedback systems - surveys, performance reviews, open forums, etc. - give feedback on what is performing and what needs improvement to initiate continuous work on the sustainability agenda (Paille et al., 2020).

Figure 2. Strategies for Cultivating Green Leadership and Culture

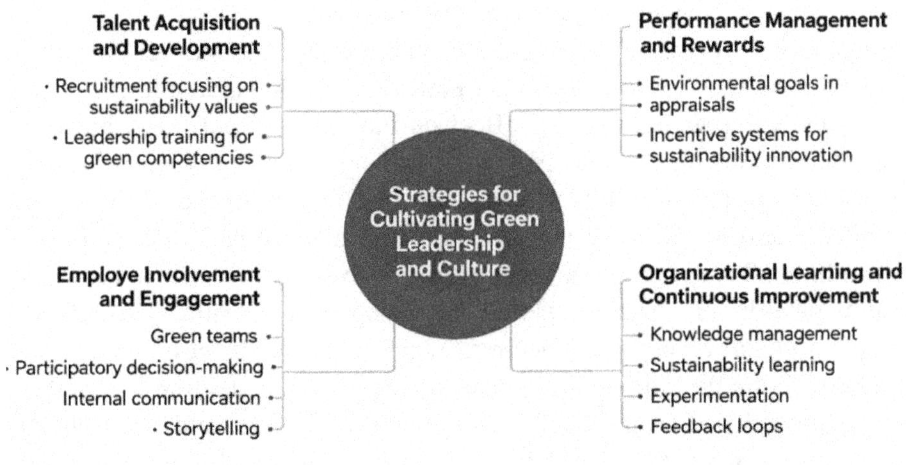

Author, (2025)

Cultivating green leadership and a sustainability-oriented culture requires an integrated approach. By aligning talent acquisition, performance management, employee engagement, and organizational learning with environmental goals, organizations can foster eco-conscious behaviors, drive innovation, and ensure continuous improvement. Such strategies not only enhance environmental responsibility but also strengthen organizational resilience, reputation, and long-term competitiveness.

6. CASE STUDIES AND BEST PRACTICES IN GREEN LEADERSHIP AND CULTURE

Real-world examples from corporations, developing economies, and higher education institutions illustrate how green leadership and culture can drive sustainability. These cases highlight both the strategies that lead to success, and the challenges organizations face during cultural transformation.

6.1 Global Examples: Corporate Leadership in Sustainability

Patagonia stands as a benchmark for green leadership. Under founder Yvon Chouinard, Patagonia embedded environmental stewardship into its core business model, using recycled materials, promoting fair labor, and pledging 1% of sales to environmental causes. This commitment has not only enhanced Patagonia's brand reputation but also fostered deep customer loyalty and industry influence. Patagonia's leadership demonstrates how visionary, ethical, and innovative approaches can make sustainability a competitive advantage (Koh, 2024).

Once again, Unilever makes a strong argument. Launched by Unilever's previous CEO Paul Polman, the Sustainable Living Plan sought to separate the expansion of the company's operations from its effect on the environment. To achieve its objectives, the firm devoted resources to sustainable product development, supply chain management, and stakeholder engagement, in line with the SDGs as out by the United Nations. Financial performance, risk management, and employee engagement have all seen significant improvements over Unilever's history, which is defined by data-driven decision-making and transparent reporting (Koh, 2024).

Interface, a global carpet manufacturer, transformed its business through the "Mission Zero" initiative, aiming for zero negative environmental impact by 2020. Interface adopted closed-loop manufacturing, invested in renewable energy, and fostered a culture of innovation. The corporate leadership inspired employees to embrace sustainability, resulting in significant reductions in waste and carbon emissions, and positioning Interface as a global leader in green manufacturing (Koh, 2024).

6.2 Regional and Sectoral Insights

Sustainability is becoming a concern of the public and private organizations in developing economies. An organization study within Qatar discovered that green leadership and HRM practices had a significant impact related to the evolution of the green organizational culture and staff behavior, which resulted in an enhancement of environmental performance in both the government and the private sectors (Al-Swidi et al., 2021). The results of this study support the significance of leadership commitment and combined HRM strategies in settings where the regulatory frameworks might be less developed.

The institutions of higher learning (HEIs) are also leading the transition towards sustainability. The University of Maribor (Slovenia) and the University of Economics-Varna (Bulgaria) have strategically integrated sustainability in their missions, operations and stakeholder interaction. Such universities took a step-by-step approach: pre-awareness to continual improvement, with leadership on every level, and promoting the sustainability culture among employees and the student body (Strukelj et al., 2023). The University of Minho (Portugal) showed that the combination of top-down and bottom-up strategies, constant monitoring, and partnership networks are the main aspects of successful cultural change (Ramisio et al., 2019).

A global survey of university leaders revealed that inclusive leadership, systemic thinking, and long-term vision are critical for embedding sustainability in higher education. Challenges such as limited resources, lack of expertise, and resistance from some academic staff persist. Successful institutions address these barriers through partnerships, public relations, and ongoing education and communication (Leal Filho et al., 2020).

6.3 Insights Gained: Success Factors and Challenges

Key success factors for embedding sustainability include *visionary and ethical leadership*, which sets a clear direction and integrates sustainability into the organization's core strategy. Such leaders inspire trust, model responsible behavior, and drive long-term commitment to environmental goals, as seen in global companies like Patagonia and Unilever, where leadership's ethical orientation and long-term vision have been pivotal for sustainable transformation.

Alignment of HRM practices - such as green recruitment, targeted training, and sustainability-linked rewards - ensures that environmental values are consistently reinforced throughout the employee lifecycle. Research shows that when HRM is strategically integrated with sustainability objectives, it amplifies green innovation and employee engagement, making sustainability a shared organizational priority.

Employee engagement is further strengthened through participatory initiatives and transparent communication. Mechanisms like green teams, suggestion schemes, and open dialogue foster a sense of ownership and collective responsibility, which are essential for overcoming resistance and embedding green behaviors.

Continuous learning and adaptation, supported by data-driven decision-making and feedback loops, enable organizations to respond to evolving sustainability challenges. Corporates that invest in organizational learning, knowledge management, and regular performance reviews are better equipped to innovate and sustain green practices over time.

The lack of resources, the incompatibility of cultures, and the unwillingness to adapt are still obstacles. Intentional leadership development, value-driven decision-making, and strategic resource allocation are necessary to overcome these hurdles and make sustainability a basic concept of the business.

Common challenges involve overcoming cultural inertia, securing buy-in from all organizational levels, and ensuring adequate resources and expertise. Both corporations and HEIs highlight the need for holistic, inclusive approaches and the importance of leadership in modeling and championing sustainability.

7. CHALLENGES AND BARRIERS TO GREEN CULTURAL TRANSFORMATION

Green cultural transformation within organizations faces a multitude of challenges, ranging from leadership resistance and short-termism to employee disengagement, institutional constraints, and the inherent paradox between growth and sustainability objectives.

7.1 Leadership Resistance, Short-Termism, and Conflicting Priorities

A significant barrier to green transformation is resistance or lack of commitment from leadership. When top management prioritizes short-term financial gains over long-term sustainability, green initiatives are often deprioritized or inadequately resourced. This short-termism is particularly problematic in contexts where immediate profitability is valued more than environmental or social outcomes, leading to a misalignment between stated sustainability goals and actual business practices (Kodua et al., 2022). Conflicting corporate priorities - such as the tension between operational efficiency and environmental stewardship - can result in fragmented or superficial sustainability efforts, impeding meaningful cultural change (Oliveira et al., 2024).

7.2 Lack of Employee Buy-In and Psychological Barriers

Employee engagement is crucial for successful green transformation, yet many organizations struggle to secure genuine buy-in. Psychological barriers such as fear of change, lack of awareness, and perceived threats to job security can foster resistance at all levels (Silvestre et al., 2025). Employees may view sustainability as an additional burden or irrelevant to their roles, especially if leadership communication is unclear or inconsistent (Alemu, 2025). Organizational inertia and established workplace norms further inhibit the adoption of sustainable behaviors, while high levels of trust in the organization can paradoxically reduce individual responsibility for sustainability, as employees may delegate these concerns to management (Silvestre et al., 2025).

7.3 Institutional, Financial, and Policy-Related Constraints

Resource constraints, particularly financial limitations, are among the most critical barriers to implementing green initiatives, especially in developing countries and small- to medium-sized enterprises (Kodua et al., 2022). The lack of access to funding, high initial costs of green technologies, and limited government incentives can stifle progress. Political and regulatory barriers, such as inconsistent policies and weak enforcement, further complicate sustainability efforts (Kodua et al., 2022). Institutional inertia, bureaucratic structures, and a lack of clear strategy or long-term planning can also slow the pace of change (Kiesnere & Baumgartner, 2019).

7.4 Managing Paradoxes Between Growth Objectives and Sustainability Commitments

Organizations often face a fundamental paradox: balancing economic growth with sustainability commitments. This tension is especially pronounced in emerging economies, where growth imperatives may overshadow environmental concerns (Oliveira et al., 2024). Short-term orientation, risk aversion, and reluctance to make trade-offs between profitability and sustainability are common internal barriers. Externally, market pressures and customer expectations can create additional obstacles, particularly when sustainability is perceived as costly or incompatible with business objectives. Addressing these paradoxes requires integrating sustainability into strategic planning, aligning incentives, and fostering a culture of innovation and adaptability (Kiesnere & Baumgartner, 2019).

Table 4. Challenges and Barriers to Green Cultural Transformation

Category of Challenge	Core Issues and Barriers	Organizational Implications
Leadership Resistance, Short-Termism, and Conflicting Priorities	Lack of leadership commitment and prioritization of short-term financial gains over long-term sustainability objectives. Tension between operational efficiency and environmental stewardship leads to fragmented implementation of green initiatives.	Weak strategic alignment between sustainability goals and business operations; limited allocation of resources; superficial or symbolic sustainability programs that fail to achieve cultural transformation.
Lack of Employee Buy-In and Psychological Barriers	Fear of change, low awareness, unclear communication, and perceived job insecurity discourage active employee participation in sustainability efforts. Cultural inertia and overreliance on management to handle environmental issues further exacerbate disengagement.	Resistance to behavioral change; low participation in green initiatives; sustainability seen as peripheral to core job responsibilities; limited diffusion of pro-environmental norms.
Institutional, Financial, and Policy-Related Constraints	Financial limitations, high initial costs of green technologies, and inadequate access to funding. Weak government incentives, inconsistent regulatory frameworks, and bureaucratic inertia hinder progress.	Slow adoption of green technologies and practices; sustainability treated as non-essential or unaffordable; stagnation in long-term transformation efforts, especially in SMEs and developing contexts.
Paradoxes Between Growth and Sustainability Objectives	Conflict between economic growth imperatives and sustainability commitments. Short-term orientation, risk aversion, and reluctance to trade off profitability for sustainability gains. External pressures such as market competition and customer expectations reinforce the tension.	Strategic contradictions undermine consistent sustainability practices; perception of sustainability as a cost center rather than a value driver; missed opportunities for innovation and long-term res

Overcoming the barriers to green cultural transformation demands visionary leadership, strategic alignment, employee engagement, and supportive institutional frameworks. Organizations that successfully navigate these challenges are those that embed sustainability into their core values, invest in capacity-building, and embrace the complexities of balancing growth with environmental stewardship.

8. MEASURING THE IMPACT OF GREEN LEADERSHIP AND CULTURE

Assessing the effectiveness of green leadership and a sustainability-oriented culture is essential for organizations aiming to achieve and demonstrate progress toward environmental, social, and economic goals. This process involves the use of key performance indicators (KPIs), a blend of qualitative and quantitative assessment tools, and increasingly, HR analytics to track cultural and leadership outcomes.

8.1 Key Performance Indicators (KPIs) for Sustainability-Oriented Culture

The key in the impact measurement of green leadership and culture is the KPIs. They offer measurable indices that are indicative of an organization on the way to integrating sustainability into its operations and culture. The contemporary models, including those based on the Global Reporting Initiative (GRI), suggest a full range of indicators that encompass the environmental, social, and economic aspects. In a recent framework, 91 strong performance indicators, 36 environmental, and 55 social indicators were identified, and organizations are able to measure and report their sustainability performance in the supply chain and within their organization (Okay et al., 2024).

In reality, corporates usually come up with their personalized KPIs to suit their respective sustainability policies. An example of an Italian food company provided a case study of the development of a dashboard consisting of approximately 70 environmental KPIs, which were directly connected with employee rewards and management by objectives. These KPIs fell under performance, commitment, control, reactivity, economic, consent, and multifunctional categories as such that sustainability would be incorporated at both corporate and departmental levels. Another popular tool is The Sustainable Development Balanced Scorecard, implementing sustainability KPIs into the enterprise management systems and aligning them with the United Nations Sustainable Development Goals (Chmutova et al., 2024).

Common sustainability KPIs include:

- Energy and resource consumption
- Greenhouse gas emissions
- Waste generation and recycling rates
- Water usage
- Employee engagement in sustainability initiatives
- Training hours on environmental topics
- Number of sustainability-related innovations or projects (Gönczi, 2023)

8.2 Qualitative and Quantitative Assessment Tools

A robust assessment of green leadership and culture requires both quantitative and qualitative tools. Quantitative tools, such as sustainability indices, balanced scorecards, and dashboards, provide objective data on performance metrics. The Composite Performance Indicator (CPI) model aggregates environmental, social, economic, and governance KPIs into a single value, allowing for benchmarking and detailed analysis of strengths and weaknesses (Pavláková Dočekalová & Kocmanová, 2016).

To round out quantitative data, qualitative techniques might pick up on subtleties in leadership and company culture. Methods such as this include self-evaluation checklists, questionnaires, focus groups, and interviews. To provide a comprehensive picture of sustainability integration, technologies such as Sustainability Tracking, Assessment & Rating System (STARS) and Green Metrics integrate quantitative indicators with qualitative input from stakeholders in higher education (Alghamdi et al., 2017). Isensee et al. (2020) noted that when evaluating intangibles like leadership commitment, employee attitudes, and the efficacy of internal communication, qualitative methodologies are very useful.

Process-oriented performance indicators and maturity models are also gaining traction. These tools assess not only outcomes but also the processes and capabilities that underpin sustainable practices, such as leadership development, stakeholder engagement, and continuous improvement (Medne & Lapiņa, 2019).

8.3 Role of HR Analytics in Tracking Cultural and Leadership Outcomes

HR analytics becomes more critical in determining the effect of green leadership and culture. Using information in HR systems, organizations may monitor the data regarding the increased employee engagement, training attendance, turnover, and performance associated with sustainability efforts. HR analytics will be able to disclose the relationships between the green training programs and the subsequent positive outcome in the pro-environmental conduct or the acceptance of the green HRM practices and the increased organizational culture and employee performance (Subrahmanyam, 2025).

Complex HR analytics systems can real-time track KPIs, benchmark and make decisions grounded in data. They are also useful in assisting organizations to seek gaps in leadership development, the effectiveness of reward systems linked to sustainability, and the process of achievement in cultural transformation goals (Okay et al., 2024). The combination of HR analytics and sustainability evaluation instruments will help organizations to adjust strategies on time and provide the culture of development.

The evaluation of the influence of the green leadership and culture should be based on the multidimensional approach which incorporates KPIs, qualitative and quantitative measurement tools, and HR analytics. This combined approach can help the organizations monitor the developments, focus on the areas that require enhancement, and make it one of the organization values to make sustainability a reality.

9. FUTURE DIRECTIONS AND RESEARCH IMPLICATIONS IN GREEN LEADERSHIP AND CULTURE

The incorporation of circular economy ideas, digital transformation, and artificial intelligence (AI) are fast changing the environment of green leadership and corporate culture. Emerging trends, cross-cultural viewpoints, and research gaps provide many opportunities for future scholarly investigation as enterprises globally work towards sustainability.

9.1 Emerging Trends: Digital Sustainability Leadership, AI-Driven Green HRM, and Circular Economy Integration

Digital sustainability leadership is gaining prominence as organizations leverage digital transformation to drive green innovation and sustainable performance. Recent research demonstrates that digital leadership not only enhances operational efficiency but also acts as a catalyst for long-term sustainability by fostering green innovation and adaptive learning cultures (Zada et al., 2025). Top management innovativeness further amplifies these effects, ensuring that sustainability initiatives are embedded into core business processes rather than remaining peripheral (Zada et al., 2025). In manufacturing and other sectors, digital transformation - when guided by sustainability-oriented leadership - facilitates green knowledge acquisition and innovation, ultimately improving sustainable corporate performance (Asif et al., 2024).

Another new development is GHRM, or green human resource management powered by artificial intelligence. According to Mahat et al. (2025), artificial intelligence is playing a more and more mediating role in the connection between GHRM practices and sustainable workplace outcomes. This means that training, employee engagement, and recruiting all benefit from AI more than before. With the use of AI, HRM operations may be made more effective and eco-friendlier by automating applicant screening, monitoring employee green habits, and supporting continuous learning.

Circular economy integration is also reshaping leadership roles. Leaders are now expected to champion resource productivity, waste elimination, and the restoration of natural systems, aligning business strategies with circular economy principles (Ratić et al., 2024). Policy frameworks such as the EU's Circular Economy Action Plan and green taxation are driving this shift, requiring leaders to adapt to new regulatory landscapes and foster innovation for long-term sustainability.

9.2 Cross-Cultural Perspectives on Green Leadership and Transformation

Cross-cultural research reveals that the effectiveness of green leadership and cultural transformation varies significantly across regions and organizational contexts. While green transformational leadership and R&D investment significantly influence green performance management in Belgium, the impact of cross-cultural perspectives is less pronounced, suggesting the need for more nuanced strategies that consider both collectivist and individualist cultural elements (Hemmet, 2024). In China and South Korea, digital leadership and transformational leadership have been shown to positively affect sustainability outcomes, but the mechanisms and boundary conditions - such as innovation capability and environmental dynamism - differ by context (Asif et al., 2024).

According to Huang et al. (2023), there is a need to align leadership approaches with local cultural values because cultural self-representation and man-nature orientation moderate the relationship between environmentally specific transformational leadership and employees' green behavior. In order to make the most of green leadership on a global scale, these results highlight the need of conducting comparative research and implementing leadership development programs that take context into account.

9.3 Research Gaps and Theoretical Contributions

Although a great step has already been made, some gaps in research are still present. To begin with, longitudinal research is also necessary to investigate the long-term effects of digital leadership and AI-based GHRM on sustainability performance since the existing research is based on cross-sectional data. Second, there is no industry-specific and comparative analysis; further studies are necessary in understanding the functioning of digital sustainability leadership in various industries and regulatory settings.

Such theoretical contributions as the carryover of organizational learning theory and dynamic capabilities theory to the concept of digital sustainability are provided to show how the concept of digital leadership leads to the development of adaptive learning cultures and an enhanced ability of firms to embrace sustainable practices (Zada et al., 2025). Low-carbon transformational leadership (LCTL) conceptualization further enhances the current knowledge of the role that leadership plays in carbon reduction and green innovation, particularly in carbon-intensive industries (Zhang and Huang, 2024).

The areas that have not been explored well are the integration of stakeholders, reverse effect of followers on the environmental behaviors of leaders, and the innovation of sophisticated tools to assess results of sustainability (Zhong et al., 2023). It is also suggested that there is a need to conduct further studies on the incorporation of AI and digital technologies into GHRM, the role of leadership in green transitions into the circulatory economy, and the impact that cultural and institutional variables have on the effectiveness of green leadership.

Figure 3. Future Directions and Research Implications

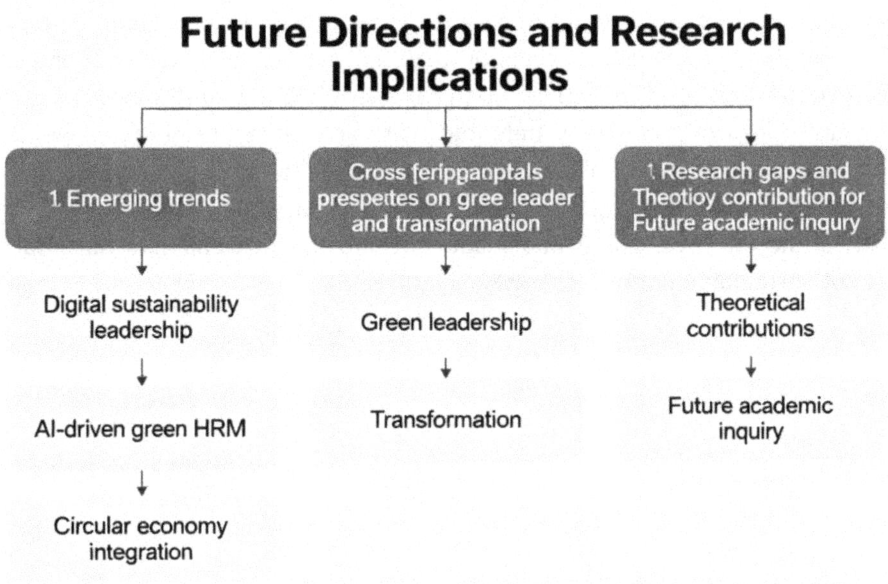

Author, (2025)

New theoretical frameworks and assessment techniques, together with longitudinal, cross-sectoral, and cross-cultural studies, should be the focuses of future study. Scholars can boost green leadership and culture by filling these knowledge gaps, which will help businesses achieve sustainable change.

10. CONCLUSION

The synthesis of the recent research reveals that green human resource management (GHRM) and sustainability-oriented leadership play the central role in ensuring the introduction of sustainability in the organizational culture and stimulation of high-quality environmental, social, and economic performance. GHRM practices: green recruitment, training, management of performance, and rewards have a direct effect on green innovation and commitment of employees and transformational leadership, and ethical leadership enhance the effect of these practices; by creating the example of sustainability and encouraging workforce participation in sustainability programs. HRMs and leaders strategic integration is not only the best approach to making organizations more resilient and adaptive to changes but also establishing a culture in which the notion of sustainability becomes not only a shared value but also a standard of organizational operation.

For practitioners, these findings underscore the need to align HR systems and leadership development with sustainability goals, ensuring that green values are reflected in daily practices and decision-making. Policymakers are encouraged to support frameworks and incentives that promote GHRM adoption and leadership accountability for sustainability outcomes. For academia, the research points to the importance of further exploring the mechanisms linking HRM, leadership, and sustainability, as well as the contextual factors that shape their effectiveness across industries and cultures. HRM and leadership are not just enablers but strategic drivers of sustainable cultural change.

REFERENCES

Ahmad, J., Al Mamun, A., Masukujjaman, M., Makhbul, Z. K. M., & Ali, K. A. M. (2023). Modeling the workplace pro-environmental behavior through green human resource management and organizational culture: Evidence from an emerging economy. *Heliyon*, *9*(9), e19134. DOI: 10.1016/j.heliyon.2023.e19134

Al-Swidi, A., Gelaidan, H., & Saleh, R. (2021). The joint impact of green human resource management, leadership and organizational culture on employees' green behaviour and organisational environmental performance. *Journal of Cleaner Production*, *316*, 128112. DOI: 10.1016/j.jclepro.2021.128112

Alabdali, M. A., Yaqub, M., Agarwal, R., Alofaysan, H., & Mohapatra, A. K. (2024). Unveiling green digital transformational leadership: Nexus between green digital culture, green digital mindset and green digital transformation. *Journal of Cleaner Production*, *450*, 141670. DOI: 10.1016/j.jclepro.2024.141670

Alemu, B. A. (2025). Cultivating a culture of sustainability: The role of organizational values and leadership in driving sustainable practices. *Business Ethics and Leadership*, *9*(1), 1–15. DOI: 10.61093/bel.9(1).79-94.2025

Alghamdi, N., den Heijer, A., & de Jonge, H. (2017). Assessment tools' indicators for sustainability in universities: An analytical overview. *International Journal of Sustainability in Higher Education*, *18*(1), 84–115. DOI: 10.1108/IJSHE-04-2015-0071

Ali, M., Shujahat, M., Fatima, N., Lopes de Sousa Jabbour, A. B., Vo-Thanh, T., Salam, M. A., & Latan, H. (2024). Green HRM practices and corporate sustainability performance. *Management Decision*, *62*(11), 3681–3703. DOI: 10.1108/MD-05-2023-0787

Ali, Z., Qin, Z., Raza-Ullah, T., Ali, A., & Sherwani, M. (2025). Leading the way to greening: Aligning leadership styles and market learning for radical and incremental green innovation. *Business Strategy and the Environment*. Advance online publication. DOI: 10.1002/bse.70039

Althnayan, S., Alarifi, A., Bajaba, S., & Alsabban, A. S. (2022). Linking environmental transformational leadership, environmental organizational citizenship behavior, and organizational sustainability performance: A moderated mediation model. *Sustainability*, *14*(14), 8779. DOI: 10.3390/su14148779

Amjad, F., Abbas, W., Zia-ur-Rehman, M., Baig, S., Hashim, M., Khan, A., & Rehman, H. U. (2021). Effect of green human resource management practices on organizational sustainability: The mediating role of environmental and employee performance. *Environmental Science and Pollution Research International*, *28*(22), 28191–28206. DOI: 10.1007/s11356-020-11307-9

Asif, M., Yang, L., & Hashim, M. (2024). The role of digital transformation, corporate culture, and leadership in enhancing corporate sustainable performance in the manufacturing sector of China. *Sustainability*, *16*(7), 2651. DOI: 10.3390/su16072651

Begum, S., Ashfaq, M., Xia, E., & Awan, U. (2022). Does green transformational leadership lead to green innovation? The role of green thinking and creative process engagement. *Business Strategy and the Environment*, *31*(1), 580–597. DOI: 10.1002/bse.2911

Biswas, S. R., Uddin, M. A., Bhattacharjee, S., Dey, M., & Rana, T. (2022). Ecocentric leadership and voluntary environmental behavior for promoting sustainability strategy: The role of psychological green climate. *Business Strategy and the Environment*, *31*(4), 1705–1718. DOI: 10.1002/bse.2978

Boeske, J., & Murray, P. A. (2022). The intellectual domains of sustainability leadership in SMEs. *Sustainability*, *14*(4), 1978. DOI: 10.3390/su14041978

Chmutova, I., Ho, S., Zhang, Ts., Mazhuta, V., & Nimets, A. (2024). Development of Key Performance Indicators of an Integrated Enterprise Management System Using a Balanced System of Sustainable Development Indicators. *Development Service Industry Management*, (2), 270–278. DOI: 10.31891/dsim-2024-6(43)

de Oliveira, U. R., Menezes, R. P., & Fernandes, V. A. (2024). A systematic literature review on corporate sustainability: Contributions, barriers, innovations and future possibilities. *Environment, Development and Sustainability*, *26*(2), 3045–3079. DOI: 10.1007/s10668-023-02933-7

Gazi, M. A. I., Dhali, S., Masud, A., Ahmed, A., Amin, M. B., Chaity, N., Senathirajah, A. R. B. S., & Abdullah, M. (2024). Leveraging Green HRM to Foster Organizational Agility and Green Culture: Pathways to Enhanced Sustainable Social and Environmental Performance. *Sustainability*, *16*(20), 8751. DOI: 10.3390/su16208751

Gönczi, J. (2023). Sustainability performance indicators. *The Annals of the University of Oradea. Economic Sciences, 32*(1), 126. https://anale.steconomiceuoradea.ro/en/wp-content/uploads/2024/02/AUOES.July_.20239.pdf?utm_source=chatgpt.com

Hasanein, A., & Metwally, A. (2025). Hospitality leading green for a sustainable scene: The mediating role of green organizational culture in the relationship between green servant leadership and environmental performance. *Geojournal of Tourism and Geosites, 60*(2 spl), 1243-1253. https://doi.org/DOI: 10.30892/gtg.602spl21-1497

Hemmet, A. (2024). Investigating the impact of green R&D, transformational leadership, and cross-cultural approaches on performance management in the manufacturing sector. *ARPHA Conference Abstracts, 7*, e129588. https://doi.org/DOI: 10.3897/aca.7.e129588

Huang, L., Guo, Z., Deng, B., & Wang, B. (2023). Unlocking the relationship between environmentally specific transformational leadership and employees' green behaviour: A cultural self-representation perspective. *Journal of Cleaner Production, 382*, 134857. DOI: 10.1016/j.jclepro.2022.134857

Isensee, C., Teuteberg, F., Griese, K., & Topi, C. (2020). The relationship between organizational culture, sustainability, and digitalization in SMEs: A systematic review. *Journal of Cleaner Production, 275*, 122944. DOI: 10.1016/j.jclepro.2020.122944

Jamal, T., Zahid, M., Martins, J. M., Mata, M. N., Rahman, H. U., & Mata, P. N. (2021). Perceived green human resource management practices and corporate sustainability: Multigroup analysis and major industries perspectives. *Sustainability, 13*(6), 3045. DOI: 10.3390/su13063045

Khaddage-Soboh, N., Yunis, M., Imran, M., & Zeb, F. (2024). Sustainable practices in Malaysian manufacturing: The influence of CSR, transformational leadership, and green organizational culture on environmental performance. *Economic Analysis and Policy, 82*, 753–768. DOI: 10.1016/j.eap.2024.04.001

Kiesnere, A. L., & Baumgartner, R. J. (2019). Sustainability management in practice: Organizational change for sustainability in smaller large-sized companies in Austria. *Sustainability, 11*(3), 572. DOI: 10.3390/su11030572

Kodua, L. T., Xiao, Y., Adjei, N., Asante, D., Ofosu, B., & Amankona, D. (2022). Barriers to green human resources management (GHRM) implementation in developing countries: Evidence from Ghana. *Journal of Cleaner Production, 340*, 130671. DOI: 10.1016/j.jclepro.2022.130671

Koh, H. W. (2024). Global leadership and sustainability. [IJSRM]. *International Journal of Scientific Research and Management, 12*(8), 7015–7026. DOI: 10.18535/ijsrm/v12i08.em03

Leal Filho, W., Eustachio, J. H. P. P., Caldana, A. C. F., Will, M., Lange Salvia, A., Rampasso, I., Anholon, R., Platje, J., & Kovaleva, M. (2020). Sustainability leadership in higher education institutions: An overview of challenges. *Sustainability*, *12*(9), 3761. DOI: 10.3390/su12093761

Mahat, D., Shrestha, S. K., Neupane, D., Karki, T. B., & Dongol, P. (2025). Green human resource management and sustainable workplace: Artificial Intelligence as mediating variable. *Edelweiss Applied Science and Technology*, *9*(6), 656–667. DOI: 10.55214/25768484.v9i6.7863

Medne, A., & Lapiņa, I. (2019). Sustainability and continuous improvement of organization: Review of process-oriented performance indicators. *Journal of Open Innovation*, *5*(3), 49. DOI: 10.3390/joitmc5030049

Mushtaq, S., & Akhtar, S. (2024). Sustainable HRM strategies, enhancing organizational resilience and advancing sustainability goals. *Journal of Management Development*, *43*(5), 712–726. DOI: 10.1108/jmd-02-2024-0072

Nakra, N., & Kashyap, V. (2024). Responsible leadership and organizational sustainability performance: Investigating the mediating role of sustainable HRM. *International Journal of Productivity and Performance Management*, *74*(2), 409–426. DOI: 10.1108/IJPPM-03-2023-0115

Okay, N. C., Sencer, A., & Taskin, N. (2024). Quantitative indicators for environmental and social sustainability performance assessment of the supply chain. *Environment, Development and Sustainability*. Advance online publication. DOI: 10.1007/s10668-024-05210-3

Paillé, P., Valeau, P., & Renwick, D. (2020). Leveraging green human resource practices to achieve environmental sustainability. *Journal of Cleaner Production*, *260*, 121137. DOI: 10.1016/j.jclepro.2020.121137

Pavláková Dočekalová, M., & Kocmanová, A. (2016). Composite indicator for measuring corporate sustainability. *Ecological Indicators*, *61*, 612–623. DOI: 10.1016/j.ecolind.2015.10.012

Piwowar-Sulej, K. (2020). Pro-environmental organizational culture: Its essence and a concept for its operationalization. *Sustainability*, *12*(10), 4197. DOI: 10.3390/su12104197

Priyadarshini, C., Chatterjee, N., Srivastava, N. K., & Dubey, R. K. (2023). Achieving organizational environmental citizenship behavior through green transformational leadership: A moderated mediation study. *Journal of Asia Business Studies*, *17*(6), 1088–1109. DOI: 10.1108/JABS-05-2022-0185

Ramísio, P. J., Pinto, L. M. C., Gouveia, N., Costa, H., & Arezes, D. (2019). Sustainability strategy in higher education institutions: Lessons learned from a nine-year case study. *Journal of Cleaner Production, 222*, 300–309. DOI: 10.1016/j.jclepro.2019.02.257

Ratić, M., Ćulibrk, J., Tasić, N., Petrović, M., & Medić, N. (2024). Integrating sustainability and circular economy principles in leadership amidst uncertainty. In *Proceedings of the International Conference on Business, Management, and Economics Engineering (FUTURE-BME 2024)* (Paper 015). https://doi.org/DOI: 10.24867/FUTURE-BME-2024-015

Roscoe, S., Subramanian, N., Jabbour, C., & Chong, T. (2019). Green human resource management and the enablers of green organisational culture: Enhancing a firm's environmental performance for sustainable development. *Business Strategy and the Environment, 28*(5), 737–750. DOI: 10.1002/bse.2277

Sajjad, A., Eweje, G., & Raziq, M. M. (2024). Sustainability leadership: An integrative review and conceptual synthesis. *Business Strategy and the Environment, 33*(4), 2849–2867. DOI: 10.1002/bse.3631

Sánchez-García, E., Montalvo-Falcón, J. V., Marco-Lajara, B., & Martínez-Falcó, J. (2024). Guiding organizations toward sustainable success: The strategic role of leadership in environmental corporate governance in the wine industry. *Corporate Social Responsibility and Environmental Management, 31*(6), 6438–6456. DOI: 10.1002/csr.2925

Silvestre, W., Begnini, S., & Abreu, I. (2025). The paradox of trust: How leadership, commitment, and inertia shape sustainability behavior in the workplace. *Administrative Sciences, 15*(7), 254. Advance online publication. DOI: 10.3390/admsci15070254

Srivastava, S., Pathak, D., Soni, S., & Dixit, A. (2024). Does green transformational leadership reinforce green creativity? The mediating roles of green organizational culture and green mindfulness. *Journal of Organizational Change Management, 37*(3), 619–640. DOI: 10.1108/JOCM-09-2023-0364

Subrahmanyam, S. (2025). Analytics: Trends, Challenges, and Prospects of Adopting Artificial Intelligence in Human Resource Management. In Kapoor, T., Singh, A., Sahay, K., Kaur, H., & Verma, R. (Eds.), *Sustainable Management Practices for Employee Retention and Recruitment* (pp. 51–72). IGI Global Scientific Publishing., DOI: 10.4018/979-8-3693-8679-8.ch003

Subrahmanyam, S. (2025). Role of Artificial Intelligence in Re-Inventing Human Resource Management (HRM) 2.0. In Kapoor, T., Singh, A., Sahay, K., Kaur, H., & Verma, R. (Eds.), *Sustainable Management Practices for Employee Retention and Recruitment* (pp. 243–264). IGI Global Scientific Publishing., DOI: 10.4018/979-8-3693-8679-8.ch012

Subrahmanyam, S. (2025). Using AI-Driven Analysis for Training and Leadership Growth. In Mrabet, J., Boudagga, S., Sharma, R., & Bansal, R. (Eds.), *Harnessing Emotion AI for Customer Support and Employee Wellbeing* (pp. 321–352). IGI Global Scientific Publishing., DOI: 10.4018/979-8-3373-3658-9.ch013

Subrahmanyam, S., Aishwaryalaxmi, N. S., Khalife, D., Shaikh, I. A. K., Faldu, R., & Asthana, N. (2024, April). Impact of Knowledge Management and Big Data Analytics Capabilities on Firm Performance. In *2024 Ninth International Conference on Science Technology Engineering and Mathematics (ICONSTEM)* (pp. 1-5). IEEE. https://doi.org/DOI: 10.1109/ICONSTEM60960.2024.10568874

Zada, M., Zada, S., Dhar, B. K., Ping, C., & Sarkar, S. (2025). Digital leadership and sustainable development: Enhancing firm sustainability through green innovation and top management innovativeness. *Sustainable Development*, *33*(5), 6704–6716. DOI: 10.1002/sd.3488

Zhang, H., & Huang, H. (2024). Low-Carbon Transformational Leadership: Conceptualization, Measurement, and Its Impact on Innovation Outcomes. *Sustainability*, *16*(24), 10844. Advance online publication. DOI: 10.3390/su162410844

Zhang, J., Ul-Durar, S., Akhtar, M. N., Zhang, Y., & Lu, L. (2021). How does responsible leadership affect employees' voluntary workplace green behaviours? A multilevel dual process model of voluntary workplace green behaviours. *Journal of Environmental Management*, *296*, 113205. DOI: 10.1016/j.jenvman.2021.113205

Zhong, J., Shao, X., Xiao, H., Yang, R., & An, X. (2023). The research on the green leadership: A systematic review and theoretical framework. *Environment, Development and Sustainability*, *27*(1), 377–408. DOI: 10.1007/s10668-023-03960-0

KEY TERMS AND DEFINITIONS

Circular Economy Integration: Circular economy integration refers to adopting business models and operational practices that minimize waste and maximize resource efficiency through reuse, recycling, and regeneration. It supports the transition from a linear "take-make-dispose" model to a sustainable system aligned with green leadership principles.

Cultural Transformation: Cultural transformation is the process of altering an organization's underlying values, norms, and behaviors to align with new strategic goals or ethical imperatives. In the green context, it involves shifting from profit-driven mindsets to sustainability-driven orientations across all levels of the organization.

Ecocentric Leadership: Ecocentric leadership is a philosophy that places the natural environment at the center of organizational decision-making. It emphasizes interdependence between human and ecological systems, promoting policies and practices that respect planetary boundaries and biodiversity.

Employee Engagement in Sustainability: Employee engagement in sustainability refers to the active participation of employees in environmental initiatives and decision-making processes. Engaged employees act as change agents, promoting eco-friendly practices and contributing to a culture of shared environmental responsibility.

Environmental Vision: Environmental vision is the long-term strategic outlook of an organization that integrates ecological sustainability into its mission, goals, and operational processes. It guides leadership actions and organizational strategies toward achieving environmental and social impact.

Green Human Resource Management (Green HRM): Green HRM refers to human resource policies and practices designed to foster environmentally responsible behavior among employees. It includes green recruitment, sustainability-focused training, eco-based performance appraisals, and incentives that reinforce environmental values.

Green Innovation: Green innovation involves developing new products, processes, or business models that reduce environmental impact while creating economic and social value. It represents a key driver of sustainable competitive advantage and organizational transformation.

Green Leadership: Green leadership refers to a leadership approach that integrates environmental sustainability into organizational vision, strategy, and practices. Green leaders promote ecological balance, ethical responsibility, and innovation to ensure long-term environmental and social well-being while maintaining organizational performance.

Sustainability-Oriented Organizational Culture: A sustainability-oriented organizational culture is a set of shared values, beliefs, and behaviors within an organization that prioritize environmental stewardship, social responsibility, and economic viability. It guides employee actions and decision-making toward achieving sustainable outcomes.

Transformational Leadership: Transformational leadership is a style where leaders inspire and motivate followers through a shared vision, intellectual stimulation, and individualized consideration. It serves as a foundation for green leadership by encouraging change and innovation toward sustainability.

Chapter 8
Ethical Leadership in the Age of Green Crisis

Martin Sposato
https://orcid.org/0000-0001-9260-9961
Zayed University, UAE

ABSTRACT

This chapter examines how ethical leadership drives environmental sustainability in modern organizations. It explores the convergence of ethical leadership, CSR, and environmental stewardship, providing frameworks for sustainable practice implementation. Drawing from organizational behavior and stakeholder perspectives, it demonstrates how ethical leaders foster environmental responsibility cultures. The analysis covers employee engagement in sustainability initiatives and knowledge management's role in environmental performance. The chapter addresses crisis management challenges and emphasizes education's importance in developing future environmental leaders. It offers practical guidance for business leaders, organizations, and policymakers, providing a roadmap for sustainable business practices in an eco-conscious global economy.

INTRODUCTION

In the face of escalating environmental challenges, the role of ethical leadership in steering organizations towards sustainable practices has become increasingly crucial. This chapter explores the intersection of ethical leadership, corporate social responsibility (CSR), and environmental stewardship in the context of the current ecological crisis in modern organizations. It examines how ethical leaders can drive

DOI: 10.4018/979-8-3373-2479-1.ch008

meaningful change within their organizations and industries, fostering a culture of environmental responsibility and sustainable business practices.

The environmental crisis presents a complex set of challenges for business leaders, requiring them to balance economic imperatives with ecological considerations and social responsibilities (Vig & Datta, 2020). Ethical leadership in this context goes beyond mere compliance with environmental regulations; it demands a proactive approach to addressing environmental issues and a commitment to long-term sustainability goals. This chapter argues that ethical leaders play a pivotal role in shaping organizational values, decision-making processes, and corporate behavior in ways that prioritize environmental sustainability.

Furthermore, the intersection of ethical leadership with emerging technologies and their potential to drive environmental sustainability is explored in this chapter. As organizations increasingly adopt artificial intelligence, the Internet of Things, and other advanced technologies, ethical leaders face new opportunities and challenges in leveraging these tools for environmental stewardship. The ethical implications of these technologies are examined, alongside their potential to create more sustainable business practices. This exploration builds upon recent research by Sposato (2024a) and ElSayary (2024), which highlights the need for leadership development to adapt to the age of artificial intelligence and its implications for sustainability efforts.

Building on these foundational perspectives, this chapter makes several distinctive contributions to both theory and practice. This chapter makes several important theoretical and practical contributions to our understanding of ethical leadership in environmental crisis management. By integrating stakeholder theory and systems theory perspectives, it develops a comprehensive framework for understanding how ethical leaders can effectively drive environmental sustainability initiatives while managing complex stakeholder relationships. The chapter examines how ethical leaders navigate the interconnected nature of organizational and ecological systems, showing how leadership decisions ripple through both domains. It provides novel insights into the role of knowledge management in environmental leadership, demonstrating how ethical leaders can leverage organizational knowledge to enhance environmental performance. The chapter also breaks new ground in exploring how ethical leaders can harness emerging technologies, particularly artificial intelligence, to advance environmental sustainability goals while maintaining ethical principles. Through detailed analysis of employee engagement, crisis management, and educational approaches, it offers practical guidance for developing and implementing ethical leadership practices that drive meaningful environmental change. The chapter's integrated theoretical framework and practical recommendations provide a roadmap for organizations seeking to enhance their environmental performance through ethical leadership, while contributing to the broader academic discourse on leadership in the age of environmental crisis.

These contributions are particularly significant in the global context, where environmental challenges transcend national boundaries. As environmental challenges transcend national boundaries, ethical leaders must be equipped to navigate complex international regulations, diverse cultural perspectives on sustainability, and global stakeholder expectations. Insights are provided into how leaders can foster cross-cultural collaboration and knowledge sharing to address environmental issues on a global scale. Through an examination of case studies and best practices from various industries and regions, a comprehensive view is offered of how ethical leadership can drive meaningful environmental change in an increasingly interconnected world.

THE ROLE OF ETHICAL LEADERSHIP IN ENVIRONMENTAL STEWARDSHIP

Ethical leadership is fundamental in driving organizations towards sustainable practices and environmental stewardship. As Northouse (2021) suggests, ethical leadership is characterized by demonstrating normatively appropriate conduct through personal actions and interpersonal relationships, and promoting such conduct among followers through two-way communication, reinforcement, and decision-making. In the context of environmental crisis, this translates to leaders who not only advocate for sustainable practices but also embody these principles in their actions and decisions.

Ethical leadership in environmental stewardship inherently involves managing diverse stakeholder interests and expectations. Leaders must balance the environmental concerns of various stakeholders, from employees and shareholders to local communities and environmental advocacy groups, while maintaining organizational performance. This multi-stakeholder approach requires leaders to develop sophisticated engagement strategies that acknowledge and address different stakeholder perspectives on environmental issues.

Recent research by Wu et al. (2021) demonstrates that ethical leadership, coupled with CSR initiatives, significantly influences employees' engagement in environmentally friendly practices. This finding underscores the importance of leadership in creating a corporate culture that values and promotes environmental stewardship at all levels of the organization. Ethical leaders set the tone for organizational behavior, influencing not only the strategic direction of the company but also the day-to-day actions of employees.

The impact of ethical leadership extends beyond the immediate organizational boundaries. As Sposato (2024b) notes, in the age of artificial intelligence and rapidly evolving technologies, leadership training and development must adapt to incorporate ethical considerations in decision-making processes. This is particularly relevant when addressing environmental challenges, as leaders must navigate the complex interplay between technological advancements, business imperatives, and ecological responsibilities.

CORPORATE SOCIAL RESPONSIBILITY AND ENVIRONMENTAL SUSTAINABILITY

Corporate Social Responsibility (CSR) has emerged as a critical framework through which organizations can address their environmental impact. López Jiménez et al. (2021) argue that CSR and ethical considerations are increasingly important in the digital environment, suggesting that ethical leadership must evolve to address new challenges posed by technological advancements. In the context of environmental crisis, CSR initiatives provide a structured approach for organizations to implement sustainable practices and contribute positively to environmental conservation efforts.

Ethical leaders play a crucial role in championing CSR initiatives that prioritize environmental sustainability. They recognize that environmental responsibility is not just a moral imperative but also a strategic necessity in an era where consumers, investors, and regulators are increasingly scrutinizing corporate environmental practices. As Hatum (2017) suggests, organizational creativity is essential for developing innovative solutions to environmental challenges. Ethical leaders foster a culture of innovation that encourages the development of sustainable technologies and practices.

Moreover, the integration of CSR into core business strategies requires a systemic approach. Drawing from systems theory, organizations can be viewed as complex adaptive systems interacting with both social and ecological environments. This perspective reveals how organizational actions create ripple effects through multiple interconnected systems - economic, social, and environmental. Ethical leaders must understand these system dynamics to effectively implement comprehensive CSR programs that address environmental issues holistically, rather than through piecemeal initiatives. This systems thinking approach enables leaders to anticipate both direct and indirect impacts of their sustainability initiatives across different stakeholder groups and ecological systems

ETHICAL LEADERSHIP AND EMPLOYEE ENGAGEMENT IN ENVIRONMENTAL INITIATIVES

One of the key roles of ethical leadership in addressing the environmental crisis is fostering employee engagement in sustainability initiatives. As Wu et al. (2021) demonstrate, ethical leadership significantly influences employees' pro-environmental behavior, particularly in the context of Industry 4.0. This finding highlights the importance of leadership in creating a corporate culture that not only values environmental stewardship but also actively encourages and empowers employees to participate in sustainability efforts.

Ethical leaders achieve this by setting clear expectations, providing resources and support for environmental initiatives, and recognizing and rewarding pro-environmental behaviors. They create an organizational climate where sustainability is not seen as an additional burden but as an integral part of the company's mission and values. This approach aligns with the principles of organizational behavior theory, as discussed by Katz & Kahn (2015), which emphasizes the role of leadership in shaping organizational culture and norms.

Furthermore, ethical leaders recognize the importance of knowledge management in driving environmental sustainability. As Martins et al. (2019) argue, knowledge management plays a crucial role in the context of sustainability. Ethical leaders facilitate the creation, sharing, and application of environmental knowledge within their organizations. They invest in employee education and training programs to build environmental awareness and competencies throughout the organization. This approach not only enhances the organization's capacity to address environmental challenges but also empowers employees to become active participants in the company's sustainability journey.

The role of ethical leadership in fostering a sense of environmental responsibility extends beyond the confines of the organization. Employees who are engaged in sustainability initiatives at work are more likely to adopt environmentally friendly behaviors in their personal lives, creating a ripple effect that amplifies the impact of organizational efforts. This phenomenon, often referred to as the 'spillover effect', has been documented in recent studies such as those by Muster & Schrader (2021), which highlight how workplace sustainability practices can influence employees' home behaviors. Ethical leaders play a crucial role in facilitating this spillover effect by consistently modeling sustainable behaviors and articulating the broader societal importance of environmental stewardship.

Moreover, the impact of ethical leadership on employee engagement in environmental initiatives is not uniform across different organizational contexts. Factors such as industry type, organizational size, and cultural setting can significantly influence the effectiveness of leadership approaches. For instance, Lee, Lee & Lee (2022), found that in knowledge-intensive industries, ethical leaders who emphasize innovation and creativity in addressing environmental challenges tend to elicit higher levels of employee engagement. Conversely, in more traditional industries, leaders who focus on the long-term benefits of sustainability for job security and community well-being often achieve better results. These findings underscore the importance of adopting a nuanced, context-specific approach to ethical leadership in driving employee engagement in environmental initiatives.

ETHICAL LEADERSHIP IN CRISIS MANAGEMENT: NAVIGATING ENVIRONMENTAL CHALLENGES

The current environmental crisis often presents leaders with complex, high-stakes situations that require effective crisis management skills. Dhamija (2023) discusses how ethical leadership can be effectively applied in crisis situations, emphasizing the need for leaders to make difficult decisions that balance short-term business needs with long-term environmental sustainability. This perspective is particularly relevant in the context of the current environmental crisis, where leaders must often make tough choices that may impact their organization's immediate profitability but are necessary for long-term ecological viability.

Crisis management in environmental contexts particularly benefits from a systems theory approach. Environmental crises often emerge from complex interactions between organizational operations, natural systems, and stakeholder behaviors. Ethical leaders must understand these systemic relationships to effectively anticipate, prevent, and respond to environmental crises. This includes recognizing how different parts of the system influence each other and identifying leverage points for intervention

Ethical leaders approach crisis management with a focus on transparency, accountability, and stakeholder engagement. They recognize that environmental crises often have far-reaching implications that extend beyond the immediate organizational boundaries. As such, they prioritize open communication with all stakeholders, including employees, customers, investors, and local communities. This approach aligns with the principles of stakeholder theory, as discussed by Jeffrey et al. (2015), which emphasizes the importance of considering the interests of all parties affected by organizational decisions.

Moreover, ethical leaders understand the importance of preparedness and proactive planning in mitigating environmental risks. They invest in developing robust environmental management systems and crisis response protocols. This proactive approach not only enhances the organization's resilience in the face of environmental challenges but also demonstrates a commitment to environmental stewardship that can enhance the company's reputation and stakeholder trust.

THE ROLE OF EDUCATION IN DEVELOPING ETHICAL LEADERS FOR ENVIRONMENTAL STEWARDSHIP

The development of ethical leaders capable of navigating the complexities of the environmental crisis is a critical challenge for management education. As highlighted by Coldwell (2017), university education plays a crucial role in equipping future leaders with the ethical framework and critical thinking skills necessary to address complex environmental challenges. By integrating sustainability and ethics into leadership education, a more robust foundation can be laid for the next generation of business leaders to tackle the environmental crisis head-on.

Educational institutions bear a significant responsibility in shaping the mindset and skills of future leaders. The integration of environmental ethics and sustainability into core curricula is essential for developing leaders who can effectively balance economic imperatives with ecological considerations. This integration should not be limited to standalone courses on sustainability but should permeate various disciplines, including finance, marketing, operations, and strategy. By doing so, students can develop a holistic understanding of how environmental considerations intersect with different aspects of business operations.

Moreover, the educational approach to developing ethical leaders for environmental stewardship must evolve to keep pace with technological advancements. As suggested by ElSayary (2024), integrating advanced technologies like generative AI into active learning environments can enhance metacognition and technological skills, which are essential for addressing complex environmental challenges. This approach to education should not only focus on theoretical knowledge but also provide practical experiences that allow students to grapple with real-world environmental issues. Case studies, simulations, and project-based learning can play a vital role in this regard, offering students opportunities to apply ethical decision-making in scenarios that mirror the complexities of environmental crisis management.

Furthermore, interdisciplinary collaboration in education is crucial for developing well-rounded ethical leaders. Environmental challenges often require solutions that draw from multiple fields of study. Therefore, educational programs should foster collaboration between business schools and departments of environmental science, engineering, and public policy. This interdisciplinary approach can help future leaders develop a more comprehensive understanding of environmental issues and the diverse stakeholder perspectives involved in addressing them.

The role of education extends beyond formal university settings. Leadership development programs within organizations should also prioritize environmental ethics and sustainability. As noted by Sposato (2024b), leadership training in the age of artificial intelligence must adapt to include ethical considerations in decision-making processes. This is particularly relevant for environmental leadership, where leaders must navigate the complex interplay between technological advancements, business imperatives, and ecological responsibilities. Continuous learning and professional development programs can help current leaders stay abreast of emerging environmental challenges and innovative solutions.

Global perspective is another critical aspect that education must address in developing ethical leaders for environmental stewardship. As environmental issues often transcend national boundaries, leaders need to understand diverse cultural perspectives on sustainability and be equipped to navigate complex international regulations. Educational programs should incorporate global case studies, international exchange programs, and cross-cultural projects to foster this global mindset.

Additionally, education plays a crucial role in developing leaders' ability to communicate effectively about environmental issues. Ethical leaders must be able to articulate the importance of sustainability initiatives to various stakeholders, including employees, investors, and the broader community. Communication skills, including the ability to translate complex environmental data into actionable insights, should be a key component of leadership education programs.

Finally, the development of ethical leaders for environmental stewardship requires a focus on long-term thinking and systems understanding. Educational programs should cultivate leaders' ability to consider the long-term consequences of business decisions on the environment and society. This includes training in systems thinking, scenario planning, and sustainability metrics. By equipping leaders with these tools, education can foster a new generation of decision-makers who are better prepared to balance short-term business needs with long-term environmental sustainability.

In conclusion, the role of education in developing ethical leaders for environmental stewardship is multifaceted and crucial. It requires a comprehensive approach that integrates sustainability and ethics across disciplines, leverages advanced technologies, fosters interdisciplinary collaboration, emphasizes global perspectives, develops effective communication skills, and cultivates long-term, systems-level thinking. By rising to this educational challenge, we can better prepare the next generation of leaders to effectively address the environmental crisis and drive meaningful change towards a more sustainable future.

THEORETICAL CONTRIBUTION

This chapter makes several significant contributions to the existing body of knowledge on ethical leadership and environmental sustainability. By integrating perspectives from organizational behavior, systems theory, and stakeholder theory, a comprehensive framework is provided for understanding the role of ethical leadership in addressing the environmental crisis. This interdisciplinary approach offers a more nuanced understanding of how ethical leadership can drive environmental sustainability in complex organizational settings.

A significant theoretical contribution of this chapter lies in its novel integration of stakeholder theory with environmental ethical leadership. By examining how ethical leaders navigate the complex web of stakeholder interests in environmental decision-making, the chapter extends traditional stakeholder theory beyond its conventional applications. It demonstrates how ethical leaders must balance and align diverse stakeholder perspectives on environmental issues, from employees and shareholders to local communities and environmental advocacy groups. This theoretical extension provides a framework for understanding how ethical leaders can create value for all stakeholders while pursuing environmental sustainability, moving beyond the traditional trade-off view between environmental and economic interests.

The chapter also makes a substantial contribution by applying systems theory to environmental ethical leadership in a way that illuminates the interconnected nature of organizational and ecological systems. Drawing on the work of scholars like Lai & Lin (2017), the chapter develops a theoretical framework that explains how ethical leaders can navigate the complex feedback loops between organizational actions and environmental impacts. This systems perspective reveals how ethical leadership decisions ripple through both organizational and environmental systems, creating multiple pathways for influence and change. The chapter's theoretical contribution lies in showing how systems thinking enables ethical leaders to anticipate and manage the unintended consequences of environmental decisions, while identifying leverage points for positive environmental change.

Furthermore, the chapter advances theory by synthesizing systems theory and stakeholder theory to create a more comprehensive model of environmental ethical leadership. This integration reveals how ethical leaders can use systems thinking to better understand and manage stakeholder relationships in the context of environmental challenges. The theoretical framework demonstrates how feedback loops in organizational systems influence and are influenced by stakeholder relationships, creating dynamic patterns of interaction that ethical leaders must navigate. This synthesis provides new insights into how ethical leaders can create positive environmental outcomes by understanding and leveraging the interconnections between organizational systems and stakeholder networks.

The chapter advances the theoretical understanding of the relationship between ethical leadership and employee engagement in environmental initiatives. Drawing on recent research, such as the work of Wu et al. (2021), it demonstrates how ethical leadership can foster a culture of environmental responsibility that permeates all levels of the organization. This contribution extends beyond simple cause-and-effect relationships, exploring the complex interplay between leadership behaviors, organizational culture, and individual employee actions. By doing so, it provides a more sophisticated model for understanding how leadership influences organizational outcomes in the context of environmental sustainability.

Furthermore, the chapter offers a theoretical basis for understanding the role of knowledge management in environmental leadership. By integrating insights from Martins et al. (2019) and others, it highlights how ethical leaders can leverage knowledge management practices to enhance their organization's environmental performance. This contribution bridges the gap between leadership theory and knowledge management literature, providing a new perspective on how these two domains intersect in the pursuit of sustainability goals. The chapter proposes that effective knowledge management, guided by ethical leadership, can serve as a catalyst for environmental innovation and adaptive organizational responses to ecological challenges.

The theoretical framework presented in this chapter also contributes to the understanding of ethical leadership in crisis management, particularly in the context of environmental challenges. By drawing on the work of Dhamija (2023) and others, it provides a conceptual model for how ethical leaders can navigate the complex trade-offs between short-term business imperatives and long-term environmental sustainability. This model emphasizes the importance of systems thinking in ethical leadership, highlighting how leaders must consider the broader ecological and social systems within which their organizations operate.

Moreover, the chapter contributes to the theoretical discourse on the role of ethical leadership in driving organizational change towards sustainability. It proposes that ethical leaders act as change agents, not only by setting the strategic direction but also by shaping the values and norms that guide organizational behavior. This perspective builds upon and extends existing change management theories by explicitly incorporating the ethical dimension of leadership in the context of environmental sustainability.

The chapter also makes a theoretical contribution by exploring the intersection of ethical leadership, environmental sustainability, and emerging technologies. By examining how ethical leaders can navigate the challenges and opportunities presented by artificial intelligence, the Internet of Things, and other advanced technologies, the chapter extends leadership theory into new domains. It proposes a framework for understanding how ethical considerations in technology adoption can be balanced with environmental goals, contributing to the emerging field of digital ethics in organizational leadership.

Finally, the chapter contributes to the theoretical understanding of ethical leadership in a global context. By considering how ethical leaders navigate diverse cultural perspectives on sustainability and complex international regulations, the chapter extends leadership theory beyond national boundaries. It proposes a model of global ethical leadership that emphasizes cross-cultural competence, stakeholder engagement, and collaborative problem-solving in addressing transnational environmental challenges.

In summary, this chapter makes multiple theoretical contributions that advance our understanding of ethical leadership in the age of environmental crisis. By integrating diverse theoretical perspectives, proposing new conceptual models, and extending existing theories into new domains, it provides a robust foundation for future research and practice in this critical area.

PRACTICAL IMPLICATIONS

The insights presented in this chapter have several significant practical implications for business leaders, organizations, educational institutions, and policymakers. These implications offer actionable guidance for implementing ethical leadership practices that drive environmental sustainability across various sectors and organizational contexts. For business leaders, the chapter underscores the critical importance of integrating environmental considerations into all aspects of organizational decision-making. Ethical leaders are called upon to go beyond mere compliance with environmental regulations and instead proactively address environmental issues through strategic initiatives. This involves developing a comprehensive understanding of the organization's environmental impact and implementing measures to mitigate negative effects while enhancing positive contributions to ecological sustainability.

In practice, business leaders should consider conducting regular environmental audits to assess their organization's ecological footprint. This process can provide valuable insights into areas where improvements can be made and help prioritize sustainability initiatives. Leaders should also focus on implementing sustainable supply chain management practices, recognizing that a significant portion of an organization's environmental impact often lies within its supply chain. Investing in clean technologies and renewable energy sources is another crucial step, demonstrating a commitment to reducing the organization's carbon footprint and fostering innovation in sustainable practices.

Ethical leaders must also prioritize the development of products and services with reduced environmental impact. This not only contributes to sustainability goals but can also create new market opportunities and enhance brand reputation. Establishing clear, measurable environmental performance targets and regularly reporting on progress is essential for maintaining accountability and demonstrating commitment to stakeholders. These targets should be ambitious yet achievable, pushing the organization to continually improve its environmental performance.

Moreover, ethical leaders must focus on fostering a culture of environmental responsibility within their organizations. This involves not only setting the strategic direction but also modeling sustainable behaviors and decision-making processes. Leaders should prioritize open communication about environmental goals and challenges, encouraging employee participation in sustainability initiatives. Recognizing and rewarding environmentally responsible behaviors can help reinforce the importance of sustainability throughout the organization and motivate employees to contribute to these efforts.

For organizations as a whole, the chapter highlights the need for a systemic approach to environmental sustainability. It suggests that companies should view environmental responsibility not as a separate initiative but as an integral part of their core business strategy. This implies the need for organizations to invest in developing robust environmental management systems, employee training programs, and knowledge management practices that support sustainable business operations. Integrating environmental considerations into strategic planning processes ensures that sustainability is considered in all major business decisions.

Organizations should consider developing cross-functional sustainability teams to drive initiatives across different departments. This approach can help break down silos and ensure that environmental considerations are integrated into all aspects of the business. Implementing environmental management systems (EMS) aligned with international standards such as ISO 14001 can provide a structured approach to managing environmental impacts and ensuring continuous improvement.

Establishing green innovation programs can drive the development of environmentally friendly products and processes, potentially creating new revenue streams and competitive advantages. Creating knowledge sharing platforms to disseminate best practices in environmental sustainability across the organization can help ensure that lessons learned and successful strategies are widely adopted.

Educational institutions can draw on the insights presented in this chapter to enhance their leadership development programs. The chapter emphasizes the need for integrating environmental ethics and sustainability into business education curricula. This goes beyond offering standalone courses on sustainability to embedding environmental considerations across various disciplines such as finance, marketing, operations, and strategy. Redesigning curricula to incorporate environmental sustainability across all business disciplines can help ensure that future leaders are equipped to address environmental challenges in all aspects of business operations.

Developing experiential learning opportunities that allow students to engage with real-world environmental challenges can provide valuable practical experience. This could include internships with sustainability-focused organizations, consulting projects for local businesses on environmental issues, or participation in environmental impact assessments. Fostering interdisciplinary collaboration between business schools and environmental science departments can help students develop a more holistic understanding of environmental issues and potential solutions.

The chapter also suggests the potential of leveraging advanced technologies in leadership education to better prepare future leaders for the complexities of environmental crisis management. This could involve the use of AI-powered simulations, virtual reality experiences, and big data analytics to enhance students' understanding of complex environmental systems and decision-making processes. Establishing partnerships with businesses and NGOs can provide students with practical exposure to environmental leadership challenges and opportunities.

For policymakers and regulatory bodies, the chapter implies the need for policies that incentivize ethical leadership and environmental responsibility in the corporate sector. This could include developing regulations that require greater transparency in environmental reporting, ensuring that stakeholders have access to accurate information about companies' environmental impacts and mitigation efforts. Creating incentives for companies that demonstrate strong environmental performance, such as tax benefits or preferential treatment in government contracts, can encourage proactive environmental stewardship.

Support for research and development in sustainable technologies can help drive innovation and create new solutions to environmental challenges. Policymakers should also consider implementing carbon pricing mechanisms or cap-and-trade systems to internalize the environmental costs of business activities. Finally, promoting public-private partnerships for environmental conservation and sustainable development can leverage the strengths of both sectors to address complex environmental challenges.

In conclusion, the practical implications of this chapter span across various stakeholders and emphasize the need for a concerted effort to address environmental challenges through ethical leadership. By implementing these practices, organizations can not only contribute to environmental sustainability but also position themselves for long-term success in an increasingly eco-conscious global economy.

FURTHER RESEARCH AVENUES

While this chapter provides a comprehensive overview of ethical leadership in the age of environmental crisis, several areas warrant further research: The impact of cultural differences on ethical leadership and environmental sustainability: Future research could explore how cultural factors influence the effectiveness of ethical leadership in driving environmental initiatives across different global contexts.

The application of stakeholder theory to environmental crisis management: Future research could explore how different stakeholder engagement strategies affect the success of environmental initiatives and crisis response efforts. Systems dynamics in environmental leadership: Research is needed to better understand how different system elements interact in environmental crisis situations and how leaders can effectively intervene in these systems. The role of technology in supporting ethical leadership for environmental sustainability: As noted by ElSayary (2024) and Sposato (2024), advanced technologies like AI are reshaping leadership practices. Further research could investigate how these technologies can be leveraged to enhance ethical decision-making and environmental management.

The long-term effects of ethical leadership on organizational environmental performance: Longitudinal studies could provide insights into how ethical leadership impacts environmental outcomes over time, and how these effects are sustained through leadership transitions. The intersection of ethical leadership, environmental sustainability, and social justice: Further research could explore how ethical leaders can address environmental challenges while also promoting social equity and justice.

The role of ethical leadership in driving circular economy initiatives: As organizations increasingly adopt circular economy principles, research could investigate how ethical leaders can facilitate this transition and overcome associated challenges. The effectiveness of different leadership development approaches in cultivating ethical and environmentally conscious leaders: Comparative studies could assess various educational and training methods to identify best practices in developing leaders capable of addressing complex environmental challenges.

INTEGRATING SYSTEMS THEORY AND ETHICAL LEADERSHIP FOR ENVIRONMENTAL SUSTAINABILITY

The application of systems theory to ethical leadership in the context of environmental sustainability offers a comprehensive framework for understanding the complex interactions between organizations, their stakeholders, and the natural environment. As Lai and Lin (2017) explain, systems theory views organizations as interconnected entities that interact with and adapt to their environment. This perspective is particularly relevant when addressing environmental challenges, as it emphasizes the need for holistic, integrated approaches to sustainability.

Ethical leaders, armed with a systems thinking approach, can more effectively navigate the complexities of environmental stewardship. They recognize that environmental issues cannot be addressed in isolation but must be considered within the broader context of organizational operations, stakeholder expectations, and societal needs. This aligns with the work of Kühl (2020), who emphasizes the importance of understanding social systems as existing between interaction and society.

Musheke & Phiri (2021) highlight the role of effective communication in organizational performance based on systems theory. In the context of ethical leadership for environmental sustainability, this underscores the importance of clear, consistent communication about environmental goals, strategies, and progress. Ethical leaders must ensure that sustainability messages are not only disseminated throughout the organization but also integrated into the company's culture and decision-making processes.

Furthermore, the application of systems theory to ethical leadership in environmental contexts emphasizes the importance of feedback loops and adaptive management. As Scott and Davis (2015) discuss in their exploration of organizations as open systems, organizations must be responsive to changes in their environment. In the face of rapidly evolving environmental challenges, ethical leaders must foster organizational agility and adaptability, continuously refining sustainability strategies based on new information and changing circumstances.

THE ROLE OF KNOWLEDGE MANAGEMENT IN ETHICAL LEADERSHIP FOR SUSTAINABILITY

Knowledge management plays a crucial role in enabling ethical leaders to drive environmental sustainability initiatives effectively. As Martins et al. (2019) argue, knowledge management is integral to sustainability efforts, facilitating the creation, sharing, and application of environmental knowledge within organizations. Ethical leaders recognize the value of this knowledge as a strategic asset in addressing environmental challenges and driving sustainable innovation.

The work of Shahzad et al. (2020) explores the influence of knowledge management processes on corporate sustainable performance through green innovation. This research underscores the importance of ethical leaders in fostering a culture of learning and innovation focused on environmental sustainability. By promoting knowledge sharing and collaborative problem-solving, ethical leaders can enhance their organization's capacity to develop and implement sustainable solutions.

Moreover, as Chopra et al. (2021) discuss, the evolution of knowledge management for business sustainability requires a forward-looking approach. Ethical leaders must not only leverage current knowledge but also anticipate future sustainability challenges and opportunities. This proactive stance aligns with the principles of ethical leadership, which emphasize long-term thinking and responsibility to stakeholders, including future generations.

The concept of green knowledge management, as explored by Abbas & Khan (2023), offers a specialized framework for ethical leaders to integrate environmental considerations into organizational knowledge processes. By fostering a green organizational culture and promoting green innovation, ethical leaders can create a virtuous cycle where environmental knowledge drives sustainable practices, which in turn generate new insights and opportunities for improvement.

ETHICAL LEADERSHIP AND TECHNOLOGICAL INNOVATION FOR ENVIRONMENTAL SUSTAINABILITY

In the age of rapid technological advancement, ethical leaders must navigate the complex interplay between innovation, sustainability, and ethical considerations. As Sposato (2024b) notes, leadership training and development in the age of artificial intelligence must adapt to incorporate ethical decision-making processes, particularly when addressing environmental challenges.

The work of Su et al. (2020) highlights the relationship between environmental leadership, green innovation practices, and environmental knowledge learning. Ethical leaders play a crucial role in fostering an organizational climate that supports green innovation, encouraging the development and adoption of technologies that can enhance environmental performance while maintaining ethical standards.

However, the integration of new technologies in sustainability efforts also presents ethical challenges. As Ore & Sposato (2022) discuss in the context of AI in recruitment, technological innovations can bring both opportunities and risks. Ethical leaders must carefully consider the potential environmental and social impacts of new technologies, ensuring that their implementation aligns with the organization's sustainability goals and ethical principles

The concept of the triple helix model of innovation, as explored by Jiménez et al. (2021), offers insights into how ethical leaders can foster collaboration between industry, academia, and government to drive sustainable innovation. By leveraging these partnerships, ethical leaders can access diverse knowledge sources, share resources, and accelerate the development of sustainable solutions.

CONCLUSION

Ethical leadership in the age of environmental crisis demands a multifaceted approach that integrates systems thinking, knowledge management, and technological innovation. By embracing these perspectives, ethical leaders can drive meaningful change within their organizations and contribute to broader societal efforts to address environmental challenges. The application of systems theory provides ethical leaders with a framework for understanding the complex interactions between their organizations and the natural environment. This holistic perspective enables more effective decision-making and strategy development in the pursuit of sustainability goals.

Knowledge management emerges as a critical tool for ethical leaders in fostering environmental sustainability. By promoting the creation, sharing, and application of environmental knowledge, leaders can enhance their organization's capacity for sustainable innovation and adaptive management in the face of evolving ecological challenges. Technological innovation, while offering significant potential for addressing environmental issues, also requires careful ethical consideration. Ethical leaders must navigate the opportunities and risks associated with new technologies, ensuring that their implementation aligns with sustainability goals and ethical principles.

As we move forward, the role of ethical leadership in addressing environmental challenges will only grow in importance. By combining moral integrity with a deep understanding of systems dynamics, knowledge management, and technological innovation, ethical leaders can drive the transformational changes needed to create more environmentally responsible and sustainable organizations. The path ahead is challenging, but it also offers opportunities for ethical leaders to make a lasting positive impact on their organizations, society, and the planet. By embracing the principles and practices outlined in this chapter, leaders can not only navigate the current environmental crisis but also pave the way for a more sustainable and ethically grounded future.

REFERENCES

Abbas, J., & Khan, S. M. (2023). Green knowledge management and organizational green culture: An interaction for organizational green innovation and green performance. *Journal of Knowledge Management*, *27*(7), 1852–1870. DOI: 10.1108/JKM-09-2022-0641

Chopra, M., Saini, N., Kumar, S., Varma, A., Mangla, S. K., & Lim, W. M. (2021). Past, present, and future of knowledge management for business sustainability. *Journal of Cleaner Production*, *328*, 129592. DOI: 10.1016/j.jclepro.2021.129592

Coldwell, D. A. (2017). Ethical leadership in crisis management: The role of university education. In *Contemporary Leadership Challenges*. IntechOpen., DOI: 10.5772/65439

Dhamija, P. (2023). Emerging ethical leadership in crisis management. In *Emotional intelligence for leadership effectiveness* (pp. 221–235). Apple Academic Press., DOI: 10.1201/9781003367635-14

ElSayary, A. (2024). Integrating generative AI in active learning environments: Enhancing metacognition and technological skills. *Journal of Systemics, Cybernetics and Informatics*, *22*(3), 34–37. DOI: 10.54808/JSCI.22.03.34

Hatum, A. (2017). The role of talent for organizational creativity. In *Mastering creativity in organizations* (pp. 15–41). Edward Elgar Publishing., DOI: 10.4337/9781786436870.00008

Jeffrey, H. L., Beswick, E., & Meade, J. (2015). Learning and unlearning sexism in the workplace: Two important barriers prevent a more inclusive environment. *Human Resource Management International Digest*, *23*(5), 18–20. DOI: 10.1108/HRMID-03-2015-0035

Jiménez, D. L., Dittmar, E. C., & Portillo, J. P. V. (2021). Cooperation and relationship in the triple helix model of innovation. *International Journal of Knowledge and Learning*, *14*(1), 1–9. DOI: 10.1504/IJKL.2021.113210

Katz, D., & Kahn, R. (2015). The social psychology of organizations. In *Organizational behavior 2* (pp. 152–168). Routledge., DOI: 10.4324/9781315702001

Kühl, S. (2020). Groups, organizations, families and movements: The sociology of social systems between interaction and society. *Systems Research and Behavioral Science*, *37*(3), 496–515. DOI: 10.1002/sres.2701

Lai, C. H., & Lin, S. H. (2017). Systems theory. The international encyclopedia of organizational communication, 41, 1-18. https://doi.org/DOI: 10.1002/9781118955567.wbieoc122

Lee, D., Lee, M., & Lee, J. (2022). The effects of ethical leadership and knowledge sharing on employee engagement in knowledge-intensive business firms. *Sustainability*, *14*(18), 11358. DOI: 10.3390/su141811358

López Jiménez, D., Dittmar, E. C., & Vargas Portillo, J. P. (2021). New directions in corporate social responsibility and ethics: Codes of conduct in the digital environment. *Journal of Business Ethics*, •••, 1–11. DOI: 10.1007/s10551-021-04929-2

Martins, V. W. B., Rampasso, I. S., Anholon, R., Quelhas, O. L. G., & Leal Filho, W. (2019). Knowledge management in the context of sustainability: Literature review and opportunities for future research. *Journal of Cleaner Production*, *229*, 489–500. DOI: 10.1016/j.jclepro.2019.04.354

Musheke, M. M., & Phiri, J. (2021). The effects of effective communication on organizational performance based on the systems theory. *Open Journal of Business and Management*, *9*(2), 659–671. DOI: 10.4236/ojbm.2021.92035

Muster, V., & Schrader, U. (2021). Green work life balance: A new perspective for green HRM and environmental management. *German Journal of Human Resource Management*, *35*(1), 72–97. DOI: 10.1177/2397002220957396

Northouse, P. G. (2021). *Leadership: Theory and practice*. Sage Publications., DOI: 10.4135/9781071800539

Ore, O., & Sposato, M. (2022). Opportunities and risks of artificial intelligence in recruitment and selection. *The International Journal of Organizational Analysis*, *30*(6), 1771–1782. DOI: 10.1108/IJOA-11-2021-2933

Scott, W. R., & Davis, G. (2015). *Organizations and organizing: Rational, natural and open systems perspectives*. Routledge., DOI: 10.4324/9781315663371

Shahzad, M., Qu, Y., Zafar, A. U., Rehman, S. U., & Islam, T. (2020). Exploring the influence of knowledge management process on corporate sustainable performance through green innovation. *Journal of Knowledge Management*, *24*(9), 2079–2106. DOI: 10.1108/JKM-12-2019-0655

Sposato, M. (2024a). Are leaders born or made? Asking the right question. *Development and Learning in Organizations*, *38*(3), 1–3. DOI: 10.1108/DLO-01-2024-0001

Sposato, M. (2024b). Leadership training and development in the age of artificial intelligence. *Development and Learning in Organizations*, *38*(4), 4–7. DOI: 10.1108/DLO-02-2024-0002

Su, X., Xu, A., Lin, W., Chen, Y., Liu, S., & Xu, W. (2020). Environmental leadership, green innovation practices, environmental knowledge learning, and firm performance. *SAGE Open*, *10*(2), 2158244020922909. DOI: 10.1177/2158244020922909

Vig, S., & Datta, M. (2020). Failure of corporate governance or business ethics: The case of ICICI Bank. Business, economics and sustainable development, the emerging issues, 109-119. DOI: DOI: 10.1007/978-981-15-6232-4_8

Wu, Q., Cherian, J., Samad, S., Comite, U., Hu, H., Gunnlaugsson, S. B., & Sial, M. S. (2021). The role of CSR and ethical leadership to shape employees' pro-environmental behavior in the era of industry 4.0. A case of the banking sector. *Sustainability*, *13*(17), 9773. DOI: 10.3390/su13179773

ADDITIONAL READING

Busquets, A., & Gómez, M. (2013). Instructional Design and Quality: Learning Strategies for the Course Plan and Formative Activities-A Practical Case of the Program of East Asian Studies. In Governance, Communication, and Innovation in a Knowledge Intensive Society (pp. 236-250). IGI Global Scientific Publishing.

Jiménez, D. L., Dittmar, E. C., & Portillo, J. P. V. (2025). Political, Regulatory, and Ethical Concerns of AI, Blockchain, and Cloud Computing in Public Administration. In Harnessing AI, Blockchain, and Cloud Computing for Enhanced e-Government Services (pp. 33-62). IGI Global Scientific Publishing.

Muenjohn, N., Montague, A., Hoare, L., & Zhang, J. (2017). Managerial competencies related to soft skills: An investigation of hotel management in Asia. Transcontinental Strategies for Industrial Development and Economic Growth, 266-278.

Muniapan, B. (2015). The Bhagavad-Gita and business ethics: A leadership perspective. In *Asian business and management practices: trends and global considerations* (pp. 232–253). IGI Global.

Muniapan, B., & Low, P. K. C. (2013). Asian leadership wisdom: The relevance of Kautilyan and Confucian values for leadership development. In Business, Technology, and Knowledge Management in Asia: Trends and Innovations (pp. 156-165). IGI Global.

Sposato, M. (2025). Beyond Leader-Centricity Embracing Followership in Global Leadership Dynamics. In Developing Managerial Skills for Global Business Success (pp. 45-64). IGI Global Scientific Publishing.

Sposato, M. (2025). Developing Digital Leadership Skills for AI-Driven Research Environments. In Impacts of Generative AI on the Future of Research and Education (pp. 235-254). IGI Global.

Sposato, M. (2025). Digital Leadership in Organizations for Digital Transformation. In Multidisciplinary Organizational Training of Human Capital in the Digital Age (pp. 77-96). IGI Global Scientific Publishing.

Sposato, M., & Dittmar, E. C. (2025). Leadership Strategies for Implementing Environmental Management Systems. In Insights Into Digital Business, Human Resource Management, and Competitiveness (pp. 277-300). IGI Global Scientific Publishing.

Sposato, M., Dittmar, E. C., & Vargas Portillo, J. P. (2025). New technologies in HR: Bridging efficiency and ethical considerations. *The International Journal of Organizational Analysis*.

KEY TERMS AND DEFINITIONS

Corporate Social Responsibility (CSR): A business approach that integrates social and environmental concerns into company operations, considering the impact of corporate decisions and activities on society, the environment, and stakeholders.

Environmental Crisis Management: The strategic approach to handling environmental challenges and emergencies, involving preparedness, response protocols, and balanced decision-making between short-term business needs and long-term environmental sustainability.

Environmental Performance: The measurable outcomes of an organization's environmental management efforts, including metrics related to resource usage, emissions, waste management, and ecological impact.

Environmental Stewardship: The responsible management and protection of the natural environment through conservation, sustainable practices, and careful consideration of environmental impact in organizational decision-making.

Ethical Leadership: A leadership approach characterized by demonstrating normatively appropriate conduct through personal actions and interpersonal relationships, while promoting such conduct among followers through two-way communication, reinforcement, and decision-making.

Green Innovation: The development and implementation of new products, processes, or practices that contribute to environmental sustainability while maintaining or enhancing business performance.

Knowledge Management: The process of creating, sharing, using, and managing environmental knowledge within an organization to enhance its capacity for sustainable innovation and adaptive management.

Stakeholder Theory: A framework that emphasizes the importance of considering and balancing the interests of all parties affected by organizational decisions, including employees, customers, investors, communities, and the environment.

Systems Theory: A theoretical framework that views organizations as complex, interconnected systems that interact with and adapt to their environment, emphasizing the relationships and feedback loops between different components.

Triple Helix Model: A framework for understanding the collaboration between industry, academia, and government in driving sustainable innovation and environmental solutions.

Chapter 9
Green Policies and Practices Across Industries

Muhammad Saeed Ahmad
https://orcid.org/0009-0008-4248-5858
The University of Faisalabad, Pakistan

Muhammad Ashfaq
The University of Faisalabad, Pakistan

Usman Ali
https://orcid.org/0009-0009-9751-0376
The University of Faisalabad, Pakistan

Tariq Jamil
https://orcid.org/0009-0009-5760-1476
The University of Faisalabad, Pakistan

ABSTRACT

This chapter examines how industries worldwide are adopting green policies and sustainable practices to combat climate change, reduce environmental impacts, and enhance organizational resilience. Focusing on key sectors like manufacturing, energy, transportation, agriculture, and services, it explores how green strategies are integrated into core operations. The role of Human Resource Management (HRM) is emphasized, highlighting its contribution in fostering green behaviors, delivering sustainability-focused training, and aligning HR practices with environmental goals. The chapter also addresses industry-specific challenges, showcases success stories, and distills lessons learned, offering actionable insights to support global climate resilience through strengthened green initiatives.

DOI: 10.4018/979-8-3373-2479-1.ch009

1. INTRODUCTION

1.1 Importance of Green Policies in Combating Climate Change

Climate change has emerged as one of the most pressing global challenges of the 21st century, posing severe threats to ecosystems, economies, and human societies. Rising global temperatures, extreme weather events, loss of biodiversity, and environmental degradation underscore the urgency of addressing climate change. Industrial activities, deforestation, and fossil fuel consumption have significantly contributed to greenhouse gas (GHG) emissions, exacerbating global warming (Islam et al., 2021). The depletion of natural resources and increasing pollution levels further highlight the dire consequences of inaction. Governments, organizations, and individuals must recognize the critical importance of adopting green policies to combat these environmental threats, ensuring a sustainable future for generations to come.

Green policies refer to strategic regulations, initiatives, and practices aimed at reducing environmental impact, promoting sustainability, and mitigating climate change. These policies encompass various measures such as renewable energy adoption, waste reduction, resource efficiency, carbon trading mechanisms, and environmental governance frameworks. By implementing green policies, industries and governments can transition toward a low-carbon economy, ensuring ecological balance while fostering economic growth. These policies also emphasize the role of sustainable development, corporate social responsibility, and environmental stewardship in creating a more resilient global economy (Zhang et al., 2022).

Recent research underscores the growing importance of green policies in combating climate change. Studies highlight that nations adopting stringent environmental regulations have witnessed significant reductions in carbon emissions. For example, the European Union's Green Deal has been instrumental in promoting sustainability across various sectors, leading to measurable improvements in air quality and energy efficiency. Research also indicates that industries investing in green technologies experience long-term cost savings and enhanced competitive advantage (Dechezleprêtre & Sato, 2017; Goubran et al., 2023).

However, challenges persist in the widespread implementation of green policies. A key issue is the lack of global consensus on regulatory standards, with some nations prioritizing economic growth over environmental sustainability. Furthermore, financial constraints and technological limitations hinder the adoption of green innovations, especially in developing countries. Studies have also identified gaps in corporate compliance with environmental regulations, with many organizations failing to integrate sustainability into their core business strategies.

Another crucial area of research focuses on the role of policy enforcement and governance in ensuring the effectiveness of green initiatives. Findings suggest that weak institutional frameworks and corruption can undermine environmental policies, leading to ineffective implementation. Additionally, public awareness and engagement play a pivotal role in driving green policies forward. Researchers emphasize the need for educational programs, incentives, and stakeholder collaboration to enhance environmental consciousness and compliance.

The intersection of green policies and economic growth has also been a topic of extensive research. While some argue that stringent environmental regulations may impose financial burdens on industries, others highlight the potential for job creation and innovation in the green economy. Studies suggest that green policies can drive sustainable business practices, fostering resilience against climate-related risks and economic downturns. In conclusion, the importance of green policies in combating climate change cannot be overstated. While significant progress has been made, continued research, policy refinement, and collaborative efforts are essential to addressing existing challenges and ensuring the long-term sustainability of our planet (Umar et al., 2024; Sachdeva & Singh, 2024).

1.2 Role of Industries in Environmental Sustainability

Industries are at the forefront of both environmental challenges and solutions. As primary drivers of economic growth, industries are simultaneously major contributors to environmental degradation and key players in promoting sustainability. The industrial sector is responsible for a significant proportion of global greenhouse gas emissions, pollution, and resource depletion, making it one of the main sources of environmental harm. Issues such as deforestation, air and water pollution, and excessive energy consumption further compound the environmental burden. However, industries also possess the capability to reverse these negative impacts by integrating sustainable practices and investing in green innovation. By acknowledging their environmental responsibility, industries can transition from being primary contributors to ecological crises to becoming leaders in environmental restoration and sustainability.

Industries can drive meaningful change through the adoption of green technologies, circular economy practices, and corporate sustainability initiatives. The integration of renewable energy sources, waste management strategies, and energy-efficient production techniques can significantly reduce industrial carbon footprints. Circular economy models, which prioritize recycling, reusing, and minimizing waste, allow industries to optimize resource consumption while reducing environmental degradation. Moreover, corporate sustainability initiatives focusing on water conservation, responsible sourcing, and eco-friendly manufacturing practices enhance industries' long-term viability and contribution to environmental protection. By embedding sustainability principles into their operations, industries can effectively balance profitability with ecological responsibility, ensuring that economic progress does not come at the cost of environmental destruction (Satrovic et al., 2025).

Companies that integrate environmental, social, and governance (ESG) criteria into their business models not only contribute to sustainability but also enhance their competitive advantage, regulatory compliance, and brand reputation. The ESG framework encourages businesses to operate with social responsibility, ethical governance, and environmental consciousness. Firms that prioritize ESG principles gain better access to investment opportunities, improve stakeholder trust, and demonstrate long-term financial resilience. Additionally, regulatory compliance with environmental policies reduces the risk of legal penalties, while a strong brand reputation for sustainability attracts environmentally conscious consumers. By embedding ESG criteria into strategic decision-making, businesses can foster sustainable growth while maintaining their market relevance in an increasingly eco-conscious world (Camilleri, 2025).

Various factors further accelerate the shift toward green industrial practices, including consumer preferences for eco-friendly products, investor pressure for sustainable business strategies, and international climate agreements such as the Paris Accord. Today's consumers are more informed and environmentally conscious, demanding transparency and sustainability from the brands they support. This shift in consumer behavior has compelled industries to adopt greener practices, such as sustainable packaging, carbon-neutral production, and ethical supply chains. Similarly, investors are placing greater emphasis on sustainable business models, favoring companies that demonstrate a commitment to environmental and social responsibility (Saleh et al., 2025). Additionally, international frameworks and climate agreements mandate industries to reduce their environmental impact, pushing them toward sustainability-driven transformations. The confluence of these factors underscores the necessity for industries to proactively integrate green policies, not only to meet regulatory expectations but also to remain competitive and relevant in a rapidly evolving global economy.

1.3 Connection Between Green Policies and Human Resource Management (HRM)

The integration of green policies within industries is not solely a technical or operational concern; it also necessitates a strategic approach to Human Resource Management (HRM). The concept of Green HRM has gained prominence in recent years, highlighting the role of HR practices in promoting environmental sustainability within organizations. Green HRM involves embedding sustainability principles into recruitment, training, performance management, and employee engagement strategies. By incorporating sustainability into HRM, organizations can create a workforce that is environmentally conscious, motivated, and aligned with corporate sustainability objectives.

Organizations that prioritize green HR practices can foster a culture of environmental responsibility, enhance employee motivation, and improve organizational commitment to sustainability goals. Green HR initiatives, such as sustainability training programs, eco-friendly workplace policies, and incentives for sustainable behavior, encourage employees to adopt environmentally responsible practices. Moreover, companies that engage employees in sustainability efforts often experience higher levels of job satisfaction, productivity, and overall corporate reputation (Khan et al., 2019; Lu et al., 2023).

Furthermore, HRM plays a critical role in facilitating the transition to sustainable business practices by promoting green leadership, developing environmentally conscious corporate policies, and ensuring compliance with environmental regulations. HR leaders can drive change by integrating sustainability into corporate values, encouraging green leadership behaviors, and aligning business strategies with environmental goals. By embedding sustainability within HRM, organizations can effectively bridge the gap between policy development and practical implementation, ensuring long-term success in their green initiatives.

1.4 Scope of the Chapter

This chapter provides a comprehensive exploration of green policies and their evolving role in shaping industrial sustainability. It examines how businesses across various sectors are adopting environmentally responsible practices, balancing economic growth with ecological preservation. By investigating industry-specific strategies, the chapter highlights how sustainable innovation, resource optimization, and policy compliance contribute to long-term environmental goals. Furthermore, it emphasizes the transformative potential of green policies in fostering resilient business ecosystems that align with global sustainability frameworks.

Beyond traditional environmental management, this chapter delves into the integration of sustainability within human resource management (HRM), emphasizing the role of Green HRM in facilitating corporate environmental responsibility. By embedding sustainability into recruitment, employee engagement, training, and leadership development, organizations can cultivate a workforce that actively contributes to green initiatives. The chapter examines how HRM acts as a catalyst in promoting eco-conscious work cultures, ensuring that sustainability goals are embedded in organizational values rather than remaining isolated as external compliance measures.

The discussion also considers the regulatory landscape, emerging trends, and technological advancements influencing green industrial transformation. It presents insights from recent research, case studies, and global best practices, offering practical guidance for businesses aiming to implement effective sustainability strategies. By bridging the gap between policy frameworks and real-world applications, this chapter provides a multidisciplinary perspective on the future of green policies, equipping organizations with the knowledge to drive impactful and lasting change.

2. OVERVIEW OF GREEN POLICIES AND PRACTICES

2.1 Definition and Scope of Green Policies

Green policies refer to structured initiatives, regulations, and strategies that governments, organizations, and businesses implement to reduce environmental harm and promote sustainability (Van Rooijen & Van Wees, 2006). These policies focus on various aspects such as energy efficiency, pollution control, conservation of natural resources, waste management, and carbon footprint reduction. At the governmental level, green policies include regulatory frameworks like carbon taxes, emission reduction targets, and incentives for renewable energy adoption. Many international agreements, such as the Kyoto Protocol and the Paris Agreement, guide national governments in formulating and enforcing these policies. In the corporate sector, green policies involve eco-friendly business practices, responsible sourcing, and adherence to environmental compliance standards. The overarching objective of these policies is to ensure environmental sustainability while balancing economic and social considerations.

Green policies have expanded significantly over the years, encompassing various sectors such as agriculture, transportation, and manufacturing. In the energy sector, green policies promote the transition from fossil fuels to renewable sources like solar, wind, and hydropower. In the transportation industry, policies encourage the adoption of electric vehicles, improved fuel efficiency standards, and enhanced public transport systems. Similarly, in the manufacturing sector, regulations mandate the reduction of industrial waste, proper disposal of hazardous materials, and adherence to sustainable supply chain practices. Environmental governance also plays a crucial role, ensuring that businesses and industries comply with environmental laws and adopt sustainable production techniques. These policies not only mitigate environmental degradation but also drive innovation and long-term economic growth.

Recent research highlights the effectiveness of green policies in reducing environmental damage and fostering economic benefits. A 2025 study published in the Journal of Environmental Management found that countries with strict environmental regulations experience lower levels of pollution and higher investments in green technologies. Similarly, a report by the International Energy Agency (IEA) in 2023 revealed that global carbon emissions declined by 3% in regions with strong regulatory frameworks on emissions control. Corporate research also shows that businesses integrating sustainability into their policies enjoy increased brand loyalty and financial performance. For instance, companies like Unilever and IKEA have adopted aggressive sustainability policies, leading to reduced waste production and increased operational efficiency (Zhou et al., 2023; Lu et al., 2025).

Globally, countries are adopting diverse green policy models to address climate change and environmental concerns. The European Union's Green Deal aims to make Europe the first climate-neutral continent by 2050, while China's aggressive investment in renewable energy has positioned it as a leader in solar and wind power production. In the United States, federal and state policies are pushing for increased electric vehicle usage and reduced dependency on coal. In Pakistan, initiatives such as the Ten Billion Tree Tsunami and Clean Green Pakistan have demonstrated a commitment to environmental sustainability. However, enforcement challenges, insufficient infrastructure, and industrial pollution continue to hinder progress. Strengthening institutional frameworks and increasing public awareness can help Pakistan align its green policies with global best practices.

2.2 Key Principles of Sustainability

Sustainability is built on three principles: environmental protection, social responsibility, and economic viability, ensuring resource conservation and long-term well-being. The "reduce, reuse, recycle" approach helps minimize waste and optimize resource use. Reducing lowers energy and resource consumption, reusing extends product life, and recycling turns waste into new products. Sustainability also involves energy efficiency, water conservation, and biodiversity protection. Many firms adopt circular economy models to reduce resource input and waste. In agriculture, organic farming and efficient irrigation promote soil health, while green buildings and renewable energy cut emissions in construction and energy sectors.

Research shows that sustainable practices improve employee satisfaction, efficiency, and profits (Sustainability, 2024). Green urban planning boosts health and economic growth (WEF, 2022). Stronger policies and technologies like AI-driven energy management and carbon capture aid sustainability (Feng et al., 2024; Bressane et al., 2024). Globally, the EU's Circular Economy Plan and corporate carbon neutrality pledges drive change. China leads in solar energy, while Pakistan promotes sustainability via the Green Stimulus Package and climate-resilient agriculture. However, challenges like deforestation, poor waste management, and low renewable energy use remain. Enhancing policies, enforcement, and public involvement is key to meeting environmental goals.

2.3 Corporate Social Responsibility (CSR) and Environmental Stewardship

Corporate Social Responsibility (CSR) is vital for environmental sustainability, requiring businesses to adopt eco-friendly practices beyond legal obligations. Companies integrate CSR by reducing emissions, ensuring sustainable supply chains, and supporting environmental initiatives. Environmental stewardship—responsible resource management—is central to CSR, with firms investing in reforestation, carbon offsets, and sustainable sourcing (Remmer & Gilbert, 2025).

Globally, businesses like Patagonia and Tesla lead in sustainability, reaping benefits such as customer loyalty, employee engagement, and profitability. A 2024 Journal of Business Ethics study found firms with strong CSR policies gain market advantage and face fewer regulatory risks (Kumar et al., 2024; Agu et al., 2024). Governments now mandate carbon reporting, while ESG investing pressures firms to prioritize sustainability. Technologies like blockchain and AI also support CSR implementation. In Pakistan, firms like Engro and Nestlé Pakistan are adopting CSR, focusing on renewable energy and water management. However, many businesses lack formal CSR policies, contributing to pollution and deforestation. Strengthening regulations, offering incentives, and raising awareness are crucial for promoting CSR. Overall, green policies, sustainability, and CSR are key to environmental well-being, requiring stronger enforcement, corporate accountability, and green technology investments for a sustainable future.

3. INDUSTRY-SPECIFIC GREEN POLICIES AND PRACTICES

3.1 Manufacturing Sector

3.1.1 Adoption of Energy-Efficient Production Processes

The manufacturing industry is a major energy consumer and a significant contributor to environmental pollution. To mitigate its impact, many companies are adopting energy-efficient production processes powered by advanced machinery, automation, and artificial intelligence (AI). These technologies optimize energy use and minimize waste. Lean manufacturing approaches, such as Six Sigma and Kaizen, help eliminate inefficiencies, boost productivity, and lower carbon emissions. Additionally, manufacturers are investing in energy-efficient tools like smart sensors and heat recovery systems, while transitioning to renewable energy sources, such as solar and wind, for powering operations.

Governments and international organizations support energy-efficient manufacturing through policies and incentives. The European Union's Energy Efficiency Directive mandates a 32.5% improvement in industrial energy efficiency by 2030. Similarly, the U.S. Department of Energy's Better Plants Program encourages manufacturers to enhance efficiency and cut greenhouse gas emissions. Research highlights the benefits of such initiatives; a 2023 study in the Journal of Cleaner Production found that companies adopting green manufacturing strategies reduced energy costs by 25% while gaining a competitive edge (Mady et al., 2023). Globally, industry leaders like Toyota and Siemens have embraced smart manufacturing, significantly reducing their carbon footprint. Tesla's Gigafactories run on 100% renewable energy, setting new standards for sustainability. In China, strict environmental regulations have driven industries to adopt energy-efficient practices. In Pakistan, sectors such as textiles and cement are gradually transitioning, but challenges like outdated machinery, high energy costs, and weak enforcement slow progress. Promoting investment in modern technology and offering financial incentives can accelerate energy-efficient manufacturing across Pakistan.

3.1.2 Use of Sustainable Materials and Waste Management

The manufacturing sector is increasingly prioritizing sustainable materials and waste management to reduce its environmental footprint. Biodegradable plastics, recycled metals, and organic fabrics are replacing traditional, resource-intensive materials. Many companies are adopting circular economy models, where materials are reused, repurposed, or recycled, minimizing waste and promoting resource efficiency. Strategies such as zero-waste production and eco-friendly packaging—like compostable materials and reduced plastic usage—are also gaining traction.

Recent studies underscore the advantages of these practices. A 2023 World Economic Forum report revealed that embracing circular economy principles could cut global waste production by up to 50% by 2030. Companies sourcing sustainable materials not only lower their carbon footprint but also attract environmentally conscious consumers, gaining a competitive advantage. Furthermore, a study published in Sustainability found that firms implementing green supply chain practices achieved a 20% reduction in raw material costs (Pinto, 2020).

Globally, leading companies are making significant strides. Apple has committed to using 100% recycled materials in its products, while Adidas produces eco-friendly footwear from ocean-recycled plastic. In Pakistan, sustainable manufacturing is emerging, particularly in the textiles and packaging industries. The textile sector is adopting organic cotton and water-saving dyeing methods. However, challenges such as limited access to sustainable materials and weak recycling infrastructure persist. To promote green manufacturing, Pakistan must strengthen waste management policies and offer incentives to businesses embracing sustainable practices.

3.2 Energy Sector

3.2.1 Transition to Renewable Energy Sources

The energy sector is at the forefront of global sustainability efforts, with a major shift toward renewable energy sources such as solar, wind, hydro, and geothermal power. Renewable energy adoption is crucial in reducing dependence on fossil fuels, lowering carbon emissions, and mitigating climate change. Governments worldwide are implementing policies and financial incentives to promote clean energy, including subsidies for solar panel installation, tax benefits for wind energy projects, and funding for research in green technologies. Many countries have set ambitious targets to achieve net-zero emissions by mid-century, accelerating the transition to renewable energy.

Recent research highlights the rapid expansion of renewable energy. According to a 2023 report by the International Renewable Energy Agency (IRENA), renewable energy accounted for over 80% of new power capacity additions globally. Solar and wind energy, in particular, have seen significant cost reductions, making them more competitive with fossil fuels. Another study in Nature Energy found that large-scale renewable energy deployment could reduce global CO2 emissions by 70% by 2050. However, challenges such as energy storage, grid integration, and initial investment costs remain barriers to full-scale adoption. Advancements in battery technology and smart grids are helping to address these issues, ensuring a more stable and efficient renewable energy supply (Koutoudjian et al., 2021; Gielen et al., 2019).

Globally, countries are making substantial investments in renewable energy. The European Union's Green Deal aims to generate 45% of its energy from renewables by 2030. China, the world's largest producer of solar panels and wind turbines, is expanding its renewable energy infrastructure at an unprecedented rate. The U.S. Inflation Reduction Act of 2022 allocated billions of dollars to clean energy projects, further driving renewable adoption. In Pakistan, renewable energy development is gaining momentum, with initiatives such as the Alternative and Renewable Energy Policy 2019 aiming to increase renewable energy capacity to 30% by 2030 (Raza et al., 2022). However, challenges such as inconsistent government policies, lack of investment, and energy storage limitations need to be addressed to ensure long-term sustainability.

3.2.2 Implementation of Carbon Capture and Storage (CCS) Technologies

Carbon capture and storage (CCS) technology is a critical component of reducing greenhouse gas emissions from the energy sector. CCS involves capturing CO2 emissions from power plants and industrial processes, transporting them, and storing them underground in geological formations. This technology is particularly important for countries still reliant on fossil fuels, as it provides a way to reduce emissions while transitioning to renewable energy. Governments and energy companies are investing in CCS projects to meet climate targets and ensure energy security.

Recent research shows that CCS has significant potential in mitigating climate change. A 2023 study published in Journal of Environmental Management found that large-scale CCS deployment could cut global CO2 emissions by 15-20% by 2050. The Intergovernmental Panel on Climate Change (IPCC) has also emphasized that achieving net-zero emissions without CCS will be challenging, especially for industries like cement and steel production. Advances in CCS technology, such as direct air capture (DAC) and bioenergy with carbon capture and storage (BECCS), are further enhancing the efficiency and scalability of carbon removal (Moreno et al., 2024).

Countries around the world are investing in CCS to complement their clean energy strategies. Norway's Northern Lights Project is one of the most advanced CCS initiatives, capturing emissions from industrial sources and storing them under the North Sea. The U.S. has allocated significant funding for CCS research, with projects such as the Petra Nova plant demonstrating the feasibility of large-scale carbon capture. In Pakistan, CCS technology is still in its early stages, with limited pilot projects in the cement and energy sectors. The country's reliance on fossil fuels for energy production makes CCS a viable option for reducing emissions. However, high costs, lack of infrastructure, and regulatory gaps remain major obstacles. To accelerate CCS adoption, Pakistan must establish supportive policies, attract foreign investment, and enhance collaboration with international climate initiatives.

3.3 Transportation Sector

3.3.1 Promoting Electric Vehicles (EVs) and Sustainable Logistics

The transportation sector is a major contributor to global carbon emissions, accounting for nearly 25% of total greenhouse gas emissions. To combat this, governments and industries are promoting electric vehicles (EVs) and sustainable logistics. EVs produce zero tailpipe emissions and have significantly lower lifecycle emissions compared to traditional internal combustion engine (ICE) vehicles.

Advancements in battery technology, improvements in charging infrastructure, and government incentives are driving the rapid adoption of EVs. Sustainable logistics, including optimizing transportation routes, using eco-friendly packaging, and switching to low-emission transport modes, further reduce environmental impacts.

Recent research underscores the benefits of EV adoption. A 2023 study published in Alexandria Engineering Journal found that widespread EV adoption could reduce global transportation emissions by up to 60% by 2050. Battery costs have declined by nearly 89% over the past decade, making EVs more affordable for consumers. Additionally, studies indicate that integrating EVs into urban logistics can lower air pollution levels and improve public health. Logistics companies like DHL and UPS have already incorporated electric delivery fleets, while ride-sharing platforms such as Uber and Lyft are incentivizing drivers to switch to EVs (Veza et al., 2023).

Globally, countries are investing in EV infrastructure. Norway leads in EV adoption, with over 80% of new car sales being electric. The European Union's Fit for 55 package includes stringent emissions regulations and support for EV expansion. In the U.S., the Inflation Reduction Act provides tax incentives for EV purchases and charging stations. In Pakistan, the Electric Vehicle Policy 2019 aims to have 30% of total vehicle sales as EVs by 2030, but adoption remains slow due to high import costs and limited charging infrastructure. Expanding charging networks and offering subsidies can accelerate EV uptake in Pakistan.

3.3.2 Reducing Carbon Emissions through Green Supply Chains

Green supply chains focus on minimizing environmental impacts by integrating sustainable practices across procurement, production, transportation, and distribution. This includes using alternative fuels, optimizing delivery routes, adopting energy-efficient warehousing, and utilizing recycled packaging materials. Many companies are transitioning to carbon-neutral logistics by investing in green fleet technologies, such as hydrogen-powered trucks and autonomous electric delivery vehicles.

A 2023 report by the World Economic Forum revealed that businesses implementing green supply chain strategies could cut carbon emissions by 40% while reducing operational costs. Research from the Sustainability Journal found that optimizing transport routes using AI and big data analytics led to a 20% decrease in fuel consumption. Companies like Amazon and Walmart are setting ambitious net-zero emission targets by investing in eco-friendly transportation and reducing reliance on fossil-fuel-based shipping (Chen et al., 2024).

Globally, the push for sustainable supply chains is accelerating. The European Union's Corporate Sustainability Reporting Directive (CSRD) requires businesses to disclose their environmental impact, encouraging companies to implement greener logistics practices. China has mandated stricter emissions standards for freight transport. In Pakistan, green supply chain initiatives are limited, but industries such as textiles and retail are exploring sustainable transportation options. Policies promoting low-emission freight transport and public-private partnerships can help Pakistan develop greener supply chains.

3.4 Agriculture Sector

3.4.1 Practices such as Sustainable Farming, Crop Rotation, and Organic Farming

Agriculture is both a driver and a victim of environmental change, contributing to deforestation, soil degradation, and water pollution. Sustainable farming practices, such as crop rotation, organic farming, and precision agriculture, help reduce environmental damage while improving food security. Crop rotation enhances soil fertility and prevents pest outbreaks, while organic farming eliminates chemical fertilizers and pesticides, promoting biodiversity. Precision agriculture, which uses data analytics and IoT-based monitoring, optimizes resource use and minimizes waste.

A 2023 study published in Agricultural Sustainability Journal found that crop rotation improved soil quality by 30% and reduced fertilizer dependency. Organic farming has been linked to lower greenhouse gas emissions and improved ecosystem health. The Food and Agriculture Organization (FAO) states that sustainable farming can increase yields while reducing environmental impact. Companies like Nestlé and Unilever are supporting regenerative agriculture initiatives to ensure sustainable food production (Zou et al., 2024).

Globally, sustainable agriculture is gaining traction. The European Union's Farm to Fork Strategy promotes organic farming, aiming for 25% of agricultural land to be organic by 2030. In the U.S., programs such as the Sustainable Agriculture Research and Education (SARE) initiative provide funding for eco-friendly farming projects. In Pakistan, sustainable farming adoption is slow but growing. The country's National Food Security Policy promotes organic farming, but limited farmer awareness and financial constraints hinder progress. Expanding subsidies for organic agriculture and improving farmer education can enhance sustainable farming in Pakistan.

3.4.2 Addressing Deforestation and Water Management

Deforestation for agricultural expansion is a significant environmental challenge, leading to biodiversity loss and carbon emissions. Sustainable land management strategies, such as agroforestry and reforestation, help mitigate these impacts. Additionally, water scarcity is a critical issue, necessitating improved irrigation techniques, rainwater harvesting, and wastewater recycling. Drip irrigation and precision watering reduce water wastage, ensuring efficient agricultural production.

Recent research highlights the importance of addressing deforestation and water management. A 2023 Global Land Use Study found that reforestation efforts could sequester up to 20% of annual CO2 emissions. Sustainable water management practices have been linked to a 25% increase in crop yields in arid regions. The United Nations Environment Programme (UNEP) emphasizes that integrating technology, such as satellite-based water monitoring, can enhance water conservation efforts (Liu et al., 2023).

3.5 Service Sector

3.5.1 Reducing Energy Use in Offices (e.g., Green Buildings)

The service sector, which includes businesses, IT companies, and financial institutions, is focusing on reducing energy consumption through green building initiatives. Green buildings incorporate energy-efficient lighting, smart HVAC systems, and renewable energy sources to minimize carbon footprints. Many organizations are adopting Leadership in Energy and Environmental Design (LEED) standards to enhance sustainability.

A 2023 Journal of Green Infrastructure study found that green buildings consume 30-50% less energy than conventional buildings. Smart office solutions, such as automated lighting and AI-driven climate control, further enhance energy efficiency. Tech giants like Google and Microsoft have committed to operating carbon-neutral campuses, investing in green building infrastructure (Kusi et al., 2024).

Globally, green building policies are expanding. The U.S. Green Building Council promotes LEED certification, while the European Union mandates energy efficiency standards for commercial properties. In Pakistan, green buildings are emerging, with projects like the Arfa Software Technology Park incorporating eco-friendly designs. However, high initial costs and limited regulatory enforcement slow widespread adoption. Government incentives and awareness programs can drive the development of energy-efficient buildings in Pakistan.

3.5.2 Encouraging Digital Transformation to Cut Resource Consumption

Digital transformation is revolutionizing the service sector by reducing resource consumption and promoting remote work. The adoption of cloud computing, paperless offices, and virtual meetings minimizes energy use and waste. AI-driven analytics optimize business operations, reducing redundancies and inefficiencies.

A 2023 report by the International Telecommunication Union (ITU) revealed that cloud-based operations could cut carbon emissions by 35% compared to traditional IT infrastructure. Companies adopting digital workflows experience cost savings and improved efficiency. Major corporations like IBM and SAP have fully digitized operations, reducing paper usage by over 80%.

4. ROLE OF HUMAN RESOURCE MANAGEMENT (HRM) IN GREEN PRACTICES

4.1 Green Recruitment: Hiring Individuals Aligned with Sustainability Values

Green recruitment involves attracting and hiring candidates committed to environmental sustainability and corporate social responsibility (CSR). It ensures new hires align with an organization's sustainability goals by evaluating environmental awareness and ethical decision-making. This process often includes eco-friendly job advertising, digital recruitment, and green competency assessments. By hiring sustainability-minded employees, organizations foster a culture supportive of green initiatives, enhancing employee engagement and employer branding. Green recruitment also improves retention, job satisfaction, and productivity (Jamil et al., 2023), while boosting public image and competitiveness. In Pakistan, green recruitment is emerging in sectors like textiles, banking, and manufacturing, especially among multinationals. However, limited awareness and resources hinder broader adoption. Government incentives and regulatory support could promote wider implementation.

4.2 Green Training and Development: Building Employee Capacity for Sustainable Operations

Green training and development involve equipping employees with the knowledge, skills, and attitudes necessary to implement and support sustainability initiatives within an organization. This includes training on energy conservation, waste management, sustainable supply chain practices, and environmentally friendly production methods. The goal is to enhance employees' environmental awareness and empower them to contribute to sustainability efforts in their respective roles.

Organizations use various methods, including workshops, e-learning modules, and on-the-job training, to foster a sustainability-oriented workforce.

Effective green training and development programs help employees integrate sustainable practices into their daily work routines. When employees understand the impact of their actions on the environment, they become more proactive in minimizing waste, reducing energy consumption, and supporting green business strategies. Moreover, training initiatives encourage innovation by enabling employees to explore eco-friendly alternatives in their work processes. As a result, organizations benefit from improved resource efficiency, cost savings, and compliance with environmental regulations.

Recent studies emphasize the critical role of green training in promoting sustainable business practices. Research by Aboramadan (2022) highlights that employees who undergo green training demonstrate higher engagement in sustainability initiatives and contribute to improved environmental performance. Another study by Barakat et al. (2023) found that organizations investing in green training experience long-term financial benefits due to enhanced operational efficiency and reduced environmental liabilities. Additionally, businesses with structured green training programs report increased employee commitment to corporate social responsibility.

4.3 Rewards and Incentives for Sustainable Practices

Rewards and incentives for sustainable practices refer to organizational strategies that recognize and encourage employees' contributions to environmental sustainability. These incentives can be financial, such as bonuses and salary increments, or non-financial, such as recognition programs, career advancement opportunities, and additional leave days. The purpose of these incentives is to reinforce positive environmental behaviors and motivate employees to actively participate in sustainability initiatives. Organizations use reward systems to drive eco-friendly behaviors, such as reducing paper waste, conserving energy, and supporting corporate sustainability goals.

The implementation of green rewards and incentives fosters a culture of sustainability within an organization. Employees are more likely to engage in pro-environmental behaviors when they perceive tangible benefits associated with their efforts. Furthermore, such reward mechanisms enhance employee morale, job satisfaction, and organizational commitment. Companies that integrate sustainability into their performance appraisal and reward systems benefit from increased employee participation in green initiatives, ultimately contributing to their long-term environmental and economic success.

Recent research underscores the effectiveness of green rewards and incentives in promoting sustainable practices. Studies indicate that organizations that link employee rewards to sustainability efforts witness a higher degree of environmental responsibility among their workforce. A study by Marrucci (2024) found that employees who receive sustainability-related incentives are more engaged in corporate green strategies, leading to improved environmental performance. Similarly, research by Jyothi (2016) suggests that non-monetary incentives, such as public recognition and career development opportunities, are equally effective in motivating employees to adopt eco-friendly practices.

4.4 Leadership's Role in Fostering a Green Organizational Culture

Leadership is key to embedding sustainability in an organization by setting a green vision, establishing supportive policies, and modeling eco-friendly behavior. Green leadership integrates environmental concerns into strategic decisions and fosters a culture of responsibility and ethical action. Leaders inspire employees to engage in sustainability, promoting practices like energy conservation and waste reduction while building stakeholder trust and enhancing reputation. Research shows transformational leadership is especially effective in promoting sustainability. Zameer et al. (2024) found proactive green leaders boost employee engagement, while Boeske (2023) links such leadership to better environmental performance. In Pakistan, green leadership is emerging, especially in multinationals. Local firms face awareness and resource challenges, but sustainability-focused leadership development could strengthen green practices and advance national environmental goals.

5. CHALLENGES IN IMPLEMENTING GREEN POLICIES

5.1 Financial Constraints and Cost of Adopting Green Technologies

Implementing green policies often requires substantial financial investment, particularly in eco-friendly technologies, renewable energy, and sustainable processes. Small and medium enterprises (SMEs) struggle with high upfront costs for solutions like solar systems and energy-efficient equipment. Ongoing maintenance costs and delayed financial returns further discourage adoption. Both developed and developing economies face challenges in funding green initiatives. Limited access to financing and uncertain ROI hinder investment, as many investors prioritize short-term gains. Research by Zheng et al. (2023) shows that firms with subsidies or tax incentives are more successful in adopting green technologies. Financial support from investors also plays a key role (Babiak & Trendafilova, 2011). Globally, countries like Germany and Sweden offer financial incentives for sustainability. In Pakistan, financial constraints—especially for SMEs—and limited green financing options hinder progress. Programs like the State Bank of Pakistan's Green Banking Guidelines aim to help, but more support is needed for widespread adoption.

5.2 Resistance to Change within Industries and Organizations

Resistance to change is a major challenge in implementing green policies, as employees, managers, and stakeholders often hesitate to shift from established practices due to fears of disruption, increased workload, or doubts about the benefits of sustainability. This reluctance is fueled by organizational inertia, rigid corporate cultures, and limited awareness of environmental responsibility, particularly in traditional sectors like manufacturing and construction.

Common misconceptions include fears that green initiatives harm productivity and profitability, or lead to job losses due to automation. A lack of leadership commitment further hinders progress, as sustainability efforts without top management support face skepticism. Research highlights that organizations with strong change management and employee engagement strategies experience less resistance. Transformational leadership and involving employees in sustainability decisions also reduce opposition (Burawat, 2019; Vila-Vázquez et al., 2018).

Globally, companies in Japan and Canada have seen success by embedding sustainability in their culture and engaging employees. In Pakistan, resistance is still significant, especially in textiles and manufacturing, due to low awareness and weak commitment. However, some multinational firms are leading by example, encouraging local adoption of green practices.

5.3 Lack of Technical Expertise and Infrastructure

The effective implementation of green policies requires technical expertise and strong infrastructure, which many organizations, especially in developing countries, lack. Gaps in knowledge related to energy-efficient processes, waste management, and environmental assessments hinder the adoption of sustainable practices. Inadequate infrastructure—such as unreliable energy and inefficient transport—further limits progress.

Without skilled personnel, organizations face difficulties operating green technologies, leading to inefficiencies and higher costs. They also struggle to identify suitable sustainability solutions, often continuing with outdated, harmful practices. Research shows that companies investing in employee training and accessing external technical support achieve better sustainability outcomes (Pavlova, 2018). Globally, countries like the Netherlands and South Korea have advanced sustainability by funding training programs and improving infrastructure. In Pakistan, technical expertise remains limited, especially among SMEs. Although some universities offer sustainability programs, greater collaboration between academia and industry is needed to develop skills and infrastructure for green policy success.

5.4 Regulatory and Policy Barriers

Regulatory frameworks are crucial for driving green policy adoption, but in many regions, weak or inconsistent regulations hinder sustainability efforts. Governments often lack clear policies, enforcement mechanisms, or incentives, while complex and outdated regulations create confusion and discourage investment in green practices. Poor coordination among agencies and industries further hampers implementation. Without a supportive regulatory environment, businesses see little motivation to prioritize sustainability. Research shows that strong environmental regulations and government incentives significantly boost corporate green initiatives (Ahmad, 2015). Countries like Denmark and Canada enforce carbon taxes, emissions limits, and certification programs to promote sustainability. In Pakistan, despite steps like the Pakistan Climate Change Act, enforcement is weak and regulatory clarity is lacking. Strengthening policies and enforcement, along with clear guidelines, can accelerate green adoption.

6. CASE STUDIES AND BEST PRACTICES

6.1 Real-World Examples of Successful Green Policies Across Industries

Sustainability is now a vital source of competitive advantage, with many companies adopting green policies to reduce environmental impact. Industries like manufacturing, retail, and technology are implementing eco-friendly practices—ranging from waste reduction and renewable energy use to sustainable supply chains and carbon neutrality. These efforts not only benefit the environment but also improve efficiency, reputation, and profitability. Real-world cases highlight effective strategies. Unilever's Sustainable Living Plan targets emissions and waste reduction, while Tesla promotes electric vehicles and renewable energy. IKEA aims to become climate positive by 2030 through sustainable sourcing and energy investments. These examples show how sustainability can align with business success.

Research supports this, showing that companies with embedded green policies see financial gains, stronger employee engagement, and brand loyalty. Manufacturing firms using circular economy models report better efficiency and cost savings (Galpin et al., 2015). Globally, Germany's Energiewende and Sweden's waste-to-energy initiatives set sustainability benchmarks. In Pakistan, companies like Sapphire and Nishat in the textile sector are investing in green practices, though challenges like limited funding and regulatory support persist. Strengthening collaboration and policy backing can help scale these efforts.

6.2 Insights from Organizations That Have Integrated Sustainability into Their Core Strategies

As businesses face increasing pressure to operate sustainably, many organizations have incorporated environmental responsibility into their long-term strategies. Companies that embed sustainability at the core of their operations not only comply with regulations but also gain a competitive edge by fostering innovation, reducing costs, and enhancing stakeholder trust. Sustainability-driven organizations prioritize green supply chains, carbon reduction strategies, and eco-friendly product development, ensuring long-term environmental and economic benefits.

A closer look at successful organizations reveals how sustainability integration translates into real-world impact. Patagonia, an outdoor apparel company, has positioned itself as a sustainability leader by using recycled materials, ethical sourcing, and a commitment to carbon neutrality. Microsoft has pledged to become carbon negative by 2030, focusing on reducing emissions and investing in carbon removal technologies. Additionally, Schneider Electric has integrated sustainability into its energy management solutions, helping industries reduce energy consumption and minimize waste. These companies demonstrate that environmental responsibility and profitability can go hand in hand.

Recent research underscores the benefits of sustainability-driven business models. A study found that organizations incorporating sustainability into their corporate strategies outperform competitors in terms of market valuation, customer loyalty, and risk management. Similarly, research indicates that firms with strong environmental, social, and governance (ESG) frameworks experience lower financial risks and greater resilience during economic downturns. These findings highlight the importance of sustainability as a core strategic priority (Agu et al., 2024).

Globally, best practices in sustainability integration can be observed in countries like Norway and the Netherlands, where businesses operate under stringent environmental standards. In Pakistan, sustainability integration remains a work in progress, with organizations like Engro and Packages Limited taking significant steps toward greener operations. Engro has launched renewable energy projects, while Packages Limited focuses on sustainable packaging solutions. Expanding these efforts through policy incentives and industry collaboration can further drive sustainability adoption in Pakistan.

7. MEASURING THE IMPACT OF GREEN POLICIES

7.1 Metrics for Assessing Sustainability Initiatives

Evaluating sustainability initiatives requires clear metrics to assess environmental, economic, and social impacts. Key indicators include carbon footprint reduction, energy and water efficiency, and waste management. Frameworks like the Global Reporting Initiative (GRI), Carbon Disclosure Project (CDP), and SASB standardize assessments and help organizations track progress and engage stakeholders. Assessments combine quantitative data—such as emissions reductions, energy savings, and waste diversion—with qualitative factors like employee engagement and policy effectiveness. Life cycle assessments (LCA) offer a holistic view of a product's environmental impact.

Research shows that firms using detailed sustainability metrics reduce costs and enhance reputations. Transparent reporting also attracts investors and customers (Baah et al., 2021). Globally, initiatives like the EU's Corporate Sustainability Reporting Directive (CSRD) mandate environmental disclosures. In Pakistan, sustainability reporting is limited. Companies like Engro and Fauji Fertilizer have adopted metrics, but broader adoption is hindered by weak regulation and low awareness. Strengthened policies and incentives can promote wider integration of sustainability assessments.

7.2 Role of HR Analytics in Tracking Green HRM Practices

The integration of sustainability into human resource management (HRM) requires a systematic approach to measuring its effectiveness. HR analytics play a crucial role in tracking green HRM practices, providing data-driven insights into employee engagement, sustainability training, and environmentally responsible behaviors. By leveraging HR analytics, organizations can assess how well green HRM initiatives—such as eco-friendly workplace policies, sustainability training programs, and green incentives—are being implemented. These insights help HR professionals refine their strategies, ensuring that sustainability goals align with workforce performance and organizational objectives.

Green HR analytics involve key performance indicators (KPIs) that measure employee participation in sustainability initiatives, the effectiveness of green training programs, and the impact of sustainability-related rewards. Metrics such as paper and energy savings due to digitalization, employee carbon footprint reduction, and sustainability-focused performance appraisals help organizations monitor progress. Advanced HR analytics tools also utilize artificial intelligence (AI) and machine learning to predict workforce trends and optimize sustainability efforts. By embedding

HR analytics in sustainability tracking, companies can create data-driven strategies that foster long-term environmental responsibility.

Recent studies emphasize the significance of HR analytics in enhancing sustainability performance. Research found that organizations that incorporate HR analytics into their sustainability strategies achieve higher employee engagement and commitment to green practices. Another study demonstrated that companies using HR analytics to track sustainability initiatives experience improved compliance with environmental regulations and enhanced corporate reputation. These findings underscore the growing role of HR analytics in shaping sustainable workplaces (Malik et al., 2021).

Globally, companies like Google and Unilever utilize HR analytics to measure employee involvement in sustainability initiatives, demonstrating the effectiveness of data-driven green HRM practices. In Pakistan, the adoption of HR analytics for sustainability tracking remains limited. However, forward-thinking organizations such as Telenor Pakistan and Engro have started incorporating analytics to monitor employee engagement in green practices. Expanding the use of HR analytics across industries could help Pakistani businesses enhance their sustainability efforts and align their workforce with global environmental goals.

8. RECOMMENDATIONS FOR ENHANCING GREEN PRACTICES

8.1 Strategies to Overcome Implementation Barriers

To achieve sustainability goals, organizations must address barriers like financial constraints, lack of technical expertise, resistance to change, and regulatory challenges. Overcoming these requires strategic actions, including financial incentives, capacity-building, and policy reforms. By embedding sustainability into core strategies, companies can transition to eco-friendly operations without sacrificing profitability. Targeted investment in green technologies, supported by government grants, tax incentives, and private funding, can offset high initial costs. Training programs equip employees with skills to support green initiatives, while fostering an inclusive culture encourages commitment to sustainability. Regulatory alignment and collaboration with policymakers create a supportive framework for green practices.

Research shows that firms investing in sustainability training achieve better policy adoption and efficiency. Financial incentives like carbon credits and green loans also boost sustainability efforts (Yafi et al., 2021). Globally, firms like Tesla and IKEA have overcome barriers through strategic investments and policy alignment. In Pakistan, companies like Engro and Nestlé Pakistan are adopting green initiatives despite challenges. Greater collaboration between industry and government can further drive sustainability progress.

8.2 Collaboration Between Industries, Governments, and Communities

A holistic approach to sustainability requires active collaboration among industries, governments, and communities. Each stakeholder plays a crucial role in driving environmental progress, with industries leading sustainable innovation, governments setting regulatory frameworks, and communities ensuring grassroots-level engagement. By fostering partnerships, stakeholders can share knowledge, resources, and best practices to address environmental challenges collectively.

Industries can collaborate with governments to develop policies that incentivize sustainable practices, such as subsidies for renewable energy projects and stricter environmental regulations. Meanwhile, engagement with communities ensures that sustainability efforts align with local needs and cultural contexts. Public-private partnerships (PPPs) serve as an effective model, enabling businesses to work alongside policymakers and local organizations to implement large-scale green projects. Additionally, industry alliances, such as the United Nations Global Compact and the World Business Council for Sustainable Development (WBCSD), promote cross-sector collaboration to advance global sustainability goals.

Research underscores the effectiveness of collaborative sustainability efforts. A study found that multi-stakeholder partnerships lead to more impactful and sustainable environmental policies. Similarly, a study highlighted that companies engaging in collaborative sustainability initiatives experience improved brand reputation and increased stakeholder trust. These studies demonstrate that a unified approach to sustainability yields significant environmental and economic benefits (Enechi & Pattberg, 2025).

Countries such as Germany and the Netherlands have successfully implemented multi-sector sustainability programs, leveraging government support, corporate participation, and community engagement. In Pakistan, collaborations between industries and the government are gradually gaining traction, with initiatives like the Green Pakistan Program promoting afforestation and environmental conservation. Expanding such partnerships and fostering greater community involvement can further strengthen sustainability efforts in the country.

8.3 Future Directions for HRM in Promoting Sustainability

Human Resource Management (HRM) plays a pivotal role in embedding sustainability into organizational culture and operations. The future of HRM lies in integrating green policies into workforce management, making environmental responsibility a core business priority. Green HRM practices—such as eco-friendly recruitment, sustainability training, and performance-based incentives—are becoming more common as companies seek greater environmental compliance. To enhance HRM's role in sustainability, organizations must embrace digital transformation, foster employee engagement, and invest in leadership development. HR analytics can track employee participation in green initiatives, helping refine sustainability strategies. Leadership training focused on environmental responsibility empowers managers to drive change, while remote work and digital tools offer new opportunities to reduce carbon footprints and promote sustainable HR practices.

Recent research underscores HRM's growing role in sustainability. Studies show that integrating sustainability into HRM boosts employee retention, productivity, and attracts top talent—especially as younger professionals value environmental responsibility (Suleman et al., 2024). Global firms like Google and Unilever have successfully adopted green HRM through technology and employee engagement. In Pakistan, companies such as Telenor and Lucky Cement have initiated green HRM practices, though broader adoption is still needed. Scaling these efforts and aligning them with national sustainability goals can enhance environmental performance across industries.

9. CONCLUSION

9.1 Summary of Key Insights

The discussion on green policies and sustainability initiatives highlights the crucial role of organizations in mitigating environmental challenges. Various aspects of sustainable practices, including green recruitment, training, and incentives, demonstrate how Human Resource Management (HRM) can shape environmentally responsible organizations. The integration of green policies requires overcoming challenges such as financial constraints, resistance to change, and regulatory barriers. By leveraging strategic planning and collaboration, industries can create a structured approach to sustainability that benefits both businesses and the environment.

Additionally, case studies and best practices from global and Pakistani organizations show that successful implementation of sustainability initiatives requires commitment at all levels. Companies that integrate sustainability into their core business strategies experience enhanced operational efficiency, regulatory compliance, and stakeholder trust. Furthermore, measuring the impact of green policies through key performance indicators (KPIs) such as carbon footprint reduction and resource efficiency enables organizations to track progress and refine their sustainability strategies. These insights emphasize that green practices are not only beneficial for the environment but also contribute to long-term business success.

9.2 Importance of Continuous Improvement in Green Policies and Practices

Sustainability is a dynamic process that requires continuous improvement and adaptation. As environmental challenges evolve, organizations must remain proactive in updating their green policies and practices. This involves staying informed about emerging sustainability trends, investing in innovative technologies, and fostering a culture of continuous learning. Without consistent improvements, even well-established sustainability efforts can become obsolete or ineffective in addressing new environmental concerns.

The role of HRM in sustaining green initiatives is particularly significant. Organizations must ensure that employees remain engaged in sustainability efforts through regular training, performance incentives, and leadership commitment. Continuous improvement also requires businesses to monitor and evaluate their environmental impact, adjusting strategies based on measurable outcomes. As global climate regulations become stricter, organizations that fail to enhance their sustainability initiatives risk facing legal and reputational challenges. Thus, a commitment to ongoing development in green policies is essential for long-term environmental and business sustainability.

9.3 Call to Action for Industries to Take a Leading Role in Climate Resilience

Industries play a crucial role in addressing climate change and advancing sustainability. Given their economic and environmental impact, businesses must go beyond compliance, setting new standards and leading the shift toward green practices. By championing sustainability, industries can drive innovation, create green jobs, and support global climate resilience. Collaboration with governments and communities is key to achieving large-scale environmental goals. Companies should engage in sustainability networks, advocate for eco-friendly policies, and support green initiatives locally and globally. In Pakistan, industries can lead by adopting renewable energy, reducing waste, and investing in sustainable supply chains. As climate risks grow, industries must take proactive leadership to secure a greener, more resilient future.

REFERENCES

Aboramadan, M. (2022). The effect of green HRM on employee green behaviors in higher education: The mediating mechanism of green work engagement. *The International Journal of Organizational Analysis*, *30*(1), 7–23.

Agu, E. E., Iyelolu, T. V., Idemudia, C., & Ijomah, T. I. (2024). Exploring the relationship between sustainable business practices and increased brand loyalty. *International Journal of Management & Entrepreneurship Research*, *6*(8), 2463–2475.

Ahmad, S. (2015). Green human resource management: Policies and practices. *Cogent Business & Management*, *2*(1), 1030817.

Baah, C., Opoku-Agyeman, D., Acquah, I. S. K., Agyabeng-Mensah, Y., Afum, E., Faibil, D., & Abdoulaye, F. A. M. (2021). Examining the correlations between stakeholder pressures, green production practices, firm reputation, environmental and financial performance: Evidence from manufacturing SMEs. *Sustainable Production and Consumption*, *27*, 100–114.

Babiak, K., & Trendafilova, S. (2011). CSR and environmental responsibility: Motives and pressures to adopt green management practices. *Corporate Social Responsibility and Environmental Management*, *18*(1), 11–24.

Barakat, B., Milhem, M., Naji, G. M. A., Alzoraiki, M., Muda, H. B., Ateeq, A., & Abro, Z. (2023). Assessing the impact of green training on sustainable business advantage: Exploring the mediating role of green supply chain practices. *Sustainability*, *15*(19), 14144.

Boeske, J. (2023). Leadership towards sustainability: A review of sustainable, sustainability, and environmental leadership. *Sustainability*, *15*(16), 12626.

Bressane, A., dos Santos Galvão, A. L., Loureiro, A. I. S., Ferreira, M. E. G., Monstans, M. C., & de Castro Medeiros, L. C. (2024). Valuing urban green spaces for enhanced public health and sustainability: A study on public willingness-to-pay in an emerging economy. *Urban Forestry & Urban Greening*, *98*, 128386.

Burawat, P. (2019). The relationships among transformational leadership, sustainable leadership, lean manufacturing and sustainability performance in Thai SMEs manufacturing industry. *International Journal of Quality & Reliability Management*, *36*(6), 1014–1036.

Camilleri, M. A. (2025). Sustainability accounting and disclosures of responsible restaurant practices in environmental, social and governance (ESG) reports. *International Journal of Hospitality Management*, *126*, 104051.

Chen, W., Men, Y., Fuster, N., Osorio, C., & Juan, A. A. (2024). Artificial intelligence in logistics optimization with sustainable criteria: A review. *Sustainability, 16*(21), 9145.

Dechezleprêtre, A., & Sato, M. (2017). The impacts of environmental regulations on competitiveness. *Review of Environmental Economics and Policy.*

Enechi, O., & Pattberg, P. (2025). What determines participation in multi-stakeholder partnerships? Assessing sustainability awareness and knowledge. *Discover Sustainability, 6*(1), 104.

Feng, T., Qamruzzaman, M., Sharmin, S. S., & Karim, S. (2024). Bridging environmental sustainability and organizational performance: The role of green supply chain management in the manufacturing industry. *Sustainability, 16*(14), 5918.

Galpin, T., Whitttington, J. L., & Bell, G. (2015). Is your sustainability strategy sustainable? Creating a culture of sustainability. *Corporate Governance, 15*(1), 1–17.

Gielen, D., Boshell, F., Saygin, D., Bazilian, M. D., Wagner, N., & Gorini, R. (2019). The role of renewable energy in the global energy transformation. *Energy strategy reviews, 24*, 38-50.

Goubran, S., Walker, T., Cucuzzella, C., & Schwartz, T. (2023). Green building standards and the united nations' sustainable development goals. *Journal of Environmental Management, 326*, 116552. PMID: 36372039

Islam, T., Hussain, D., Ahmed, I., & Sadiq, M. (2021). Ethical leadership and environment specific discretionary behaviour: The mediating role of green human resource management and moderating role of individual green values. *Canadian Journal of Administrative Sciences, 38*(4), 442–459.

Jamil, S., Zaman, S. I., Kayikci, Y., & Khan, S. A. (2023). The role of green recruitment on organizational sustainability performance: A study within the context of green human resource management. *Sustainability, 15*(21), 15567.

Jyothi, J. (2016). Non-monetary benefits & its effectiveness in motivating employees. *Clear International Journal of Research in Commerce & Management, 7*(5).

Khan, M. A. S., Ali, M., Usman, M., Saleem, S., & Jianquo, D. (2019). Interrelationships between ethical leadership, green psychological climate, and organizational environmental citizenship behavior: The moderating role of gender. *Frontiers in Psychology, 10*, 1977. PMID: 31555173

Koutoudjian, G., Diniz, L., Cespedes, R., & UNDP, V. G. (2021). About IRENA The International Renewable Energy Agency (IRENA) is an intergovernmental organisation that supports countries in their transition to a sustainable energy future, and serves as the principal platform for international co-operation, a centre of excellence, and a repository of policy, technology, resource and financial knowledge on renewable energy. *IRENA promotes the. IRENA promotes the, 24*.

Kumar, A., Snehvrat, S., Kumari, P., Priyadarshani, P., & Ray, P. (2024). A corporate social responsibility (CSR)-led competitiveness framework through a bibliometric analysis (2001–2023). *Journal of Asia Business Studies*, *18*(3), 737–766.

Kusi, E., Boateng, I., & Danso, H. (2024). Energy consumption and carbon emission of conventional and green buildings using building information modelling (BIM). *International Journal of Building Pathology and Adaptation*.

Liu, M., Chen, Y., Chen, K., & Chen, Y. (2023). Progress and hotspots of research on land-use carbon emissions: A global perspective. *Sustainability*, *15*(9), 7245.

Lu, H., Cai, S., Liu, Y., & Chen, H. (2023). How GHRM impacts employee OCBE: The role of emotions and value discrepancy. *International Journal of Manpower*, *44*(2), 318–333.

Lu, Q., Deng, J., Chen, S., & Hussain, Y. (2025). Managerial myopia and its barrier to green innovation in high-pollution enterprises: A machine learning approach. *Journal of Environmental Management*, *376*, 124477. PMID: 39946802

Mady, K., Battour, M., Aboelmaged, M., & Abdelkareem, R. S. (2023). Linking internal environmental capabilities to sustainable competitive advantage in manufacturing SMEs: The mediating role of eco-innovation. *Journal of Cleaner Production*, *417*, 137928.

Malik, S. Y., Hayat Mughal, Y., Azam, T., Cao, Y., Wan, Z., Zhu, H., & Thurasamy, R. (2021). Corporate social responsibility, green human resources management, and sustainable performance: Is organizational citizenship behavior towards environment the missing link? *Sustainability*, *13*(3), 1044.

Marrucci, L., Daddi, T., & Iraldo, F. (2024). Creating environmental performance indicators to assess corporate sustainability and reward employees. *Ecological Indicators*, *158*, 111489.

Moreno, D., Bang, A., Nielsen, S., & Thellufsen, J. Z. (2024). Strategic management of CO2: A scalable model for CCS in decarbonised societies. *Journal of Environmental Management*, *368*, 122175. PMID: 39178794

Pavlova, M. (2018). Fostering inclusive, sustainable economic growth and "green" skills development in learning cities through partnerships. *International Review of Education*, *64*, 339–354.

Pinto, L. (2020). Green supply chain practices and company performance in Portuguese manufacturing sector. *Business Strategy and the Environment*, *29*(5), 1832–1849.

Raza, M. A., Aman, M. M., Rajpar, A. H., Bashir, M. B. A., & Jumani, T. A. (2022). Towards achieving 100% renewable energy supply for sustainable climate change in Pakistan. *Sustainability*, *14*(24), 16547.

Remmer, S., & Gilbert, D. U. (2025). Causal Mechanisms in Corporate Social Responsibility (CSR) Management: A Meta-Synthesis of Micro-CSR Research. *Journal of Management Studies*.

Sachdeva, C., & Singh, T. (2024). Green transformational leadership and pro-environmental behavior: Unravelling the underlying mechanism in the context of hotel industry. *The International Journal of Organizational Analysis*, *32*(2), 255–271.

Saleh, I., Abu Afifa, M., & Alkhawaja, A. (2025). Environmental, social, and governance (ESG) disclosure, earnings management and cash holdings: Evidence from a European context. *Business Ethics, the Environment & Responsibility*, *34*(2), 295–308.

Satrovic, E., Usman, O., & Alola, A. A. (2025). Drivers of eco-efficiency and the criticality of environmental policy stringency in OECD countries. *Journal of Cleaner Production*, •••, 145157.

Suleman, A. R., Amponsah-Tawiah, K., & Ametorwo, A. M. (2024). The role of employee environmental commitment in the green HRM practices, turnover intentions and environmental sustainability nexus. *Benchmarking*, *31*(9), 3055–3078.

Umar, M., Ahmad, A., Sroufe, R., & Muhammad, Z. (2024). The nexus between green intellectual capital, blockchain technology, green manufacturing, and sustainable performance. *Environmental Science and Pollution Research International*, •••, 1–13. PMID: 38285260

Van Rooijen, S. N., & Van Wees, M. T. (2006). Green electricity policies in the Netherlands: An analysis of policy decisions. *Energy Policy*, *34*(1), 60–71.

Veza, I., Asy'ari, M. Z., Idris, M., Epin, V., Fattah, I. R., & Spraggon, M. (2023). Electric vehicle (EV) and driving towards sustainability: Comparison between EV, HEV, PHEV, and ICE vehicles to achieve net zero emissions by 2050 from EV. *Alexandria Engineering Journal*, *82*, 459–467.

Vila-Vázquez, G., Castro-Casal, C., Álvarez-Pérez, D., & del Río-Araújo, L. (2018). Promoting the sustainability of organizations: Contribution of transformational leadership to job engagement. *Sustainability*, *10*(11), 4109.

Yafi, E., Tehseen, S., & Haider, S. A. (2021). Impact of green training on environmental performance through mediating role of competencies and motivation. *Sustainability*, *13*(10), 5624.

Zameer, H., Wang, Y., & Yasmeen, H. (2024). Strengthening green competitive advantage through organizational learning and green marketing capabilities in a big data environment: A moderated-mediation model. *Business Process Management Journal*, *30*(6), 2047–2072.

Zhang, D., Zheng, M., Feng, G. F., & Chang, C. P. (2022). Does an environmental policy bring to green innovation in renewable energy? *Renewable Energy*, *195*, 1113–1124.

Zheng, Q., Li, J., & Duan, X. (2023). The impact of environmental tax and R&D tax incentives on green innovation. *Sustainability*, *15*(9), 7303.

Zhou, Y. (2023). Worldwide carbon neutrality transition? Energy efficiency, renewable, carbon trading and advanced energy policies. *Energy Reviews*, *2*(2), 100026.

Zou, Y., Liu, Z., Chen, Y., Wang, Y., & Feng, S. (2024). Crop rotation and diversification in China: Enhancing sustainable agriculture and resilience. *Agriculture*, *14*(9), 1465.

Chapter 10
A Study on Understanding Climate Change and Organizational Responsibility

R. Velmurugan
https://orcid.org/0000-0002-7925-9757
Karpagam Academy of Higher Education, Coimbatore, India

R. Bhuvaneswari
https://orcid.org/0009-0003-9396-6129
Dr. Mahalingam College of Engineering and Technology, Pollachi, India

J. Sudarvel
https://orcid.org/0000-0001-6656-1992
Karpagam Academy of Higher Education, Coimbatore, India

Joji Abey
https://orcid.org/0000-0003-4176-3023
Kingdom University, Bahrain

ABSTRACT

Organizational responsibility, climate change, is important to their being able to tackle the global environmental crisis. A rise in temperatures, the rise of extreme weather events, disrupted ecosystems, all these are the results of human influence on the climate. In particular, organizations, both in private and public sectors, have a vast role in lessening its effects through sustainable practices, reduction of carbon

DOI: 10.4018/979-8-3373-2479-1.ch010

footprints and advocating for policies that value environmental conservation. In context of climate change, Corporate responsibility involves taking up eco-friendly technologies, energy efficient processes and supporting climate action. Additionally, organizations are required to spread the word, teach stakeholders, and establish a sustainability based culture. Integrating climate change considerations into their operations and decision making processes has the potential of adding organization related value in the global fight against climate change and advocacy of long term environmental stewardship.

INTRODUCTION

Climate change has established itself as the utmost global challenge of the twenty-first century because it creates serious impacts on ecosystems alongside economies as well as societies. Global temperatures continue to increase because of human-made GHG emissions which cause multiple severe environmental results including sea-level rise together with extreme weather events and biodiversity decline (IPCC, 2021). Multiple disaster-proportional climate change predictions have been issued by the Intergovernmental Panel on Climate Change (IPCC in 2021. Climate change mitigation requires collective action between organizations and individual citizens and government institutions because the problem has reached a significant scale through business operations within economic structures that control environmental results.

The economic decisions organizations make serve as primary causes which determine either the maintenance or the reduction of global warming damage. Manufacturing institutions and energy production facilities together with transportation systems constitute the primary contributors to environmental deterioration and carbon emissions (Stern, 2007). Business activities cause significant GHG emissions to occur between business operations and supply chain functions as global resource usage continues to raise worldwide pollution levels according to Rockström et al. (2009). The total exploitation of natural resources together with the destructive impact of deforestation results in environmental deterioration because industry creates harmful waste that speeds up climate change (Steffen et al., 2015). Organizations maintain solution-generation power due to their innovative capabilities coupled with their policy support activities and responsible conduct. Through sustainable approaches organizations fight environmental pollution along with improving sustainable consumption practices and power efficiency (Porter & Kramer, 2011).

The study analyzes origin factors and outcome elements of climate change throughout this chapter. Fossil fuel emissions along with industrial activity and deforestation processes cause substantial increases of carbon dioxide (CO_2) and Greenhouse Gas (GHGs) concentration within the atmosphere (NASA, 2020). Deforestation causes Earth to lose its ability to absorb carbon which results in deteriorating climate change situations (Houghton, 2005). Rapid global warming develops due to substantial methane emissions generated by agricultural production and during industrial operations (Shindell et al., 2009). The extensive impacts of climate change affect natural systems as well as damage economic structures while degrading the quality of human welfare. Global warming causes water shortages despite reduced food availability and it threatens public health and leads to increased occurrences of storms alongside wildfires and floods (Field et al., 2012).

Each business organization holds an essential requirement to address environmental matters. The incorporation of sustainability through business core strategies depends on the implementation of Corporate Social Responsibility (CSR) and Environmental Social Governance (ESG) frameworks as described by (Eccles et al., 2014). The adoption of sustainability by organizations enables them to accomplish two goals through environmental protection together with enhanced corporate image and operational performance and stakeholder satisfaction (Elkington, 1997). Sustainable business practices that combine carbon footprint decrease methods with renewable power utilization alongside circular economy approaches generate extended economic value with parallel environmental benefits according to Bocken et al. (2014). Organizations must now state their intentions regarding carbon emission reductions and low-carbon economy shift because Paris Agreement under UNFCCC (2015) combined with regulatory systems have introduced these requirements.

Organizations pursuing sustainability need to apply three comprehensive methods that include technology-driven innovations and supply chain sustainability and stakeholder participation. Technology innovations that include CCS combined with energy-efficient manufacturing and electric transportation systems offer proven methods to decrease emissions (Pacala & Socolow, 2004). Impact investing together with green finance developed as two funding mechanisms which enable sustainable projects by providing support and fostering environmental sustainability (Clark et al., 2015). Businesses create environmental protection initiatives through associations between governments and NGOs and stakeholder involvement (Schaltegger & Wagner, 2011).

Climate change presents an unmanageable worldwide problem which warrants an immediate joint effort from every sector of society. Organizations operate at two levels because they both generate climate change effects and possess potential to decrease these effects. Organizational sustainable business operations enable identification of environmental impacts which leads to sustained global economic development that refuses disruption. Organizations hold essential responsibility according to the chapter which enables a presentation of enduring economic prosperity-strategic environmental sustainability methods for businesses.

THE SCIENCE OF CLIMATE CHANGE

The long-lasting global patterns of climate alteration stems primarily from human-induced activities which give rise to climate change. The climate change phenomenon exhibits heat-related temperature movements together with modifications in rain patterns as well as changes in weather extremity intensity. Rising levels of greenhouse gases function as the primary trigger for global warming that protests to create climate change (IPCC, 2021). Industrial activities together with forest clearing and fossil fuel combustion have made this process unfold much faster.

Industrialization functions as a primary factor behind rising GHG emissions patterns. Since the Industrial Revolution sectoral growth has triggered an increasing demand for fossil fuel products coal oil and natural gas (Pachauri & Meyer, 2014). The release of carbon dioxide from these energy sources stands as the most important factor behind climate change because CO2 constitutes the major greenhouse gas. Toxic gas emissions continue increasing at worrisome levels which the IPCC (2021) indicates reach extreme human historical records.

Climate change has greatly worsened because of forest clearing operations. The carbon sink function of forests happens through photosynthesis when they absorb atmospheric CO2 (Bonan, 2008). Major deforestation acts performed for agricultural needs and urban growth and timber extraction limit nature's capacity to absorb CO2 and allow atmospheric CO2 to rise in concentration. Tropical deforestation produces 10% of total human-emitted CO2 emissions according to Baccini et al. (2012). Local and global weather patterns suffer major disruptions through deforestation because it destroys the hydrological cycle and reduces rainfall (Lawrence & Vandecar 2015).

Methane (CH4) functions as a strong greenhous gas substance which plays a significant role in climate change impacts. Multiple human activities produce this gas including livestock and manure handling in agriculture and waste disposal and fossil fuel extraction according to Saunois et al. (2020). Livestock creatures including cattle emit methane as a byproduct through a digestive method known as enteric fermentation which functions as a byproduct to produce CH4. Together with wetlands and rice paddies and permafrost melting process methane levels continue to rise. The warming potential of CH4 exceeds that of CO2 significantly so this gas stands as a key factor in climate change battles (Shindell et al., 2012).

The agricultural practice of excessive synthetic fertilizer use creates the primary source of nitrous oxide (N2O) emissions in the atmosphere (Davidson, 2009). Soil microbiological processes of nitrification and denitrification which occur because of fertilizer use produce N2O gases that escape into the atmosphere. Scientific studies show that the 100-year assessment of N2O as a greenhouse gas exceeds CO2 by 300 times (IPCC, 2014). The release of N2O from industrial processes combined with biomass fires enlarges the impact of climate change (Reay et al., 2012).

Large scale environmental effects triggered by climate change remain severe across the planet. Global temperature elevation produces more frequent as well as stronger instances of extreme weather events like hurricanes plus heatwaves and droughts and heavy rainfall (Easterling et al., 2000). Environmental disruptions from these events cause damage to infrastructure and endanger human life as well as disrupt ecosystems. Prolonged droughts along with heatwaves resulted in major water shortages and destroyed farms throughout the world (Hansen et al., 2012).

Ongoing climate change causes significant elevation in sea levels which emerges as a vital issue. The process of temperature increase within the climate results in both seawater thermal expansion in addition to glacier and polar ice caps melting thus accelerating sea level rises (Church & White, 2011). Locations near the coast together with international islands situated at low elevations show special sensitivity to rising sea levels because they face increased threats of water inundation combined with shoreline destruction and population relocation. Studies from the IPCC (2019) demonstrated that sea levels increased by 15 centimeters since the early 20th century while demonstrating the path for even higher levels during the upcoming century.

Climate change leads to the direct loss of biodiversity throughout the planet. Several species experience severe challenges when they try to adjust their biology to fast-moving human-made alterations in their environment (Bellard et al., 2012). Nature destruction together with migration pattern adaptations and feeding chain disturbances cause species to become extinct. Research shows that coral reefs experience coral bleaching and major ecosystem deterioration because they strongly react to ocean temperature increases (Hoegh-Guldberg et al., 2007). Polar bears along with penguins represent polar species that face losing their habitats because of melting ice caps (Post et al., 2013).

The changing climate endangers both water and food supply systems at a dangerous level. The combination of temperature rise and modifications in precipitation reduces both agricultural output and the performance of farming operations (Lobell et al., 2011). The rising occurrence of droughts alongside floods creates farming interruptions which results in food scarcity along with increased expenses of food. The decreasing amounts of glaciers and snowpacks that supply freshwater to millions of people have created an urgent water scarcity problem (Barnett et al., 2005). Saltwater continues to spread into freshwater sources because of rising sea levels which threaten drinking water accessibility for coastal residents (Kundzewicz et al., 2008).

Climate change represents a complicated worldwide problem that stems mainly from human-made activities including manufacturing and tree clearing along with unreasonable releases of greenhouse gasses. The scientific understanding proves that we must develop quick-response strategies for mitigating and adapting to dangers from temperature increases and worsened weather patterns alongside rising coastal waters and diminishing species populations and food and water resource disruptions.

The IPCC (2021) identifies local as well as national and global action which requires quick implementation to decrease emissions and shift energy sources to sustainable ones and develop environmental protection policies against climate change threats.

THE ROLE OF ORGANIZATIONS IN CLIMATE CHANGE

Organizations serve fundamental roles in escalating as well as slowing down the impacts of climate change that has grown into a worldwide environmental emergency. Climate change receives contributions from different sectors which generate energy use as well as waste production and deplete natural resources while creating carbon dioxide emissions (IPCC, 2021). All organizations regardless of sector impact the environment either positively or negatively yet manufacturing, transportation, and agriculture prove most influential for change. Organizations can effectively fight climate change through their implementation of responsible business operations and policies along with sustainable innovation (Stern, 2007).

CONTRIBUTIONS OF ORGANIZATIONS TO CLIMATE CHANGE

1. Energy Consumption

The main factor behind organizational climate change impact stems from their energy consumption because industry sectors depend heavily on fossil fuels to produce electricity and conduct their operations. The combustion of coal oil and natural gas at power plants and heating facilities and production sites emits major amounts of greenhouse gases (GHGs) into the air thus intensifying global warming effects. Fossil fuel combustion produces carbon dioxide (CO_2) as the main greenhouse gas which creates the greenhouse effect and leads to persistent climate change (IEA, 2020).

According to the International Energy Agency industrial sector uses 37% of global energy with their consumption being the focus of their report. The sector incorporates notable energy-intensive industries which include steel production alongside cement and chemical manufacturing because they need substantial amounts of energy. Plant operations in these industries use mass amounts of heat and electricity which leads to high fossil fuel use and CO_2 emission increases (Smil, 2017).

Sustainable development faces a crucial hurdle because the world continues increasing its energy consumption while depending heavily on fossil fuels. Business entities are now researching renewable power technology and energy-efficient solutions which enable them to diminish their environmental impact. The implementation of renewable energy systems needs considerable funding together with government backing and modern technological developments to establish a sustainable industrial sector with environmentally friendly practices.

2. Carbon Emissions

The activities conducted by organizations result in direct and indirect carbon emissions that create major negative impacts on the environment. The burning or combustion of fossil fuels at physical locations generates direct emissions through industrial processes, factory operations and transportation activities and power generation facilities. Carbon dioxide emissions (CO_2) along with other pollutants enter the atmosphere when coal, oil and natural gas combustion takes place. Manufacturing facilities operate their energy-consuming machines but corporate transportation systems including trucks and ships and airplanes extensively burn fuel to produce additional emissions (Gillingham et al., 2018).

Corporate carbon emissions result from both their production processes as well as their supply chain activities and product utilization across their lifecycles. Organization emissions termed Scope 3 encompass multiple phases from raw material recovery to outsourced production operations and energy consumption in office facilities and shipping and discarding products. Due to global greenhouse gas (GHG) emissions the transportation sector produces about 14% of the total while freight and logistics operations significantly increase this environmental impact. The worldwide push for global trade and e-commerce has worsened cargo transport-related emissions since most of these emissions stem from air and maritime shipping that use fossil fuels (IPCC, 2021).

Industries engaged in mining as well as refining and chemical processing operations discharge greenhouse gases simultaneous to their CO_2 emissions. The emission of methane from mining and oil extraction creates atmospheric heating equivalent to CO_2 emissions at about 25 times the strength in its 100-year lifecycle. The industrial release of nitrous oxide (N_2O) from agricultural sources makes this gas 300 times more effective than CO_2 at capturing heat from the atmosphere which contributes substantially to climate change (Schlesinger & Bernhardt, 2020).

Organizational pressure continues to rise because both direct and indirect emissions demonstrate extensive consequences thus organizations need to establish sustainable methods including clean energy adoption along with enhanced supply chain management and carbon offset programs. The reduction of emissions needs organizations to work with governments and industries together with consumers toward establishing a low-carbon economy in order to minimize climate change impacts.

3. Waste Generation and Pollution

The waste produced by organizations creates extensive environmental pollution as well as enduring harm to ecosystems throughout the environment. Industrial operations together with manufacturing processes and commercial operations generate very high levels of solid waste along with liquid waste and hazardous materials that can pollute environmental soil and water sources while increasingly polluting the surrounding atmosphere unless proper waste management strategies exist. Earth systems face disruption and biodiversity suffers because of poor waste handling combined with industrial waste fluid movement and plastic contamination which further pushes climate change effects. Waste materials including microplastics together with heavy metals along with synthetic chemicals stay in the environment for extended time periods which create substantial dangers for marine organisms and wildlife and human health (Hoornweg & Bhada-Tata, 2012).

The improper handling of hazardous waste stands as an urgent problem among various industries which include electronics and textiles along with chemicals. E-waste disposal results in toxic substance emissions that enter groundwater and soil and hence pollute food and water sources with lead, mercury and cadmium. The textile industry produces significant wastewater amounts that include synthetic dyes and heavy metals together with microplastics which cause water pollution alongside aquatic environment destruction. The manufacturing of chemicals alongside industrial refinement operations generates VOCs and dangerous gases that reduce air quality while leading to respiratory illnesses (Geyer et al., 2017).

Municipal and industrial waste disposed in landfills produces methane (CH_4) which creates far more global warming effects than carbon dioxide (CO_2) does. When organic waste consisting of food scraps together with biodegradable materials decomposes anaerobically within landfills the process produces methane that acts as a heat-trapping substance to strengthen climate change. As reported by the United Nations Environment Programme (UNEP) waste dumpsites represent the leading human-manufactured methane emission sources so effective waste management plays a crucial role in global warming reduction. Organizations can lessen environmental damage while promoting sustainability through waste reduction strategies which must incorporate recycling along with composting and circular economy transition (UNEP, 2019).

Organizations now implement sustainable waste management approaches because they want to minimize the adverse effects of waste production. Organizations should work toward cutting down single-use plastics while improving their recycling initiatives along with introducing environmentally friendly manufacturing operations. Regulatory bodies together with governments have started implementing stricter waste disposal standards to push industries toward selecting greener approaches that include funding waste-to-energy projects. Organizations exercise a critical role in environmental conservation together with pollution reduction through their commitment to waste reduction and proper disposal methods.

4. Resource Depletion

Organizations exhaust natural resources to an extent that creates severe environmental effects including deforestation and water scarcity as well as reduction in biological diversity. Forest quick depletion together with freshwater exhaust and non-renewable resource shortages disrupts natural ecosystems while endangering wildlife areas and worsens climate problems. The agriculture sector together with manufacturing industries along with mining operations and energy producers heavily depend on natural resources as they extract them beyond sustainable levels to fulfill rising world demands. Environmental systems experience mounting stress which results in decreased capability to restore their functions supporting all life forms.

Agricultural practices drive the majority of deforestation as forests are being razed down to make space for farming land used to raise livestock together with large-scale crop plantations. Deforestation extended across vast areas causes decreased carbon capture capabilities and creates severe problems that include deteriorated soil quality and diminished biodiversity and modified climate zones in the affected regions. The extracting of trees from land reduces Earth's ability to take in carbon dioxide (CO_2) which consequently heightens global warming alongside catastrophic storm events. Deforestation creates a threat to indigenous populations and wildlife because it restricts species habitats which causes their numbers to decrease resulting in increased endangerment (Foley et al., 2011).

Industrial water usage generates intense water deficiencies in areas facing water rationing along with decreasing water flow reservoirs. The industrial sectors running textile manufacturing facilities and power plants and mining operations need very large quantities of water mainly for process cooling and waste management systems. The removal of excessive quantities of groundwater and surface water affects the supply of clean drinking water and impair irrigation for local agricultural activities. The ongoing removal of water from sources can result in desertification of land which ultimately damages crop growth and creates food-based challenges. Higher water needs for industrial operations cause increased competition over water supplies hence demanding superior long-term sustainability approaches through water management and conservation (Rockström et al., 2009).

The extraction of minerals and fossil fuels to excess speeds up the reduction of natural resources as well as environmental destruction. The extraction of minerals through mining operations creates numerous ecological damage effects by removing important geological matter while damaging natural habitats and fluid pollution followed by toxic waste accumulation. Fossil fuel extraction through methods like coal extraction and oil extraction and natural gas extraction causes depletion of limited resources while its operation produces greenhouse gas emissions that trigger climate change. The harmful actions of deep-sea mining and oil drilling coupled with fracking operations result in permanent deterioration of marine and terrestrial ecosystems which weakens biodiversity while reducing ecosystem resilience (Steffen et al., 2015).

Organizations need to establish sustainable practices that include responsible sourcing methods together with reforestation projects, water preservation approaches and circle economic management to reduce environmental damage due to wasteful resource utilization. The use of natural resources receives growing support from environmental agencies together with government bodies who promote both technological innovations and strict regulations and alternative energy solutions for responsible stewardship. Sustainable business strategies enable industries to guard ecosystems together with biodiversity and secure a long-term sustainable planet.

MITIGATION STRATEGIES FOR ORGANIZATIONS

Organizations exhaust natural resources to an extent that creates severe environmental effects including deforestation and water scarcity as well as reduction in biological diversity. Forest quick depletion together with freshwater exhaust and non-renewable resource shortages disrupts natural ecosystems while endangering wildlife areas and worsens climate problems. The agriculture sector together with manufacturing industries along with mining operations and energy producers heavily depend on natural resources as they extract them beyond sustainable levels to fulfill rising world demands. Environmental systems experience mounting stress which results in decreased capability to restore their functions supporting all life forms.

Agricultural practices drive the majority of deforestation as forests are being razed down to make space for farming land used to raise livestock together with large-scale crop plantations. Deforestation extended across vast areas causes decreased carbon capture capabilities and creates severe problems that include deteriorated soil quality and diminished biodiversity and modified climate zones in the affected regions. Deforestation reduces Earth's carbon dioxide (CO_2) absorption capacity thus it generates quicker global warming along with intensified extreme weather conditions. Deforestation creates a threat to indigenous populations and wildlife because it restricts species habitats which causes their numbers to decrease resulting in increased endangerment (Foley et al., 2011).

Industrial water usage generates intense water deficiencies in areas facing water rationing along with decreasing water flow reservoirs. The industrial sectors running textile manufacturing facilities and power plants and mining operations need very large quantities of water mainly for process cooling and waste management systems. The removal of excessive quantities of groundwater and surface water affects the supply of clean drinking water and impair irrigation for local agricultural activities. The ongoing removal of water from sources can result in desertification of land which ultimately damages crop growth and creates food-based challenges. The rising need for water-intensive manufacturing increases the battle for water resources thus requiring urgent long-term sustainable solutions in water conservation (Rockström et al., 2009).

The extraction of minerals and fossil fuels to excess speeds up the reduction of natural resources as well as environmental destruction. Mining operations extract minerals while ruining landscapes thus creating major ecological problems that poison water resources along with producing fatal toxic remnants. Fossil fuel extraction through methods like coal extraction and oil extraction and natural gas extraction causes depletion of limited resources while its operation produces greenhouse gas emissions that trigger climate change. The harmful actions of deep-sea mining and oil drilling coupled with fracking operations result in permanent deterioration of marine and terrestrial ecosystems which weakens biodiversity while reducing ecosystem resilience (Steffen et al., 2015).

Organizations who want to reduce their environmental impact need to establish sustainable practices which include responsible sourcing methods together with reforestation projects together with water conservation methods and circular economy models. Regulations backed by governments and environmental agencies have become stronger to combine with technological developments and alternative energy systems which guarantee proper natural resource management. ölutions in industrial practices lead to ecosystem protection together with biodiversity preservation that results in planetary stability for future generations.

1. Energy Efficiency and Green Technologies

The aim to lessen the industry's energy efficiency is still vital to mitigate environmental emissions, operational expenses, and environmentally protection. Although it has been reduced, the energy required of industrial operations and factories is still high since manufacturing facilities, processing centers, and production areas require significant amounts of both electricity and heat. Efficient technology adoption and process enhancement program helps in improving organizations together with lower emissions through improvement in efficiency, productivity, and market competition (IEA, 2021).

Assumption 1: An efficient approach to increase energy efficiency levels is to upgrade lighting infrastructure along with heating and cooling systems. The adoption of LED lighting is beneficial for organizations as they are ideal at lowering demand for electricity, as opposed to regular incandescents and fluorescent bulbs. Modern building facilities equipped with a digital HVAC control system and smart thermostats can efficiently manage settings temperatures, and thus use less energy by minimizing unneeded consumption. When modern insulation and new building design are combined, companies are able to reduce their building energy usage target.

With automation platforms, industrial energy management is reached by way of automation of the maintenance and optimization process of process, and automation of the prediction of when maintenance is required, and by real time monitoring of the assets. The use of industry sensor technology, data analytical models and artificial learning systems allow machines to find operational losses by means of work routine optimization and output energy control based on market demands. Automation allows industrial operation to be more precise and ultimately waste reduction results in sustainable and cost competitive business operations. According to IEA (2021), businesses using AI based energy management systems will enjoy important energy savings and at the same time maintain their operational performance.

Essential measures to halt climate change are the practice of green technologies that lead to improved efficiencies. Carbon capture and storage (CCS) is a promising advance: It allows CO_2 emissions to be pulled on the industrial processes before being buried or used in industrial processes. CCS technology provides a means of large emission reduction in, in principle, high pollution emitting industries (the cement and steel producers and chemical producers) thereby enabling these industries to be more sustainable (Hepburn et al., 2019).

This means that industrial facilities must adopt low emission logistics based transportation systems for supply chain operations for the adoption of energy efficient industrial operations. The electric trucks coupled with hydrogen vehicles that industrial enterprises are investing in do something about cutting fuel usage, and they are optimizing route planning to reduce carbon emissions by using rail transportation since it creates less emissions than surface freight. Renewable wind and solar served industrial operations that provide fuel to utility scale vehicles charged by electric power, as well as the sustainable operation of factories.

Industry operations have the potential to cut carbon emissions while continuing to maintain operational excellence and monetary efficiency when energy efficient technologies are combined with green technologies using AI optimization algorithms. Continuous implementation of these practices is the key to the future of sustainable development since industries around the world are trying to make their operations carbon-free.

2. Waste Reduction and the Circular Economy Movements are Practiced Together

Circular economy principles provide organizations the opportunity to reduce the quantity of waste generated and create sustainability by using resources for longer, and reducing the amount of waste and natural system regeneration. The circular economy works differently than traditional lines of economy because it embraces product solutions with infinite or extended use potential and recyclability in accordance with optimal resource utilization. Firstly, the circular economy principles (Ellen Mac Arthur Foundation, 2013) allow the businesses to continue pursuing environmental benefits whilst at the same time gaining cost efficiencies and supply chain stability and extended business lifetime sustainability.

Fundamental to the circular economy is a fundamental design objective that seeks to create resistant products that are recyclable. High quality materials and modular construction of products are the means to increase product longevity organizations adopt such that replacement demands are minimized. In design, breaking down the products will make it easier to fix and reuse as well as recycle the components before they become waste. Closed loop manufacturing practice can be applied by various industries to deal with waste product after its operation life, making it possible to produce new production requirements, hence both lowering of material use and generation of waste.

Sustainable packaging development is considered the major measure in reducing waste. Due to this increasing number of organizations would choose such packaging material as biodegradable and compostable or recyclable products to replace the single use plastics. Plants based bioplastics and reusable packaging systems with lightweight materials is applied by the cities in order to reduce the waste production while lowering their environmental impact. Today, packaging companies offer take back programs that allow customers to bring back their used packaging materials to recycling or reuse methods in order to increase material loop closures.

Industrial composting systems are the ideal solutions to the disposal of food business and hospitality service waste materials and the similar sorts of waste. By turning food waste and agricultural residues that would have otherwise gone to landfills into biodegradable materials, organizations can cut powerful landfill emissions of methane. Composting provides valuable nutrient resources for farming operations, land development and restoration, and gardening operations so soil amendments derived from composting are used by many firms to support agricultural operations.

Industrial symbiosis is an advanced way of going circular that brings industrial waste to benches as useful input materials for other enterprises. The cooperation between industries is based on how to use minimally leaked materials efficiently in reducing charging expenses and minimizing pollution indices. Conversely, the heat surplus heat in one industrial site can be used as community district heating service and by products of power generating facilities like fly ash can be used in cage production. Therefore, industrial symbiosis is an attempt to improve sustainability factors and create new economic opportunities through cooperative action among distinct industries (Ghisellini et al., 2016).

Digital blockchain and Internet of Things (IoT) as well as artificial intelligence (AI) technologies are used by organizations to monitor material flow and maximize operational output in order to enhance waste management systems. Smart waste sorting techniques as well as automation of ways to recycle materials and use predictive analytics boost the amount of available materials and reduce landfills.

Circular economy principles adopted in business activities minimize waste output from companies, as well as adopt lesser resources and development of a sustainable economic system. Organisations must embrace the principles of circular economy as their success in business and environmental preservation over the long run will come down to a great extent on regulatory requirement and sustainable product expectations growing stronger.

3. Carbon Offsetting and Environmental Initiatives

To make up for unavoidable emissions, organizations can take into account a carbon offsetting program, which is a mechanism that helps equalize the organization's carbon footprint by supporting projects that remove or reduce greenhouse gas (GHG) emissions from the atmosphere. These programs are important in the realisation of corporate sustainability strategies and enabling companies to offset their environmental footprint and help reach the goals of long term climate. Organizations can support global efforts to stop the effects of climate change and conserve natural ecosystems by funding reforestation, afforestation and conservation projects (Gold Standard, 2021).

Reforestation and possibly afforestation are some of, if not the best carbon offset strategies. Reforestation, in particular, involves the planting of trees in deforested areas while afforestation involves the creation of new forests in areas devoid of them as well. Trees are natural carbon sinks as they absorb this carbon dioxide (CO_2) during photosynthesis so these projects are natural carbon sinks and help decrease the levels of atmospheric carbon. Forests also have climate benefits, but other than that they provide habitat for wildlife, enhance biodiversity, prevent soil erosion and improve water cycles. Large tree-planting initiatives form part of larger companies' investments that keep down emissions and bring about wider environmental and social gains.

Organizations concerned with protecting the forests, wetlands and other ecosystems that act as carbon 'stores' on the planet usually embark on conservation projects. Conservation efforts are crucial to maintain natural carbon storage, as these practices cause their releases of large amounts of CO_2 to the atmosphere. The support of organizations can help organizations aim to prevent illegal logging, restore the degraded lands and promote sustainable land management practices. Conservation organisations and local communities have also worked together with many businesses to protect the environment for long term sustainability and the economic development of the people.

Most companies are also seeking to achieve carbon neutrality by offsetting carbon from direct projects along with an equivalent amount carbon removal. Usually accomplished by horizontal operational efficiencies, the adoption of renewable energy and buying carbon credits. Carbon credits are reductions in emission verified by certified environmental projects that issue carbon credits. Although purchasing these credits allows businesses to offset their emissions financially into supporting sustainable development and clean energy initiatives (UNEP, 2020), they also provide substantial benefits including job creation for communities close to that power source for example.

Also, in order to raise their efforts of sustainability, organizations are through the corporate social responsibility (CSR) program that benefits the environment and changing the climate. Examples of such CSR initiatives can be these investments in clean energy research, in projects of sustainable agriculture, in community driven environmental projects. This is also the case for many companies that participate in employee led programs such as volunteer tree planting event, recycling drive, eco awareness campaign and so on. Moreover, these initiatives not only help to achieve environmental goals but also add to corporate reputation, stakeholder trust and employee engagement (Carroll and Shabana, 2010).

Organizations that want to make the most out of their carbon offsetting and sustainability efforts need to guarantee that those efforts are transparent, accountable and in line with international climate standards. Independent certifications, like Gold Standard and the Verified Carbon Standard (VCS), are the way of measuring if offset projects really provide real environmental impacts, lasting benefits. When businesses weave carbon offsetting into the larger realm of the company's sustainability strategies, it can be a game changer for their actions in general regarding climate responsibility as well as for the future that will be greener and more resilient.

On the one hand, organizations are contributors to climate change, and on the other hand, organizations are also climate change mitigators. Energy consumption, waste generation and resource depletion have serious environmental footprints on the industries having manufacturing, transportation and agriculture. Yet organizations also have the ability to embrace sustainable practices that positively affect them. Transitioning to renewable energy, improving energy efficiency with environmentally friendly suppliers, reducing waste production, commit to carbon offsetting, and investing in arrays of renewable energy to downplay global climate action. Considering the importance of climate change, corporate sustainability activity which is not a responsibility but a must for long-term ecological and economic durability (Rockström et al., 2009; Steffen et al., 2015).

ORGANIZATIONAL RESPONSIBILITY TOWARDS CLIMATE CHANGE

Organizational responsibility is asserted on two types: regulatory compliance and voluntary action which attempt to minimize environmental impact. This responsibility has the key aspects including:

1. Corporate Sustainability and ESG (Environmental, Social, and Governance) Initiatives

However, nowadays, more and more organizations are starting to integrate sustainability into their business planning by aligning the strategies with the concepts of Environmental, Social, and Governance (ESG). The shift indicates that the result of such a recognition is a greater focus on sustainable business practices which can reduce environmental and social risks, increase long term profits, customer trust and regulatory compliance. Responsible corporate governance and sustainable economic growth (Eccles et al., 2012) include ESG focused initiatives such as reducing carbon footprints, ethic sourcing and providing transparent on reporting on sustainability.

It's a key high priority for organizations who have picked up ESG principles because energy efficiency, renewable energy investments and emissions reduction strategies are all ways to minimize carbon footprint. Companies are yet to be carbon neutral but they have been actively working on providing low carbon technology, reshuffling supply chain logistics, and carbon offset programs. Additionally, many firms have science based emissions reduction targets in accordance with the Paris Agreement target to limit global temperature rise (CDP, 2020). There is research that shows businesses that adjust carbon liabilities are indeed more financially resilient and better market valued than their peers (Clark et al., 2015).

Another critical aspect of integrating sustainability in business model is promoting ethical sourcing. Ethical sourcing strives to provide each and every part sourced by a company in a socially and environmentally responsible manner. It entails having quality suppliers with fair labor practices, low impact on the environment, as well as transparency in their supply chain. In particular, fashion, electronics, agriculture, among others have placed special emphasis on reducing exploitative labor practices, preventing deforestation and limiting the pollution from sourcing activities

(Soundararajan et al., 2018). For instance, certifications like Fair Trade, Rainforest Alliance and Forest Stewardship Council (FSC) enable companies of prove their fair sourcing and right production way techniques.

Sustainability reporting with transparency is paramount for organizations that aim to gain credibility and establish accountability in their ESG initiatives. There are increasing demands of all form of corporate environmental and social impact disclosures by investors, regulators, and consumers. A large number of companies (many of which use the Global Reporting Initiative (GRI), Sustainability Accounting Standards Board (SASB) and Task Force on Climate-related Financial Disclosures (TCFD)) publish sustainability reports. Transparency gives businesses to ability to track progress on their ESG commitments, highlighted areas for improvement, and enhance stakeholder confidence (Ioannou & Serafeim, 2017).

In this regard, financial sector is also to a large extent carrying the torch by integrating ESG principles into business models through which institutional investors and asset managers' increasingly incorporate ESG issues in investment decisions. Businesses have access to funding through sustainable finance and green bonds to support its economic viability of sustainability initiatives (Friede et al., 2015).

In general, organizations that are able to integrate sustainability into their operation would benefit from such an integration by becoming less risky and more efficient at operations, as well as improving brand reputation. With the increasing global demand for ESG aligned companies to survive in ever changing business landscape, they are going to be better equipped to thrive in such an environment and coexist with a more sustainable and just future.

2. Carbon Footprint Reduction

Together, energy efficiency improvement, the adoption of renewable energy and waste reduction strategies offer companies the means to significantly lessen their carbon footprint and firmly support global climate targets (Paris Agreement and UN Sustainable Development Goals) (CDP, 2020). By automating industrial operations into a smart system with AI driven energy management and upgrading industrial equipment, the energy efficiency can be improved, resulting in massive reduction on the consumption of energy and greenhouse gas emissions (IEA, 2021). Further transitioning away from fossil fuels is also achieved through taking on 100% renewable energy through power purchase agreements (PPAs) and on-site generation (REN21, 2021). Energy use is only one area that organizations can reduce emissions by applying waste reduction strategies aligned with the principles of a circular economy, like sustainable packaging, material recycling and industrial symbiosis (where one industry's waste is used as raw material by another) (Geissdoerfer et al., 2017).

These strategies reduce and mitigate the waste related emissions, namely methane (CH_4) which is more potent than carbon dioxide (UNEP. 2019).

Besides cutting down emissions, many companies have set to reduce carbon emissions to carbon neutrality and net zero targets by investing in carbon offset project like reforestation, afforestation and carbon capture and storage (CCS) (Gold Standard 2021). The most common net zero strategies include direct emission reductions and through schemes such as the restoration of ecosystems and through removing carbon from the atmosphere (UNEP, 2020). Further, organizations are integrating low carbon supply chains, sustainable transportation and responsible sourcing in their value chains as well (BloombergNEF, 2022). With an increase in governmental restriction of climate regulations and an increase in the consideration of Environmental, Social, and Governance (ESG) performance for the investors, proactive businesses that reduce their emissions are granted with a competitive advantage (as they offer and facilitate regulatory compliance, financial resilience and brand reputation). Aligning corporate strategies with global decarbonization efforts not only supports climate change mitigation, but also sets a company up for longevity and taking the market leadership in what is at once a changing low carbon economy.

3. Sustainable Supply Chain Management

Assessment of environmental impact of the supply chain is also observed in an organization where those organizations should evaluate the carbon footprint associated with sourcing, production, and distribution of products as well as the resource consumption and waste generation (Seuring & Müller, 2008). Green logistics (adopting of green logistics) is a critical aspect of sustainable supply chain as it makes it possible to optimize the transportation network, reduce the fuel consumption and utilize the low emission vehicles in minimizing the greenhouse gas (GHG) emissions (McKinnon et al., 2010). Intermodal transportation, route optimization algorithm, and electric or hydrogen powered delivery fleets are becoming more popular among companies when cutting emissions in logistics operations (Rogers et al., 2012). Moreover, digital supply chain management tools, particularly blockchain and AI based tracking systems allow companies to track the stream of sustainability performance of their suppliers and ensure that their suppliers comply with environmental regulations and corporate sustainability goals (Montabon et al., 2016).

An environmentally responsible supply chain also relies on sourcing the sustainable materials. Organisations can threaten to impact less on the environment by giving best to renewable, recycled or ethically brought material sources, diminishing dependence on virgin supplies and lower all round discharges (Geissdoerfer et al., 2017). Closed loop supply chains are spreading in many industries especially in fashion, electronics and consumer goods for collecting end-of-life products that are reconditioned and reintegrated to production cycles (Ghisellini et al., 2016). Additionally, suppliers need to be made to conform with eco friendly practices so as to ensure sustainability throughout the value chain. As companies are increasingly setting supplier sustainability criteria, these criteria are often set with environmental, social, and governance (ESG) standards and with the ISO 14001, the FSC (Forest Stewardship Council) and Fair Trade (Tachizawa & Wong. 2015. Businesses which put emphasis on sustainable supply chain management not only diminish environmental risks but also improve the brand reputation, regulatory compliance and long term profitability in a market that encourages sustainability driven business model (Pagell & Shevchenko, 2014).

4. Regulatory Compliance and Policy Advocacy

Environmental regulations are implemented to change climate, minimize pollution and use of natural resources in a sustainable way, all across the globe. The regulations comprising these regulations include carbon taxes, emission trading systems (ETS), and extremely strict environmental standards as they have requirements on companies' environmental impacts (OECD, 2019). Incentives to reduce greenhouse gas (GHG) output from companies can be accomplished through the use of carbon taxes that put a price on emissions (Nordhaus, 2018). Emission trading systems (ETS) also facilitate the trading of emission allowances within an overseen cap in order to facilitate cost effective reductions of carbon footprints (Ellerman et al., 2016). Moreover, governments mandate mandatory sustainability reporting, increased tightening of pollution controls, and extended producer responsibility (EPR) programs which impose on industries to manage waste and emissions to better standards (UNEP, 2020). If these regulations are not adhered to, then contractual penalties, possible lawsuits, and damaged reputation would all be a result of their noncompliance, and integrated environmentally compliant business practices within these organizations are essential (Porter and Van der Linde, 1995).

Organizations go beyond compliance in corporate advocacy and industry collaboration (Delmas & Montes-Sancho, 2010) to actively play a role in shaping the sustainability focused policies. Kindly support a green policy framework by government and the legacy of NGOs and international agencies to run initiatives like renewable energy incentives, deforestation prevention and energy policy of circular economy (CDP, 2021). Those businesses that actively align with sustainable regulations and call for fresh policies that embrace progress for climate gain an advantage: through incentives by securing them, confidence from investors, and relationships with stakeholders (Eccles et al., 2012). Besides, corporate sustainability leadership in terms of public commitments to net zero targets, responsible sourcing and transparent environmental reporting further influence other regulatory trends and industry standards (Friede et al., 2015). Their ability to remain competitive and focused on long term success in a low carbon economy is dependent on their support and ability to comply with those regulations as well as be able to support and advocate sustainable business models (Sullivan & Gouldson, 2017).

5. Corporate Social Responsibility (CSR) and Stakeholder Engagement

Organizations are actively working with stakeholders, including employees, consumers, investors, and local communities, through promoting environmental awareness and sustainability initiative (Freeman et al., 2010). Companies that promote the culture of environmental responsibility will not only encourage the practices at individual level but also at organizational level (Doppelt, 2017). Green workplace programs, sustainability training and Eco friendly policies constitute a corporate sustainability effort for the reduction of its environmental footprint (Hart, 1995). At the same time, businesses teach consumers the behavior of such consumption by providing eco-labeled product, responsible packaging and carbon footprint disclosure (Ottman, 2011). Other key aspect is community engagement in which organizations join local governments, schools and NGOs to promote environmental initiatives (Carroll & Shabana, 2010). The presence of these efforts helps brands' reputation, builds stakeholder confidence and helps in meeting long term sustainability goals (Elkington, 1997).

Many corporate social responsibility (CSR) programs are reforestation projects, water preservation efforts, climate education efforts, and strive to mitigate or even reverse many of the impacts on the environment (Matten & Moon, 2008). The efforts of tree planting initiatives assist in offsetting carbon emissions, in restoring degraded ecosystems and further in supporting biodiversity conservation (Gold Standard, 2021). Gleick (2014) also covers how many organizations invest in conservation projects for global water scarcity and community resilience, for example rainwater harvesting, watershed restoration, and clean water access programs. There are such things as climate education initiatives like public awareness campaigns, workshops on sustainability, outreach to schools, that give communities the knowledge and tools to fight back climate change (UNESCO, 2020). What organizations achieve by using stakeholder engagement and by integrating CSR driven environmental programs is not just improved corporate sustainability but significant long term environmental impact (Porter & Kramer 2011). With the increasing trends of sustainability expectations, businesses that would engage with their stakeholders as well as their advocacy in helping the environment will remain competitive and foster a greener more responsible future, as suggested by Bansal and Roth (2000).

CHALLENGES AND FUTURE DIRECTIONS

More and more organizations are realizing their responsibility in fighting climate change, but it is very difficult for them to act on this. As such, investments into renewable energy, energy efficient infrastructure as well as low carbon technologies are often constrained by financial policies especially so among small and medium sized enterprises (SMEs) (Lash & Wellington, 2007). Furthermore, corporate sustainability efforts are further complicated by regulatory complexities, as businesses have to deal with dissimilar environmental policies and compliance requirements and also carbon pricing mechanism in individual jurisdictions (OECD, 2019). Further, the adoption of sustainable practices is further hindered by organizational resistance to change, which can be attributed mainly to short term profit pressure, lack of expertise and existing business models (Kotter & Schlesinger, 2008). Developing these solutions requires leadership, collaboration of stakeholders and incentives to move towards low carbon operations (Eccles & Klimenko, 2019).

Green technologies have improved; consumers demand and policies develop at the international level for sustainability offer opportunities to businesses to improve their environmental responsibility despite all these challenges. Carbon capture and storage (CCS), bio based materials and also ai driven energy optimisation are some industrial sustainability innovations (Hepburn et al., 2019). Switching technologies is also spurred by consumers, who are making conscious efforts to buy eco products, supporting ethical brands and demanding transparency for companies on efforts. Among other things. (White et al., 2019) On top of that, agreements regarding global climate such as the Paris Agreement and net zero commitment is compelling corporations to set sciencebased targets and to transform their supply chain as sustainable as possible (UNFCCC, 2015). According to Geissdoerfer et al. (2017), circular economy models will be the trend to look for as one to aid in resource efficiency and waste minimization, while carbon capture technologies (IEA, 2021) and digital innovations for sustainably tracking will be needed (Saberi et al., 2019). Adopting these developments allows organizations to move beyond the sustainability challenges they face under the conditions of long term resilience, competitiveness and stewardship (Bansal and DesJardine, 2014).

CONCLUSION

Knowledge about climate change and the role of organization responsibility is necessary in standing for a sustainable future. Consequently, with increasing complexities of climate risks, businesses ought to abandon traditional corporate models in favor of sustainability – driven strategies by integrating ESG principles (Eccles et al., 2012). Reduction of carbon footprint, conservation of natural resources and boosting of circular economy principles for minimization of emissions and waste (Geissdoerfer et al., 2017) are vital for organizations. A systems thinking approach to business and transition to low carbon economies is required with businesses working with governments, stakeholders and consumers towards effective change (Rockström et al., 2009). This transformation can be achieved through clear sustainability commitments, investments in renewable energy as well as transparent climate risk reporting so there can be accountability and long term environmental resilience (Carney, 2015).

To support the balance of economic growth and ecological preservation, organizations need to use innovation, alignment of policy, and responsible business (Porter and Kramer, 2011). Artificial intelligence for energy efficiency and carbon capture solutions as well as digitalization in the sustainable supply chain, among others, are technological advancements that can speed up the corporate sustainability efforts (Hepburn et al., 2019). Moreover, complying with the international climate policies like Paris Agreement and UN Sustainable Development Goals (SDGs) allows businesses to navigate their path while complying with the regulatory landscape as well as to foster global climate action (UNFCCC, 2015). Sustainability embedded in core corporate strategies helps companies to manage climate risks, build a better brand reputation, attract ethical investors and future proofing of operations (Grewal & Serafeim, 2020). Therefore, a commitment to responsible environmental stewardship helps organizations become part of a resilient, low carbon future by ensuring livability of the planet for our generations and for the future generations (Bansal & DesJardine, 2014).

REFERENCES

Baccini, A.. (2012). Estimated carbon dioxide emissions from tropical deforestation improved by carbon-density maps. *Nature Climate Change*.

Bansal, P., & DesJardine, M. R. (2014). Business sustainability: It is about time. *Strategic Organization*, *12*(1), 70–78.

Bansal, P., & Roth, K. (2000). Why companies go green: A model of ecological responsiveness. *Academy of Management Journal*, *43*(4), 717–736.

Barnett, T. P.. (2005). Potential impacts of a warming climate on water availability in snow-dominated regions. *Nature*.

Bellard, C.. (2012). Impacts of climate change on the future of biodiversity. *Ecology Letters*.

BloombergNEF. (2022). *Corporate Clean Energy Buying Surges to New Record in 2021*.

Bocken, N. M. P., Short, S. W., Rana, P., & Evans, S. (2014). A literature and practice review to develop sustainable business model archetypes. *Journal of Cleaner Production*, *65*, 42–56.

Bonan, G. B. (2008). Forests and climate change: Forcings, feedbacks, and the climate benefits of forests. *Science*.

Carney, M. (2015). *Breaking the tragedy of the horizon–climate change and financial stability. Speech at Lloyd's of London, 29 September 2015*. Bank of England.

Carroll, A. B., & Shabana, K. M. (2010). The business case for corporate social responsibility: A review of concepts, research and practice. *International Journal of Management Reviews*, *12*(1), 85–105.

CDP. (2020). *The Carbon Disclosure Project: Global Climate Change Report 2020*. Retrieved from www.cdp.net

CDP. (2021). *Engaging the chain: Driving speed and scale*. CDP Worldwide.

Church, J. A., & White, N. J. (2011). Sea-level rise from the late 19th to the early 21st century. *Surveys in Geophysics*.

Clark, G. L., Feiner, A., & Viehs, M. (2015). *From the stockholder to the stakeholder: How sustainability can drive financial outperformance*. University of Oxford & Arabesque Partners.

Davidson, E. A. (2009). The contribution of manure and fertilizer nitrogen to atmospheric nitrous oxide since 1860. Nature Geoscience.

Delmas, M. A., & Montes-Sancho, M. J. (2010). Voluntary agreements to improve environmental quality: Symbolic and substantive cooperation. *Strategic Management Journal, 31*(6), 575–601.

Doppelt, B. (2017). *Leading change toward sustainability: A change-management guide for business, government and civil society*. Routledge.

Easterling, D. R.. (2000). Climate extremes: Observations, modeling, and impacts. *Science*.

Eccles, R. G., Ioannou, I., & Serafeim, G. (2012). The impact of corporate sustainability on organizational processes and performance. *Management Science, 60*(11), 2835–2857.

Eccles, R. G., & Klimenko, S. (2019). The investor revolution: Shareholders are getting serious about sustainability. *Harvard Business Review, 97*(3), 106–116.

Elkington, J. (1997). *Cannibals with forks: The triple bottom line of 21st century business*. Capstone.

Ellen MacArthur Foundation. (2013). *Towards the Circular Economy: Economic and Business Rationale for an Accelerated Transition*. Ellen MacArthur Foundation.

Ellerman, A. D., Convery, F. J., & De Perthuis, C. (2016). *Pricing carbon: The European Union emissions trading scheme*. Cambridge University Press.

Field, C. B., Barros, V. R., & Dokken, D. J.. (2012). *Managing the risks of extreme events and disasters to advance climate change adaptation: A special report of the Intergovernmental Panel on Climate Change (IPCC)*. Cambridge University Press.

Foley, J. A.. (2011). Solutions for a cultivated planet. *Nature, 478*(7369), 337–342.

Freeman, R. E., Harrison, J. S., & Zyglidopoulos, S. (2010). *Stakeholder theory: Concepts and strategies*. Cambridge University Press.

Friede, G., Busch, T., & Bassen, A. (2015). ESG and financial performance: Aggregated evidence from more than 2000 empirical studies. *Journal of Sustainable Finance & Investment*, 5(4), 210–233.

Geissdoerfer, M., Savaget, P., Bocken, N. M., & Hultink, E. J. (2017). The circular economy–A new sustainability paradigm? *Journal of Cleaner Production, 143*, 757–768.

Geyer, R., Jambeck, J. R., & Law, K. L. (2017). Production, use, and fate of all plastics ever made. *Science Advances*, *3*(7), e1700782.

Ghisellini, P., Cialani, C., & Ulgiati, S. (2016). A review on circular economy: The expected transition to a balanced interplay of environmental and economic systems. *Journal of Cleaner Production*, *114*, 11–32.

Gleick, P. H. (2014). The world's water volume 8: The biennial report on freshwater resources. Island Press.

Gold Standard. (2021). *Carbon Market Report: Ensuring Integrity in Carbon Offsetting*. Retrieved from www.goldstandard.org

Gold Standard. (2021). *Investing in climate security and sustainable development.* Gold Standard Foundation.

Grewal, J., & Serafeim, G. (2020). Research on corporate sustainability: Review and directions for future research. *Foundations and Trends in Accounting*, *14*(3-4), 73–127.

Hansen, J.. (2012). Perception of climate change. *Proceedings of the National Academy of Sciences.*

Hart, S. L. (1995). A natural-resource-based view of the firm. *Academy of Management Review*, *20*(4), 986–1014.

Hepburn, C., Adlen, E., Beddington, J., Carter, E. A., Fuss, S., Mac Dowell, N., & Wilcox, J. (2019). The technological and economic prospects for CO_2 utilization and removal. *Nature*, *575*(7781), 87–97.

Hoegh-Guldberg, O.. (2007). Coral reefs under rapid climate change and ocean acidification. *Science.*

Hoornweg, D., & Bhada-Tata, P. (2012). *What a waste: A global review of solid waste management.* World Bank.

Houghton, R. A. (2005). Tropical deforestation as a source of greenhouse gas emissions. Tropical deforestation and climate change, 13-21.

IEA. (2020). *Energy Technology Perspectives 2020.* International Energy Agency.

IEA. (2021). *Net zero by 2050: A roadmap for the global energy sector.* International Energy Agency.

Ioannou, I., & Serafeim, G. (2017). The consequences of mandatory corporate sustainability reporting. *Harvard Business School Working Paper, 11-100.*

IPCC. (2014, 2019, 2021). Climate Change Reports. Intergovernmental Panel on Climate Change.

IPCC. (2021). [The Physical Science Basis.]. *Climatic Change*, •••, 2021.

IPCC. (2021). *Climate Change 2021: The Physical Science Basis. Contribution of Working Group I to the Sixth Assessment Report of the Intergovernmental Panel on Climate Change*. Cambridge University Press.

Kotter, J. P., & Schlesinger, L. A. (2008). Choosing strategies for change. *Harvard Business Review*, *86*(7/8), 130–139.

Kundzewicz, Z. W.. (2008). The implications of projected climate change for freshwater resources and their management. *Hydrological Sciences Journal*.

Lash, J., & Wellington, F. (2007). Competitive advantage on a warming planet. *Harvard Business Review*, *85*(3), 94–102.

Lobell, D. B.. (2011). Climate trends and global crop production. *Science*.

Matten, D., & Moon, J. (2008). "Implicit" and "explicit" CSR: A conceptual framework for a comparative understanding of corporate social responsibility. *Academy of Management Review*, *33*(2), 404–424.

McKinnon, A., Browne, M., Whiteing, A., & Piecyk, M. (2010). *Green logistics: Improving the environmental sustainability of logistics*. Kogan Page Publishers.

Montabon, F., Pagell, M., & Wu, Z. (2016). Making sustainability sustainable. *The Journal of Supply Chain Management*, *52*(2), 11–27.

NASA. (2020). *Scientific consensus: Earth's climate is warming*. National Aeronautics and Space Administration.

Nordhaus, W. (2018). *The climate casino: Risk, uncertainty, and economics for a warming world*. Yale University Press.

OECD. (2019). *Taxing energy use 2019: Using taxes for climate action*. Organisation for Economic Co-operation and Development.

Ottman, J. A. (2011). *The new rules of green marketing: Strategies, tools, and inspiration for sustainable branding*. Berrett-Koehler Publishers.

Pacala, S., & Socolow, R. (2004). Stabilization wedges: Solving the climate problem for the next 50 years with current technologies. *Science*, *305*(5686), 968–972.

Pagell, M., & Shevchenko, A. (2014). Why research in sustainable supply chain management should have no future. *The Journal of Supply Chain Management, 50*(1), 44–55.

Porter, M. E., & Kramer, M. R. (2011). Creating shared value. *Harvard Business Review, 89*(1/2), 62–77.

Porter, M. E., & Van der Linde, C. (1995). Toward a new conception of the environment-competitiveness relationship. *The Journal of Economic Perspectives, 9*(4), 97–118.

REN21. (2021). *Renewables 2021: Global Status Report*. Renewable Energy Policy Network for the 21st Century.

Rockström, J., Steffen, W., & Noone, K.. (2009). A safe operating space for humanity. *Nature, 461*(7263), 472–475.

Rogers, D. S., Lambert, D. M., Croxton, K. L., & García-Dastugue, S. J. (2012). Supply chain management: Processes, partnerships, performance. Supply Chain Management Institute.

Saberi, S., Kouhizadeh, M., Sarkis, J., & Shen, L. (2019). Blockchain technology and its relationships to sustainable supply chain management. *International Journal of Production Research, 57*(7), 2117–2135.

Schaltegger, S., & Wagner, M. (2011). Sustainable entrepreneurship and sustainability innovation: Categories and interactions. *Business Strategy and the Environment, 20*(4), 222–237.

Schlesinger, W. H., & Bernhardt, E. S. (2020). *Biogeochemistry: An Analysis of Global Change*. Academic Press.

Seuring, S., & Müller, M. (2008). Core issues in sustainable supply chain management – A Delphi study. *Business Strategy and the Environment, 17*(8), 455–466.

Shindell, D. T., Faluvegi, G., & Koch, D. M.. (2009). Improved attribution of climate forcing to emissions. *Science, 326*(5953), 716–718.

Smil, V. (2017). *Energy and civilization: A history*. MIT Press.

Soundararajan, V., Brown, J. A., & Wicks, A. C. (2018). Can multi-stakeholder initiatives improve global supply chains? *Business Ethics Quarterly, 29*(3), 385–412.

Steffen, W., Richardson, K., & Rockström, J.. (2015). Planetary boundaries: Guiding human development on a changing planet. *Science, 347*(6223), 1259855.

Steffen, W., Richardson, K., & Rockström, J.. (2015). Planetary boundaries: Guiding human development on a changing planet. *Science, 347*(6223), 1259855.

Stern, N. (2007). *The economics of climate change: The Stern Review*. Cambridge University Press.

Sullivan, R., & Gouldson, A. (2017). The governance of corporate responses to climate change: An international comparison. *Business Strategy and the Environment, 26*(4), 413–425.

Tachizawa, E. M., & Wong, C. Y. (2015). The performance of green supply chain management governance mechanisms: A supply network and complexity perspective. *The Journal of Supply Chain Management, 51*(3), 18–32.

UNEP. (2019). *Global Waste Management Outlook 2019*. United Nations Environment Programme.

UNEP. (2020). *Emissions Gap Report 2020*. United Nations Environment Programme.

UNEP. (2020). *Sustainability and reporting trends in 2020: Preparing for the future*. United Nations Environment Programme.

UNESCO. (2020). *Education for sustainable development: A roadmap*. United Nations Educational, Scientific and Cultural Organization.

UNFCCC. (2015). *The Paris Agreement*. United Nations Framework Convention on Climate Change.

White, K., Hardisty, D. J., & Habib, R. (2019). The elusive green consumer. *Harvard Business Review, 97*(4), 124–133.

Compilation of References

Abbas, J., & Khan, S. M. (2023). Green knowledge management and organizational green culture: An interaction for organizational green innovation and green performance. *Journal of Knowledge Management*, *27*(7), 1852–1870. DOI: 10.1108/JKM-09-2022-0641

Abbas, Q., HongXing, Y., Ramzan, M., & Fatima, S. (2024). BRICS and the climate challenge: Navigating the role of factor productivity and institutional quality in CO2 emissions. *Environmental Science and Pollution Research International*, *31*(3), 4348–4364. DOI: 10.1007/s11356-023-31321-x PMID: 38102427

Aboramadan, M. (2022). The effect of green HRM on employee green behaviors in higher education: The mediating mechanism of green work engagement. *The International Journal of Organizational Analysis*, *30*(1), 7–23.

Acquah, I. S. K., Agyabeng-Mensah, Y., & Afum, E. (2021). Examining the link among green human resource management practices, green supply chain management practices and performance. *Benchmarking*, *28*(1), 267–290. DOI: 10.1108/BIJ-05-2020-0205

Adams, C. A., Alhamood, A. M., & He, X. (2022). The development and implementation of GRI standards: practice and policy issues. In *Handbook of accounting and sustainability* (pp. 26–43). Edward Elgar Publishing., DOI: 10.4337/9781800373518.00009

Agu, E. E., Iyelolu, T. V., Idemudia, C., & Ijomah, T. I. (2024). Exploring the relationship between sustainable business practices and increased brand loyalty. *International Journal of Management & Entrepreneurship Research*, *6*(8), 2463–2475.

Ahmad, J., Al Mamun, A., Masukujjaman, M., Makhbul, Z. K. M., & Ali, K. A. M. (2023). Modeling the workplace pro-environmental behavior through green human resource management and organizational culture: Evidence from an emerging economy. *Heliyon*, *9*(9), e19134. DOI: 10.1016/j.heliyon.2023.e19134

Ahmad, S. (2015). Green human resource management: Policies and practices. *Cogent Business & Management*, *2*(1), 1030817.

Alabdali, M. A., Yaqub, M., Agarwal, R., Alofaysan, H., & Mohapatra, A. K. (2024). Unveiling green digital transformational leadership: Nexus between green digital culture, green digital mindset and green digital transformation. *Journal of Cleaner Production*, *450*, 141670. DOI: 10.1016/j.jclepro.2024.141670

Alemu, B. A. (2025). Cultivating a culture of sustainability: The role of organizational values and leadership in driving sustainable practices. *Business Ethics and Leadership*, *9*(1), 1–15. DOI: 10.61093/bel.9(1).79-94.2025

Alghamdi, N., den Heijer, A., & de Jonge, H. (2017). Assessment tools' indicators for sustainability in universities: An analytical overview. *International Journal of Sustainability in Higher Education*, *18*(1), 84–115. DOI: 10.1108/IJSHE-04-2015-0071

Alharbi, S. S., Tabash, M. I., Farooq, U., & Issa, S. S. (2025). How does climate policy uncertainty determine green innovation adoption? New perspectives from the BRICS. *Journal of Economic Asymmetries*. Advance online publication. DOI: 10.1016/j.jeca.2025.e00411

Alibašić, H. (2018). The nexus of sustainability and climate resilience planning: Embedding climate resilience policies in local governments. *The International Journal of Climate Change: Impacts and Responses*, *10*(2), 21–33.

Ali, M., Shujahat, M., Fatima, N., Lopes de Sousa Jabbour, A. B., Vo-Thanh, T., Salam, M. A., & Latan, H. (2024). Green HRM practices and corporate sustainability performance. *Management Decision*, *62*(11), 3681–3703. DOI: 10.1108/MD-05-2023-0787

Ali, Z., Qin, Z., Raza-Ullah, T., Ali, A., & Sherwani, M. (2025). Leading the way to greening: Aligning leadership styles and market learning for radical and incremental green innovation. *Business Strategy and the Environment*. Advance online publication. DOI: 10.1002/bse.70039

Al-Romeedy, B. S., & Singh, A. (2026). Emerging Green Entrepreneurship: A Path to Green Business Models. In *Incentives and Benefits for Adopting Green Entrepreneurship Practices* (pp. 79-104). IGI Global Scientific Publishing.

Al-Romeedy, B. S., Hussein, H., & Singh, A. (2025). Peaks and Valleys: Navigating Over-Tourism in Mountain Destinations. In *Balancing Mountain Tourism, Cultural Heritage, and Environmental Stability* (pp. 431-444). IGI Global Scientific Publishing.

Al-Swidi, A., Gelaidan, H., & Saleh, R. (2021). The joint impact of green human resource management, leadership and organizational culture on employees' green behaviour and organisational environmental performance. *Journal of Cleaner Production, 316*, 128112. DOI: 10.1016/j.jclepro.2021.128112

Althnayan, S., Alarifi, A., Bajaba, S., & Alsabban, A. S. (2022). Linking environmental transformational leadership, environmental organizational citizenship behavior, and organizational sustainability performance: A moderated mediation model. *Sustainability, 14*(14), 8779. DOI: 10.3390/su14148779

Amjad, F., Abbas, W., Zia-Ur-Rehman, M., Baig, S. A., Hashim, M., Khan, A., & Rehman, H. U. (2021). Effect of green human resource management practices on organizational sustainability: The mediating role of environmental and employee performance. *Environmental Science and Pollution Research International, 28*, 28191–28206. DOI: 10.1007/s11356-020-11307-9

Amoako, D. K., Zakuan, M. N., Okyere-Kwakye, E., & Tetteh, F. K. (2023). Effect of Training and Reward on Social Sustainability in Ghana's Cocoa Supply Chain: The Role of Green Buyer-Supplier Relationship. *Journal of International Food & Agribusiness Marketing, 35*(2), 212–243. DOI: 10.1080/08974438.2021.1981511

Antoni, C. H., Baeten, X., Perkins, S. J., Shaw, J. D., & Vartiainen, M. (2017). Reward management. *Journal of Personnel Psychology*. Advance online publication. DOI: 10.1027/1866-5888/a000187

Ashipala, D. O., & Singh, A. (2025). Successful Local Experience in Waste Management of Urban Destination. In *Solid Waste Management for Rural Regions* (pp. 257–274). IGI Global Scientific Publishing.

Asif, M., Yang, L., & Hashim, M. (2024). The role of digital transformation, corporate culture, and leadership in enhancing corporate sustainable performance in the manufacturing sector of China. *Sustainability, 16*(7), 2651. DOI: 10.3390/su16072651

Baah, C., Opoku-Agyeman, D., Acquah, I. S. K., Agyabeng-Mensah, Y., Afum, E., Faibil, D., & Abdoulaye, F. A. M. (2021). Examining the correlations between stakeholder pressures, green production practices, firm reputation, environmental and financial performance: Evidence from manufacturing SMEs. *Sustainable Production and Consumption, 27*, 100–114.

Babiak, K., & Trendafilova, S. (2011). CSR and environmental responsibility: Motives and pressures to adopt green management practices. *Corporate Social Responsibility and Environmental Management, 18*(1), 11–24.

Baccini, A.. (2012). Estimated carbon dioxide emissions from tropical deforestation improved by carbon-density maps. *Nature Climate Change*.

Bahadur, A. V., Ibrahim, M., & Tanner, T. (2013). Characterising resilience: Unpacking the concept for tackling climate change and development. *Climate and Development*, *5*(1), 55–65.

Baker, L., Jacobson, H., McGrosky, A., Hinz, E., & Associates. (2025). *Ambient temperature and wet bulb globe temperature outperform heat index in predicting hydration status and heat perception in a semi-arid environment.* Annals of Human Biology. Retrieved from https://www.tandfonline.com/doi/full/10.1080/03014460.2025.2456152

Bangar, V., Goyal, R., & Pandey, R. (2020). Climate change responses and sustainable development: integration of mitigation and adaptation. In *Sustainable Development Goals: An Indian Perspective* (pp. 203–214). Springer International Publishing.

Bansal, P. (2005). Evolving sustainably: A longitudinal study of corporate sustainable development. *Strategic Management Journal*, *26*(3), 197–218.

Bansal, P., & DesJardine, M. R. (2014). Business sustainability: It is about time. *Strategic Organization*, *12*(1), 70–78. DOI: 10.1177/1476127013520265

Bansal, P., & Roth, K. (2000). Why companies go green: A model of ecological responsiveness. *Academy of Management Journal*, *43*(4), 717–736.

Barakat, B., Milhem, M., Naji, G. M. A., Alzoraiki, M., Muda, H. B., Ateeq, A., & Abro, Z. (2023). Assessing the impact of green training on sustainable business advantage: Exploring the mediating role of green supply chain practices. *Sustainability*, *15*(19), 14144.

Barnett, T. P.. (2005). Potential impacts of a warming climate on water availability in snow-dominated regions. *Nature*.

Barney, J. (1991). Firm resources and sustained competitive advantage. *Journal of Management*, *17*(1), 99–120.

Bateman, K. M., McCausland, J., & Walsh, N. (2025). *Exploring human impact on the Hudson River over time and space.* Justice-Oriented Science Education. Retrieved from https://books.google.com/books?id=68xIEQAAQBAJ

Baumgärtner, S., & Quaas, M. (2010). What is sustainability economics? *Ecological Economics*, *69*(3), 445–450. DOI: 10.1016/j.ecolecon.2009.11.019

Becker, G. S. (1964). Human capital. The University of Chicago Press Chicago and London Cone Communications. (2017). 2017 Cone Communications CSR Study. http://www.conecomm.com/research-blog/2017-csrstudy

Beg, N., Morlot, J. C., Davidson, O., Afrane-Okesse, Y., Tyani, L., Denton, F., & Rahman, A. A. (2002). Linkages between climate change and sustainable development. *Climate Policy*, *2*(2-3), 129–144.

Begum, S., Ashfaq, M., Xia, E., & Awan, U. (2022). Does green transformational leadership lead to green innovation? The role of green thinking and creative process engagement. *Business Strategy and the Environment*, *31*(1), 580–597. DOI: 10.1002/bse.2911

Bellard, C.. (2012). Impacts of climate change on the future of biodiversity. *Ecology Letters*.

Bhan, M., Sharma, D., Ashwin, A. S., & Mehra, S. (2017). Policy forum: Nationally-determined climate commitments of the BRICS: At the forefront of forestry-based climate change mitigation. *Forest Policy and Economics*, *85*, 172–175. DOI: 10.1016/j.forpol.2017.09.013

Bhardwaj, M., Prakash, N., & Malhotra, R. K. (2024). Energy Sector and Carbon Trading in the Future: A Reference to India. *2024 International Conference on Smart Devices, ICSD 2024*. https://doi.org/DOI: 10.1109/ICSD60021.2024.10751388

Biswas, S. R., Uddin, M. A., Bhattacharjee, S., Dey, M., & Rana, T. (2022). Ecocentric leadership and voluntary environmental behavior for promoting sustainability strategy: The role of psychological green climate. *Business Strategy and the Environment*, *31*(4), 1705–1718. DOI: 10.1002/bse.2978

BloombergNEF. (2022). *Corporate Clean Energy Buying Surges to New Record in 2021*.

Bocken, N. M. P., Short, S. W., Rana, P., & Evans, S. (2014). A literature and practice review to develop sustainable business model archetypes. *Journal of Cleaner Production*, *65*, 42–56.

Boeske, J. (2023). Leadership towards sustainability: A review of sustainable, sustainability, and environmental leadership. *Sustainability*, *15*(16), 12626. DOI: 10.3390/su151612626

Boeske, J., & Murray, P. A. (2022). The intellectual domains of sustainability leadership in SMEs. *Sustainability*, *14*(4), 1978. DOI: 10.3390/su14041978

Boiral, O., & Paillé, P. (2012). Organizational citizenship behaviour for the environment: Measurement and validation. *Journal of Business Ethics, 109*(4), 431–445.

Bolis, I., Morioka, S. N., Leite, W. K. D. S., & Zambroni-de-Souza, P. C. (2021). Sustainability is all about values: The challenges of considering moral and benefit values in business model decisions. *Sustainability, 13*(2), 664. DOI: 10.3390/su13020664

Bonan, G. B. (2008). Forests and climate change: Forcings, feedbacks, and the climate benefits of forests. *Science*.

Bowen, F., & Aragon-Correa, J. A. (2014). Greenwashing in corporate environmentalism research and practice: The importance of what we say and do. *Organization & Environment, 27*(2), 107–112.

Bressane, A., dos Santos Galvão, A. L., Loureiro, A. I. S., Ferreira, M. E. G., Monstans, M. C., & de Castro Medeiros, L. C. (2024). Valuing urban green spaces for enhanced public health and sustainability: A study on public willingness-to-pay in an emerging economy. *Urban Forestry & Urban Greening, 98*, 128386.

Bullard, R. D. (2018). *Dumping in Dixie: Race, Class, and Environmental Quality*. Westview Press.

Burawat, P. (2019). The relationships among transformational leadership, sustainable leadership, lean manufacturing and sustainability performance in Thai SMEs manufacturing industry. *International Journal of Quality & Reliability Management, 36*(6), 1014–1036.

Burns, J. P., & Zhiren, Z. (2010). Performance management in the government of the People's Republic of China: Accountability and control in the implementation of public policy. *OECD Journal on Budgeting, 10*(2), 7.

Cameron, J. (2001). Negative effects of reward on intrinsic motivation—A limited phenomenon: Comment on Deci, Koestner, and Ryan (2001). *Review of Educational Research, 71*(1), 29–42. DOI: 10.3102/00346543071001029

Camilleri, M. A. (2025). Sustainability accounting and disclosures of responsible restaurant practices in environmental, social and governance (ESG) reports. *International Journal of Hospitality Management, 126*, 104051.

Carbon Majors Report, C. D. P. (2017). *The 100 Companies Responsible for 70% of Global Carbon Emissions*. https://www.cdp.net/en

Carney, M. (2015). *Breaking the tragedy of the horizon–climate change and financial stability. Speech at Lloyd's of London, 29 September 2015*. Bank of England.

Carroll, A. B., & Shabana, K. M. (2010). The business case for corporate social responsibility: A review of concepts, research and practice. *International Journal of Management Reviews, 12*(1), 85–105. DOI: 10.1111/j.1468-2370.2009.00275.x

CDP. (2020). *The Carbon Disclosure Project: Global Climate Change Report 2020.* Retrieved from www.cdp.net

CDP. (2021). *Engaging the chain: Driving speed and scale*. CDP Worldwide.

Center for International Environmental Law. (2019). *Plastic & climate: The hidden costs of a plastic planet.* Retrieved from https://www.ciel.org/plasticandclimate

Center for Sustainable Systems. (2021). Carbon Footprint Factsheet. University of Michigan. https://css.umich.edu/publications/factsheets/sustainability-indicators/carbon-footprint-factsheet

Chakroun, S., Salhi, B., Ben Amar, A., & Jarboui, A. (2020). The impact of ISO 26000 social responsibility standard adoption on firm financial performance: Evidence from France. *Management Research Review, 43*(5), 545–571. DOI: 10.1108/MRR-02-2019-0054

Chen, W., Men, Y., Fuster, N., Osorio, C., & Juan, A. A. (2024). Artificial intelligence in logistics optimization with sustainable criteria: A review. *Sustainability, 16*(21), 9145.

Chen, W., Zhang, Q., Wang, C., Li, Z., Geng, Y., Hong, J., & Cheng, Y. (2022). Environmental sustainability challenges of China's steel production: Impact-oriented water, carbon and fossil energy footprints assessment. *Ecological Indicators, 136*, 108660. DOI: 10.1016/j.ecolind.2022.108660

Chiang, F. (2005). A critical examination of Hofstede's thesis and its application to international reward management. *The International Journal of Human Resource Management, 16*(9), 1545-1563. https://doi.org/DOI: 10.1080/09585190500239044

Chmutova, I., Ho, S., Zhang, Ts., Mazhuta, V., & Nimets, A. (2024). Development of Key Performance Indicators of an Integrated Enterprise Management System Using a Balanced System of Sustainable Development Indicators. *Development Service Industry Management*, (2), 270–278. DOI: 10.31891/dsim-2024-6(43)

Chopra, M., Saini, N., Kumar, S., Varma, A., Mangla, S. K., & Lim, W. M. (2021). Past, present, and future of knowledge management for business sustainability. *Journal of Cleaner Production, 328*, 129592. DOI: 10.1016/j.jclepro.2021.129592

Church, J. A., & White, N. J. (2011). Sea-level rise from the late 19th to the early 21st century. *Surveys in Geophysics*.

Clark, G. L., Feiner, A., & Viehs, M. (2015). *From the stockholder to the stakeholder: How sustainability can drive financial outperformance*. University of Oxford & Arabesque Partners.

Coldwell, D. A. (2017). Ethical leadership in crisis management: The role of university education. In *Contemporary Leadership Challenges*. IntechOpen., DOI: 10.5772/65439

Congressional Budget Office. (2024, February 15). The budget and economic outlook: 2023 to 2033. Retrieved March 5, 2025, from https://www.cbo.gov/publication/60030

Cook, J., Oreskes, N., Doran, P. T., Anderegg, W. R. L., Verheggen, B., Maibach, E. W., & Rice, K. (2013). Quantifying the consensus on anthropogenic global warming in the scientific literature. *Environmental Research Letters*, *8*(2), 024024. DOI: 10.1088/1748-9326/8/2/024024

Costa, M. D., & Opare, S. (2025). Impact of corporate culture on environmental performance. *Journal of Business Ethics*, *196*(1), 61–92. DOI: 10.1007/s10551-024-05674-3

Craig, , C., & Allen, M. (2013). Sustainability information sources: Employee knowledge, perceptions, and learning. *Journal of Communication Management (London)*, *17*(4), 292–307. DOI: 10.1108/JCOM-05-2012-0035

Crane, A., Matten, D., Glozer, S., & Spence, L. J. (2014). *Business ethics: Managing corporate citizenship and sustainability in the age of globalization*. Oxford University Press., DOI: 10.1093/hebz/9780198810070.001.0001

Danciu, V. (2013). The sustainable company: New challenges and strategies for more sustainability. *Theoretical and Applied Economics*, *20*(9), 7–26.

Das, R., Singh, A., & Roy, K. (2025). Homestay Initiative: Building a Resilient Pathway for Garden Communities in Darjeeling's Tea Tourism. In *Global Practices and Innovations in Sustainable Homestay Tourism* (pp. 229-258). IGI Global Scientific Publishing.

Das, S. C., & Singh, R. K. (2016). Green HRM and organizational sustainability: An empirical review. *Kegees Journal of Social Science*, *8*(1), 227–236.

Das, S., & Dash, M. (2024). Green compensation and reward system: A novel approach towards the growth and sustainability of organization. *Journal of Advanced Zoology*, *45*(2), 1213–1219.

Datta, P. (2012). An applied organizational rewards distribution system. *Management Decision*, *50*(3), 479–501. DOI: 10.1108/00251741211216241

Davidson, E. A. (2009). The contribution of manure and fertilizer nitrogen to atmospheric nitrous oxide since 1860. Nature Geoscience.

D-CARBONIZE. (2025, February 22). *Comparison of GHG emissions by sector*. D-CARBONIZE. Retrieved March 5, 2025, from https://d-carbonize.eu/greenhouse-gases/comparison-ghg-emissions-sector/#:~:text=The%20energy%20sector%20is%20responsible,significantly%20contributing%20to%20global%20warming

De Mendonca, T. R., & Zhou, Y. (2019). Environmental performance, customer satisfaction, and profitability: A study among large US companies. *Sustainability, 11*(19), 5418. DOI: 10.3390/su11195418

de Oliveira, U. R., Menezes, R. P., & Fernandes, V. A. (2024). A systematic literature review on corporate sustainability: Contributions, barriers, innovations and future possibilities. *Environment, Development and Sustainability, 26*(2), 3045–3079. DOI: 10.1007/s10668-023-02933-7

De Palma, A., Lindsey, R., Proost, S., Riou, Y., & Trannoy, A. (2025). *Why combating climate change is so challenging*. Université de Cergy-Pontoise. Retrieved from https://thema.u-cergy.fr/IMG/pdf/2025-02.pdf

Dechezleprêtre, A., & Sato, M. (2017). The impacts of environmental regulations on competitiveness. *Review of Environmental Economics and Policy*.

Deci, E. L., Nezlek, J., & Sheinman, L. (1981). Characteristics of the rewarder and intrinsic motivation of the rewardee. *Journal of Personality and Social Psychology, 40*(1), 1. DOI: 10.1037/0022-3514.40.1.1

Delmas, M. A., & Burbano, V. C. (2011). The drivers of greenwashing. *California Management Review, 54*(1), 64–87. DOI: 10.1525/cmr.2011.54.1.64

Delmas, M. A., & Montes-Sancho, M. J. (2010). Voluntary agreements to improve environmental quality: Symbolic and substantive cooperation. *Strategic Management Journal, 31*(6), 575–601.

DesJardins, J. (2005). Business and environmental sustainability. *Business & Professional Ethics Journal, 24*(1/2), 35–59. https://www.jstor.org/stable/27801371

Dhamija, P. (2023). Emerging ethical leadership in crisis management. In *Emotional intelligence for leadership effectiveness* (pp. 221–235). Apple Academic Press., DOI: 10.1201/9781003367635-14

DiMaggio, P. J., & Powell, W. W. (1983). The iron cage revisited: Institutional isomorphism and collective rationality in organizational fields. *American Sociological Review, 48*(2), 147–160.

Ding, G. K. (2005). Developing a multicriteria approach for the measurement of sustainable performance. *Building Research and Information, 33*(1), 3–16. DOI: 10.1080/0961321042000322618

Dissanayake, D. (2021). Sustainability key performance indicators and the global reporting initiative: Usage and challenges in a developing country context. *Meditari Accountancy Research, 29*(3), 543–567. DOI: 10.1108/MEDAR-08-2019-0543

Dolganova, I., Bach, V., Rödl, A., Kaltschmitt, M., & Finkbeiner, M. (2022). Assessment of critical resource use in aircraft manufacturing. *Circular Economy and Sustainability, 2*(3), 1193–1212.

Doni, F., & Gasperini, A. (2014). The internal integrated reporting and the value creation. A case study approach. In *Knowledge and Management Models for Sustainable Growth. Proceedings. 11-13 June Matera Italy* (Vol. 9, pp. 2658-2696). Institute of Knowledge Asset Management.

Doppelt, B. (2017). *Leading change toward sustainability: A change-management guide for business, government and civil society*. Routledge.

Dubey, R., Gunasekaran, A., & Chakrabarty, A. (2015). World-class sustainable manufacturing: Framework and a performance measurement system. *International Journal of Production Research, 53*(17), 5207–5223. DOI: 10.1080/00207543.2015.1012603

Dyllick, T., & Hockerts, K. (2002). Beyond the business case for corporate sustainability. *Business Strategy and the Environment, 11*(2), 130–141. DOI: 10.1002/bse.323

Dyllick, T., & Muff, K. (2016). Clarifying the meaning of sustainable business: Introducing a typology from business-as-usual to true business sustainability. *Organization & Environment, 29*(2), 156–174. DOI: 10.1177/1086026615575176

Easterling, D. R.. (2000). Climate extremes: Observations, modeling, and impacts. *Science*.

Eccles, R. G., Ioannou, I., & Serafeim, G. (2014). The impact of corporate sustainability on organizational processes and performance. *Management Science, 60*(11), 2835–2857. DOI: 10.1287/mnsc.2014.1984

Eccles, R. G., & Klimenko, S. (2019). The investor revolution: Shareholders are getting serious about sustainability. *Harvard Business Review, 97*(3), 106–116.

Ekechukwu, D. E., & Simpa, P. (2024). A comprehensive review of renewable energy integration for climate resilience. *Engineering Science & Technology Journal, 5*(6), 1884–1908.

Ekins, P., & Zenghelis, D. (2021). The costs and benefits of environmental sustainability. *Sustainability Science*, *16*, 949–965. DOI: 10.1007/s11625-021-00910-5

Elkington, J. (1997). The triple bottom line. *Environmental management: Readings and cases, 2*, 49-66.

Elkington, J. (1999). *Cannibals with forks: The triple bottom line of 21st century business*. Capstone.

Elkington, J. (2006). Governance for sustainability. *Corporate Governance*, *14*(6), 522–529. DOI: 10.1111/j.1467-8683.2006.00527.x

Ellen MacArthur Foundation. (2013). *Towards the Circular Economy: Economic and Business Rationale for an Accelerated Transition*. Ellen MacArthur Foundation.

Ellen MacArthur Foundation. (2020). *Circular economy in detail*. https://www.ellenmacarthurfoundation.org/circular-economy/concept

Ellerman, A. D., Convery, F. J., & De Perthuis, C. (2016). *Pricing carbon: The European Union emissions trading scheme*. Cambridge University Press.

ElSayary, A. (2024). Integrating generative AI in active learning environments: Enhancing metacognition and technological skills. *Journal of Systemics, Cybernetics and Informatics*, *22*(3), 34–37. DOI: 10.54808/JSCI.22.03.34

Enechi, O., & Pattberg, P. (2025). What determines participation in multi-stakeholder partnerships? Assessing sustainability awareness and knowledge. *Discover Sustainability*, *6*(1), 104.

Epstein, M. J., & Rejc Buhovac, A. (2014). A New Day for Sustainability. *Strategic finance, 96*(1).

European Commission. (2023). *Corporate Sustainability Due Diligence Directive*. Retrieved from https://ec.europa.eu/commission/presscorner/detail/en/ip_23_1104

European Commission. (2023). Hydrogen-powered aviation: Future developments. Retrieved from https://ec.europa.eu

European Commission. (2023). The European Green Deal: A Just Transition. Retrieved from https://ec.europa.eu/info/strategy/priorities-2019-2024/european-green-deal_en

European Commission. (2023, June). *Progress report* (v.1.0). European Commission. Retrieved March 5, 2025, from https://defence-industry-space.ec.europa.eu/system/files/2023-06/Progress%20report%20v.1.0.pdf

European Commission. (n.d.). *Methane emissions*. European Commission. Retrieved March 5, 2025, from https://energy.ec.europa.eu/topics/carbon-management-and-fossil-fuels/methane-emissions_en

Ewing, J. (2017). *Faster, higher, farther: The inside story of the Volkswagen scandal*. Random House.

Feng, T., Qamruzzaman, M., Sharmin, S. S., & Karim, S. (2024). Bridging environmental sustainability and organizational performance: The role of green supply chain management in the manufacturing industry. *Sustainability*, *16*(14), 5918.

Ferguson, P., & Wollersheim, L. (2023). From sustainable development to resilience?(Dis) continuities in climate and development policy governance discourse. *Sustainable Development*, *31*(1), 67–77.

Fernández, E., Junquera, B., & Ordiz, M. (2003). Organizational culture and human resources in the environmental issue: A review of the literature. *International Journal of Human Resource Management*, *14*(4), 634–656. DOI: 10.1080/0958519032000057628

Field, C. B., Barros, V. R., & Dokken, D. J.. (2012). *Managing the risks of extreme events and disasters to advance climate change adaptation: A special report of the Intergovernmental Panel on Climate Change (IPCC)*. Cambridge University Press.

Flammer, C. (2015). Does corporate social responsibility lead to superior financial performance? A regression discontinuity approach. *Management Science*, *61*(11), 2549–2568. DOI: 10.1287/mnsc.2014.2038

Foley, J. A.. (2011). Solutions for a cultivated planet. *Nature*, *478*(7369), 337–342.

Food and Agriculture Organization (FAO). (2021). *State of the World's Forests 2021*. Rome: FAO. Retrieved from https://www.fao.org/state-of-forests

Freeman, R. E. (1999). Divergent stakeholder theory. *Academy of Management Review*, *24*(2), 233–236.

Freeman, R. E., Harrison, J. S., & Zyglidopoulos, S. (2018). *Stakeholder theory: Concepts and strategies*. Cambridge University Press.

Friede, G., Busch, T., & Bassen, A. (2015). ESG and financial performance: Aggregated evidence from more than 2000 empirical studies. *Journal of Sustainable Finance & Investment*, *5*(4), 210–233. DOI: 10.1080/20430795.2015.1118917

Gadenne, D., Mia, L., Sands, J., Winata, L., & Hooi, G. (2012). The influence of sustainability performance management practices on organisational sustainability performance. *Journal of Accounting & Organizational Change*, *8*(2), 210–235. DOI: 10.1108/18325911211230380

Gagné, M., Deci, E. L., & Ryan, R. M. (2018). Self-determination theory applied to work motivation and organizational behavior. *The SAGE handbook of industrial, work & organizational psychology: Organizational psychology, 2*, 97-121.

Galderisi, A. (2016). Nexus approach to disaster risk reduction, climate adaptation and ecosystems' management: New paths for a sustainable and resilient urban development. In *Peri-urban areas and food-energy-water nexus: Sustainability and resilience strategies in the age of climate change* (pp. 11–21). Springer International Publishing.

Galpin, T., Whitttington, J. L., & Bell, G. (2015). Is your sustainability strategy sustainable? Creating a culture of sustainability. *Corporate Governance*, *15*(1), 1–17.

Garbie, I. H. (2015). Sustainability awareness in industrial organizations. *Procedia CIRP*, *26*, 64–69. DOI: 10.1016/j.procir.2015.03.003

Gazi, M. A. I., Dhali, S., Masud, A., Ahmed, A., Amin, M. B., Chaity, N., Senathirajah, A. R. B. S., & Abdullah, M. (2024). Leveraging Green HRM to Foster Organizational Agility and Green Culture: Pathways to Enhanced Sustainable Social and Environmental Performance. *Sustainability*, *16*(20), 8751. DOI: 10.3390/su16208751

Gbadegesin, T. K. (2024). Climate change and energy dynamics: Evidence from BRICS and G7 countries. *Energy & Environment*. Advance online publication. DOI: 10.1177/0958305X241302692

Geissdoerfer, M., Savaget, P., Bocken, N. M., & Hultink, E. J. (2017). The circular economy–A new sustainability paradigm? *Journal of Cleaner Production*, *143*, 757–768.

George, R. A., Siti-Nabiha, A. K., Jalaludin, D., & Abdalla, Y. A. (2016). Barriers to and enablers of sustainability integration in the performance management systems of an oil and gas company. *Journal of Cleaner Production*, *136*, 197–212. DOI: 10.1016/j.jclepro.2016.01.097

Geyer, R., Jambeck, J. R., & Law, K. L. (2017). Production, use, and fate of all plastics ever made. *Science Advances*, *3*(7), e1700782.

Ghisellini, P., Cialani, C., & Ulgiati, S. (2016). A review on circular economy: The expected transition to a balanced interplay of environmental and economic systems. *Journal of Cleaner Production*, *114*, 11–32.

Gielen, D., Boshell, F., Saygin, D., Bazilian, M. D., Wagner, N., & Gorini, R. (2019). The role of renewable energy in the global energy transformation. *Energy strategy reviews, 24*, 38-50.

Gielen, D., Boshell, F., Saygin, D., Bazilian, M. D., Wagner, N., & Gorini, R. (2019). The role of renewable energy in the global energy transformation. *Energy Strategy Reviews, 24*, 38–50. DOI: 10.1016/j.esr.2019.01.006

Gleick, P. H. (2014). The world's water volume 8: The biennial report on freshwater resources. Island Press.

Glendinning, P. M. (2002). Performance management: Pariah or messiah. *Public Personnel Management, 31*(2), 161–178. DOI: 10.1177/009102600203100203

Gold Standard. (2021). *Carbon Market Report: Ensuring Integrity in Carbon Offsetting*. Retrieved from www.goldstandard.org

Gold Standard. (2021). *Investing in climate security and sustainable development*. Gold Standard Foundation.

Gönczi, J. (2023). Sustainability performance indicators. *The Annals of the University of Oradea. Economic Sciences, 32*(1), 126. https://anale.steconomiceuoradea.ro/en/wp-content/uploads/2024/02/AUOES.July_.20239.pdf?utm_source=chatgpt.com

Goubran, S., Walker, T., Cucuzzella, C., & Schwartz, T. (2023). Green building standards and the united nations' sustainable development goals. *Journal of Environmental Management, 326*, 116552. PMID: 36372039

Grewal, J., & Serafeim, G. (2020). Research on corporate sustainability: Review and directions for future research. *Foundations and Trends in Accounting, 14*(3-4), 73–127.

Grimm, J. H., Hofstetter, J. S., & Sarkis, J. (2016). Exploring sub-suppliers' compliance with corporate sustainability standards. *Journal of Cleaner Production, 112*, 1971–1984. DOI: 10.1016/j.jclepro.2014.11.036

Grossman, K. R. (2025). *The role of mindfulness practice in warding off despair and sustaining ecological action in the context of existential threat.* ProQuest Dissertations. Retrieved from https://search.proquest.com/openview/a8b48976572cf9b286b67a83cf19e7d4/1?pq-origsite=gscholar

Grynko, T., Krupskyi, O. P., Koshevyi, M., & Maximchuk, O. (2017). Tangible and intangible rewards in service industries: Problems and prospects.

Hameed, H., & Akram, K. (2024). Adoption of Green Human Resource Management Practices in Pakistan: Case of Leather Industry. *Pakistan Social Sciences Review, 8*(4), 314–329.

Hansen, J.. (2012). Perception of climate change. *Proceedings of the National Academy of Sciences.*

Harrison, D. A., & Klein, K. J. (2007). What's the difference? Diversity constructs are separation, variety, or disparity in organizations. *Academy of Management Review, 32*(4), 1199–1228.

Hart, S. L. (1995). A natural-resource-based view of the firm. *Academy of Management Review, 20*(4), 986–1014.

Hart, S. L., & Milstein, M. B. (2003). Creating sustainable value. *The Academy of Management Perspectives, 17*(2), 56–67. DOI: 10.5465/ame.2003.10025194

Hasanein, A., & Metwally, A. (2025). Hospitality leading green for a sustainable scene: The mediating role of green organizational culture in the relationship between green servant leadership and environmental performance. *Geojournal of Tourism and Geosites, 60*(2 spl), 1243-1253. https://doi.org/DOI: 10.30892/gtg.602spl21-1497

Hasan, M. A., Abubakar, I. R., Rahman, S. M., Aina, Y. A., Chowdhury, M. M. I., & Khondaker, A. N. (2020). The synergy between climate change policies and national development goals: Implications for sustainability. *Journal of Cleaner Production, 249*, 119369.

Hatum, A. (2017). The role of talent for organizational creativity. In *Mastering creativity in organizations* (pp. 15–41). Edward Elgar Publishing., DOI: 10.4337/9781786436870.00008

Heede, R. (2013). Tracing anthropogenic carbon dioxide and methane emissions to fossil fuel and cement producers, 1854–2010. *Climatic Change, 122*(1–2), 229–241. DOI: 10.1007/s10584-013-0986-y

Hemmet, A. (2024). Investigating the impact of green R&D, transformational leadership, and cross-cultural approaches on performance management in the manufacturing sector. *ARPHA Conference Abstracts, 7*, e129588. https://doi.org/DOI: 10.3897/aca.7.e129588

Hendry, C., Woodward, S., Bradley, P., & Perkins, S. (2000). Performance and rewards: Cleaning out the stables. *Human Resource Management Journal, 10*(3), 46–62.

Hepburn, C., Adlen, E., Beddington, J., Carter, E. A., Fuss, S., Mac Dowell, N., & Wilcox, J. (2019). The technological and economic prospects for CO_2 utilization and removal. *Nature*, *575*(7781), 87–97.

Hertwich, E. G. (2021). Increased carbon footprint of materials production driven by rise in investments. *Nature Geoscience*, *14*, 151–155. DOI: 10.1038/s41561-021-00690-8

Hoegh-Guldberg, O.. (2007). Coral reefs under rapid climate change and ocean acidification. *Science*.

Hoornweg, D., & Bhada-Tata, P. (2012). *What a waste: A global review of solid waste management*. World Bank.

Houghton, R. A. (2005). Tropical deforestation as a source of greenhouse gas emissions. Tropical deforestation and climate change, 13-21.

Huang, L., Guo, Z., Deng, B., & Wang, B. (2023). Unlocking the relationship between environmentally specific transformational leadership and employees' green behaviour: A cultural self-representation perspective. *Journal of Cleaner Production*, *382*, 134857. DOI: 10.1016/j.jclepro.2022.134857

IEA. (2020). *Energy Technology Perspectives 2020*. International Energy Agency.

IEA. (2021). *Net zero by 2050: A roadmap for the global energy sector*. International Energy Agency.

IKEA. (2021). *Sustainability Report*. Retrieved from https://about.ikea.com/en/sustainability

Intergovernmental Panel on Climate Change (IPCC). (2014). *Climate change 2014: Synthesis report*. Geneva, Switzerland: IPCC. https://doi.org/DOI: 10.1017/CBO9781107415324

Intergovernmental Panel on Climate Change (IPCC). (2021). *Climate change 2021: The physical science basis*. Retrieved from https://www.ipcc.ch/report/ar6/wg1

Intergovernmental Panel on Climate Change (IPCC). (2022). *Sixth Assessment Report: Mitigation of Climate Change*. https://www.ipcc.ch/ar6/

Intergovernmental Panel on Climate Change (IPCC). (2023). *AR6 synthesis report: Climate change 2023: Synthesis report*. Intergovernmental Panel on Climate Change. Retrieved March 5, 2025, from https://www.ipcc.ch/report/ar6/syr/downloads/report/IPCC_AR6_SYR_LongerReport.pdf

Intergovernmental Panel on Climate Change (IPCC). (2023). *Climate Change 2023: Synthesis Report.* Geneva: IPCC. DOI: https://doi.org/DOI: 10.1016/j.scitotenv.2023.164942

International Air Transport Association (IATA). (2024). *Fact sheet: Sustainable aviation fuels.* Retrieved March 5, 2025, from https://www.iata.org/en/iata-repository/pressroom/fact-sheets/fact-sheet-sustainable-aviation-fuels/

International Air Transport Association (IATA). (n.d.). *Fact sheet on climate change.* Retrieved March 5, 2025, from https://www.iata.org/contentassets/d13875e9ed784f75bac90f000760e998/fact_sheet_on_climate_change.pdf

International Energy Agency. (2020, October). Iron and steel technology roadmap. Retrieved March 5, 2025, from https://www.iea.org/reports/iron-and-steel-technology-roadmap

International Energy Agency. (n.d.). *Ammonia technology roadmap: Executive summary.* Retrieved March 5, 2025, from https://www.iea.org/reports/ammonia-technology-roadmap/executive-summary

Ioannou, I., & Serafeim, G. (2017). The consequences of mandatory corporate sustainability reporting. *Harvard Business School Working Paper, 11-100.*

IPCC. (2014). [*Synthesis Report.* Intergovernmental Panel on Climate Change.]. *Climatic Change,* •••, 2014.

IPCC. (2014, 2019, 2021). Climate Change Reports. Intergovernmental Panel on Climate Change.

IPCC. (2021). *Climate Change 2021: The Physical Science Basis. Contribution of Working Group I to the Sixth Assessment Report of the Intergovernmental Panel on Climate Change.* Cambridge University Press.

IPCC. (2021). Climate Change 2021: The Physical Science Basis. Retrieved from https://www.ipcc.ch/report/ar6/wg1

IPCC. (2023). Climate Change 2023: Synthesis Report. Intergovernmental Panel on Climate Change. Retrieved from https://www.ipcc.ch/report/ar6/syr/

Isensee, C., Teuteberg, F., Griese, K., & Topi, C. (2020). The relationship between organizational culture, sustainability, and digitalization in SMEs: A systematic review. *Journal of Cleaner Production, 275,* 122944. DOI: 10.1016/j.jclepro.2020.122944

Islam, T., Hussain, D., Ahmed, I., & Sadiq, M. (2021). Ethical leadership and environment specific discretionary behaviour: The mediating role of green human resource management and moderating role of individual green values. *Canadian Journal of Administrative Sciences*, *38*(4), 442–459.

Jabbour, C. J. C., & de Sousa Jabbour, A. B. L. (2016). Green human resource management and green supply chain management: Linking two emerging agendas. *Journal of Cleaner Production*, *112*, 1824–1833.

Jabbour, C. J. C., & Santos, F. C. A. (2008). The central role of human resource management in the search for sustainable organizations. *International Journal of Human Resource Management*, *19*(12), 2133–2154.

Jackson, S. E., Renwick, D. W. S., Jabbour, C. J. C., & Muller-Camen, M. (2011). State-of-the-art and future directions for green human resource management: Introduction to the special issue. *German Journal of Human Resource Management*, *25*(2), 99–116.

Jakobsen, M., James, O., Moynihan, D., & Nabatchi, T. (2019). JPART virtual issue on citizen-state interactions in public administration research. *Journal of Public Administration: Research and Theory*, *29*(4), e8–e15. DOI: 10.1093/jopart/muw031

Jamal, T., Zahid, M., Martins, J. M., Mata, M. N., Rahman, H. U., & Mata, P. N. (2021). Perceived green human resource management practices and corporate sustainability: Multigroup analysis and major industries perspectives. *Sustainability*, *13*(6), 3045. DOI: 10.3390/su13063045

Jamil, N. N., & Wahyuni, E. T. (2024). Do Some Sustainability Reporting Measures Enable ESG Greenwashing: A Case Study of Oil and Gas Industry in Malaysia. *Global Business and Management Research*, *16*(3s), 123–128.

Jamil, S., Zaman, S. I., Kayikci, Y., & Khan, S. A. (2023). The role of green recruitment on organizational sustainability performance: A study within the context of green human resource management. *Sustainability*, *15*(21), 15567.

Javed, S., Yingjun, Z., Kumar, N., & Meghwar, S. (2024). Green human resource management practices and organizational sustainability performance in Pakistan's healthcare sector: A descriptive analysis. *International Journal of Research in Business & Social Science*, *13*(1).

Jeffrey, H. L., Beswick, E., & Meade, J. (2015). Learning and unlearning sexism in the workplace: Two important barriers prevent a more inclusive environment. *Human Resource Management International Digest*, *23*(5), 18–20. DOI: 10.1108/HRMID-03-2015-0035

Jiménez, D. L., Dittmar, E. C., & Portillo, J. P. V. (2021). Cooperation and relationship in the triple helix model of innovation. *International Journal of Knowledge and Learning*, *14*(1), 1–9. DOI: 10.1504/IJKL.2021.113210

Johanson, U., Aboagye, E., & Yao, J. (2022). Integrating business model for sustainability and performance management to promote occupational health and safety—A discussion of value. *Frontiers in Sustainability*, *3*, 950847. DOI: 10.3389/frsus.2022.950847

Jones, P., Clarke-Hill, C., Comfort, D., & Hillier, D. (2008). Marketing and sustainability. *Marketing Intelligence & Planning*, *26*(2), 123–130. DOI: 10.1108/02634500810860584

Jyothi, J. (2016). Non-monetary benefits & its effectiveness in motivating employees. *Clear International Journal of Research in Commerce & Management*, *7*(5).

Katz, D., & Kahn, R. (2015). The social psychology of organizations. In *Organizational behavior 2* (pp. 152–168). Routledge., DOI: 10.4324/9781315702001

Kaur, J., Khanna, R., Kumar, R., & Sunil, G. (2024). Role of Blockchain Technologies in Goods and Services Tax. *Proceedings - 2024 3rd International Conference on Sentiment Analysis and Deep Learning, ICSADL 2024*, 607–612. https://doi.org/DOI: 10.1109/ICSADL61749.2024.00104

Kaur, M., Singh, A., & Kaur, A. (2025). Eco-Friendly Skies: Sustainable Marketing Strategies for the Aviation Industry. In *Sustainable Marketing Practices in the Aviation Industry* (pp. 151-182). IGI Global Scientific Publishing.

Kaur, S., & Singh, A. (2025). India's Peaks Under Pressure: The Impact of Climate Change and Gender Issues on Mountain Tourism. In *Balancing Mountain Tourism, Cultural Heritage, and Environmental Stability* (pp. 309-320). IGI Global Scientific Publishing.

Kaur, A., & Sharma, V. (2024). Nexus between climate change and mitigation approaches for sustainable development: A bibliometric review. *Global Knowledge, Memory and Communication*.

Kaur, M., Singh, A., & Kaur, A. (2025a). Navigating the environmental pitfalls of overtourism and finding sustainable solutions. In *Solutions for Managing Overtourism in Popular Destinations* (pp. 61–84). IGI Global Scientific Publishing.

Kelly, K., Liu, W., & Presslee, A. (2023). The interactive effect of organizational identification and reward type on reward valuation. *Contemporary Accounting Research*, *40*(3), 1733–1759. DOI: 10.1111/1911-3846.12868

Khaddage-Soboh, N., Yunis, M., Imran, M., & Zeb, F. (2024). Sustainable practices in Malaysian manufacturing: The influence of CSR, transformational leadership, and green organizational culture on environmental performance. *Economic Analysis and Policy*, *82*, 753–768. DOI: 10.1016/j.eap.2024.04.001

Khalid, U., Rizvi, S. A. A., Abideen, Z. U., Imran, M., & Raza, H. (2025). Exploring the Obstacles to the Implementation of Green HRM Practices for Sustainable Small and Medium Enterprises in Pakistan. *Journal of Management & Social Science*, *2*(1), 122–137.

Khan, A. J. (2022). Green human resource practices: A sustainable approach to increase employee performance. *iRASD. Journal of Management.*

Khan, K., Gogia, E. H., Shao, Z., Rehman, M. Z., & Ullah, A. (2025). The impact of green HRM practices on green innovative work behaviour: Empirical evidence from the hospitality sector of China and Pakistan. *BMC Psychology*, *13*(1), 96. PMID: 39901289

Khan, K., Khan, Q., Jalaly, S., Shams, M. S., & Shah, T. A. (2024). Green human resource management and green service behaviour in organisations: Examining underlying mechanism. *SA Journal of Human Resource Management*, *11*(1), 113–129.

Khan, M. A. S., Ali, M., Usman, M., Saleem, S., & Jianquo, D. (2019). Interrelationships between ethical leadership, green psychological climate, and organizational environmental citizenship behavior: The moderating role of gender. *Frontiers in Psychology*, *10*, 1977. PMID: 31555173

Khanna, R., Sharma, S., Czeh, T., & Adhana, K. (2025). Challenges, Interventions, and Strategies for Higher Education Transformation: Vision 2047. *Lecture Notes in Networks and Systems, 1075 LNNS*, 637 – 655. https://doi.org/DOI: 10.1007/978-981-97-6106-7_38

Khan, Z. N., Iqbal, I., & Saba, S. (2025). The Impact of Green Human Resources Management to Enhance Pro-Environmental Behaviour among Organizational Employees. *Pakistan Social Sciences Review*, *9*(2), 306–321.

Khorbaladze, E. (2025). BRICS: A comprehensive mechanism bringing changes into the world financial system and international climate diplomacy. In *The Multilateralism of the New Development Bank on the Sustainable Economic Growth in BRICS Nations and Beyond*. Emerald Publishing., DOI: 10.1108/978-1-83662-292-520251006

Kiesnere, A. L., & Baumgartner, R. J. (2019). Sustainability management in practice: Organizational change for sustainability in smaller large-sized companies in Austria. *Sustainability, 11*(3), 572. DOI: 10.3390/su11030572

Kim, H., Woo, E., Uysal, M., & Kwon, N. (2018). The effects of corporate social responsibility (CSR) on employee well-being in the hospitality industry. *International Journal of Contemporary Hospitality Management, 30*(3), 1584–1600.

Kim, Y. J., Kim, W. G., Choi, H. M., & Phetvaroon, K. (2019). The effect of green human resource management on hotel employees' eco-friendly behavior and environmental performance. *International Journal of Hospitality Management, 76*, 83–93. DOI: 10.1016/j.ijhm.2018.04.007

Kodua, L. T., Xiao, Y., Adjei, N., Asante, D., Ofosu, B., & Amankona, D. (2022). Barriers to green human resources management (GHRM) implementation in developing countries: Evidence from Ghana. *Journal of Cleaner Production, 340*, 130671. DOI: 10.1016/j.jclepro.2022.130671

Koh, H. W. (2024). Global leadership and sustainability. [IJSRM]. *International Journal of Scientific Research and Management, 12*(8), 7015–7026. DOI: 10.18535/ijsrm/v12i08.em03

Kotter, J. P., & Schlesinger, L. A. (2008). Choosing strategies for change. *Harvard Business Review, 86*(7/8), 130–139.

Koutoudjian, G., Diniz, L., Cespedes, R., & UNDP, V. G. (2021). About IRENA The International Renewable Energy Agency (IRENA) is an intergovernmental organisation that supports countries in their transition to a sustainable energy future, and serves as the principal platform for international co-operation, a centre of excellence, and a repository of policy, technology, resource and financial knowledge on renewable energy. *IRENA promotes the. IRENA promotes the, 24*.

Kramer, M. R., & Porter, M. (2011). *Creating shared value* (Vol. 17). FSG.

Kühl, S. (2020). Groups, organizations, families and movements: The sociology of social systems between interaction and society. *Systems Research and Behavioral Science, 37*(3), 496–515. DOI: 10.1002/sres.2701

Kukreti, M., Sehajpal, S., Tiwari, R., & Sood, K. (2024a). Driving global health and sustainable development goals with smart technology. In *Driving Global Health and Sustainable Development Goals With Smart Technology*. IGI Global., DOI: 10.4018/979-8-3373-0240-9

Kumar, A., Snehvrat, S., Kumari, P., Priyadarshani, P., & Ray, P. (2024). A corporate social responsibility (CSR)-led competitiveness framework through a bibliometric analysis (2001–2023). *Journal of Asia Business Studies, 18*(3), 737–766.

Kundzewicz, Z. W.. (2008). The implications of projected climate change for freshwater resources and their management. *Hydrological Sciences Journal.*

Kusi, E., Boateng, I., & Danso, H. (2024). Energy consumption and carbon emission of conventional and green buildings using building information modelling (BIM). *International Journal of Building Pathology and Adaptation.*

Kusnawan, A., Riyadi, S., Lestari, S. D., & Mudjijah, S. (2025). Green Corporate Performance Management: A Strategic Approach to Environmental and Organizational Sustainability. *Primanomics: Jurnal Ekonomi & Bisnis, 23*(1), 256–272. DOI: 10.31253/pe.v23i1.3523

Lacy, P., & Rutqvist, J. (2015). *Waste to wealth: The circular economy advantage.* Palgrave Macmillan.

Lai, C. H., & Lin, S. H. (2017). Systems theory. The international encyclopedia of organizational communication, 41, 1-18. https://doi.org/DOI: 10.1002/9781118955567.wbieoc122

Larik, K. A., Mangi, M., & Ahmed, S. I. (2024). Green Initiatives in Pakistani Universities: Nexus among Human Resource Practices, Employee Environment Behavior, and Innovative Work with Mediating Role of Work Engagement in Pakistani Universities. *Business Review, 19*(1), 68–86.

Lash, J., & Wellington, F. (2007). Competitive advantage on a warming planet. *Harvard Business Review, 85*(3), 94–102.

Lavrik, M., & Saurombe, A. (2025). BRICS and Climate Change Law: An Opportunity to Save the Planet. *Potchefstroom Electronic Law Journal, 28*, 1–51. DOI: 10.17159/1727-3781/2025/v28i0a17634

Leal Filho, W., Eustachio, J. H. P. P., Caldana, A. C. F., Will, M., Lange Salvia, A., Rampasso, I., Anholon, R., Platje, J., & Kovaleva, M. (2020). Sustainability leadership in higher education institutions: An overview of challenges. *Sustainability, 12*(9), 3761. DOI: 10.3390/su12093761

Leal-Arcas, R., Al Zarkani, M., Jbara, L., Mubwana, R. M., Margaritidou, M., & van der Berg, A. (2022). Climate change mitigation law and policy in the BRICS. In *Research Handbook on Climate Change Mitigation Law*. Edward Elgar Publishing Ltd.

Lee, D., Lee, M., & Lee, J. (2022). The effects of ethical leadership and knowledge sharing on employee engagement in knowledge-intensive business firms. *Sustainability, 14*(18), 11358. DOI: 10.3390/su141811358

Linnenluecke, M. K., & Griffiths, A. (2010). Corporate sustainability and organizational culture. *Journal of World Business, 45*(4), 357–366. DOI: 10.1016/j.jwb.2009.08.006

Liu, M., Chen, Y., Chen, K., & Chen, Y. (2023). Progress and hotspots of research on land-use carbon emissions: A global perspective. *Sustainability, 15*(9), 7245.

Liu, X., Yuan, X., Ge, X., & Jin, Z. (2024). Adaptation and innovation: How does climate vulnerability shapes corporate green innovation in BRICS. *International Review of Financial Analysis, 94*. Advance online publication. DOI: 10.1016/j.irfa.2024.103272

Li, W., Lee, H., Liu, Y., Tang, T., & Zheng, G. (2024). Reward exploitation or exploration? The effects of local government sustainability strategies and performance information on citizen evaluation. *Public Management Review*, •••, 1–22. DOI: 10.1080/14719037.2024.2399147

Lobell, D. B.. (2011). Climate trends and global crop production. *Science*.

Longoni, A., & Cagliano, R. (2016). Human resource and customer benefits through sustainable operations. *International Journal of Operations & Production Management, 36*(12), 1719–1740. DOI: 10.1108/IJOPM-11-2014-0564

López Jiménez, D., Dittmar, E. C., & Vargas Portillo, J. P. (2021). New directions in corporate social responsibility and ethics: Codes of conduct in the digital environment. *Journal of Business Ethics*, •••, 1–11. DOI: 10.1007/s10551-021-04929-2

Lu, H., Cai, S., Liu, Y., & Chen, H. (2023). How GHRM impacts employee OCBE: The role of emotions and value discrepancy. *International Journal of Manpower, 44*(2), 318–333.

Luo, L., & Tang, Q. (2023). The real effects of ESG reporting and GRI standards on carbon mitigation: International evidence. *Business Strategy and the Environment, 32*(6), 2985–3000. DOI: 10.1002/bse.3281

Lu, Q., Deng, J., Chen, S., & Hussain, Y. (2025). Managerial myopia and its barrier to green innovation in high-pollution enterprises: A machine learning approach. *Journal of Environmental Management, 376*, 124477. PMID: 39946802

Lüthi, D., Le Floch, M., Bereiter, B., Blunier, T., Barnola, J. M., & Siegenthaler, U. (2008). High-resolution carbon dioxide concentration record 650,000–800,000 years before present. *Nature*, *453*(7193), 379–382. DOI: 10.1038/nature06949 PMID: 18480821

Machalaba, C., Cybulska, L., & Fogarty, F.. (2025). Application of a One Health Approach to Translate Policies to Action. *CABI One Health*. Advance online publication. DOI: 10.1079/cabionehealth.2025.0005

Mady, K., Battour, M., Aboelmaged, M., & Abdelkareem, R. S. (2023). Linking internal environmental capabilities to sustainable competitive advantage in manufacturing SMEs: The mediating role of eco-innovation. *Journal of Cleaner Production*, *417*, 137928.

Mahat, D., Shrestha, S. K., Neupane, D., Karki, T. B., & Dongol, P. (2025). Green human resource management and sustainable workplace: Artificial Intelligence as mediating variable. *Edelweiss Applied Science and Technology*, *9*(6), 656–667. DOI: 10.55214/25768484.v9i6.7863

Maiti. (2025, January 20). *Fast fashion's detrimental effect on the environment*. Retrieved March 5, 2025, from https://earth.org/fast-fashions-detrimental-effect-on-the-environment/

Malhotra, R. K., Sharma, C. A. P., Upreti, H., Sharma, S., Prakash, N., & Mittal, S. (2024). Role of Psychological Contract in Employee Engagement in Higher Education: Empirical Study on Uttarakhand. In P. P.K. (Ed.), *E3S Web of Conferences* (Vol. 556). EDP Sciences. https://doi.org/DOI: 10.1051/e3sconf/202455601030

Malik, S. Y., Hayat Mughal, Y., Azam, T., Cao, Y., Wan, Z., Zhu, H., & Thurasamy, R. (2021). Corporate social responsibility, green human resources management, and sustainable performance: Is organizational citizenship behavior towards environment the missing link? *Sustainability*, *13*(3), 1044.

Mannan, M. M. J., Aiman, A. N., Saba, S. I., & Kiani, M. B. K. (2023). In-depth review of green human resource management along with latest initiatives. *NUST Business Review, 5*(1).

Mansaray, H. E. (2019). The role of human resource management in employee motivation and performance-An overview. *Budapest International Research and Critics Institute (BIRCI). Journal*, *2*(3), 183–194. DOI: 10.33258/birci.v2i3.405

Marrucci, L., Daddi, T., & Iraldo, F. (2024). Creating environmental performance indicators to assess corporate sustainability and reward employees. *Ecological Indicators*, *158*, 111489.

Martins, M. A. B. (2019). The BRICS commitment on climate change: process towards an effective approach in the path of sustainable development. *Contributions to Economics*, 175 – 187. https://doi.org/DOI: 10.1007/978-3-030-02662-2_9

Martins, V. W. B., Rampasso, I. S., Anholon, R., Quelhas, O. L. G., & Leal Filho, W. (2019). Knowledge management in the context of sustainability: Literature review and opportunities for future research. *Journal of Cleaner Production*, *229*, 489–500. DOI: 10.1016/j.jclepro.2019.04.354

Matten, D., & Moon, J. (2008). "Implicit" and "explicit" CSR: A conceptual framework for a comparative understanding of corporate social responsibility. *Academy of Management Review*, *33*(2), 404–424.

Matzembacher, D. E., & Meira, F. B. (2019). Sustainability as business strategy in community supported agriculture: Social, environmental and economic benefits for producers and consumers. *British Food Journal*, *121*(2), 616–632. DOI: 10.1108/BFJ-03-2018-0207

McKinnon, A., Browne, M., Whiteing, A., & Piecyk, M. (2010). *Green logistics: Improving the environmental sustainability of logistics*. Kogan Page Publishers.

Medne, A., & Lapiņa, I. (2019). Sustainability and continuous improvement of organization: Review of process-oriented performance indicators. *Journal of Open Innovation*, *5*(3), 49. DOI: 10.3390/joitmc5030049

Mehak, S. S., & Batcha, H. M. (2024). Analyzing the influence of green human resource practices on organizational sustainability: The role of green attitudes and performance of employees'. *Environment, Development and Sustainability*, •••, 1–34. DOI: 10.1007/s10668-024-05157-5

Meirinhos, G., Cardoso, A., Neves, M., Silva, R., & Rêgo, R. (2023). Leadership styles, motivation, communication and reward systems in business performance. *Journal of Risk and Financial Management*, *16*(2), 70. DOI: 10.3390/jrfm16020070

Merriman, K. K., Sen, S., Felo, A. J., & Litzky, B. E. (2016). Employees and sustainability: The role of incentives. *Journal of Managerial Psychology*, *31*(4), 820–836. DOI: 10.1108/JMP-09-2014-0285

Meuer, J., Koelbel, J., & Hoffmann, V. H. (2020). On the nature of corporate sustainability. *Organization & Environment*, *33*(4), 319–341.

Miah, M., Szabó-Szentgróti, G., & Walter, V. (2024). A systematic literature review on green human resource management (GHRM): An organizational sustainability perspective. *Cogent Business & Management*, *11*(1), 2371983. DOI: 10.1080/23311975.2024.2371983

Microsoft. (2020, January 16). *Microsoft will be carbon negative by 2030*. Microsoft. https://blogs.microsoft.com/blog/2020/01/16/microsoft-will-be-carbon-negative-by-2030/

Miller, K., & Monaghan, P. (2017). Flexible working arrangements: Green benefits and challenges. *Journal of Environmental Management, 193*, 69–78.

Molina-Azorin, J. F., López-Gamero, M. D., Tarí, J. J., Pereira-Moliner, J., & Pertusa-Ortega, E. M. (2021). Environmental management, human resource management and green human resource management: A literature review. *Administrative Sciences, 11*(2), 48. DOI: 10.3390/admsci11020048

Montabon, F., Pagell, M., & Wu, Z. (2016). Making sustainability sustainable. *The Journal of Supply Chain Management, 52*(2), 11–27.

Moreno, D., Bang, A., Nielsen, S., & Thellufsen, J. Z. (2024). Strategic management of CO2: A scalable model for CCS in decarbonised societies. *Journal of Environmental Management, 368*, 122175. PMID: 39178794

Mottaz, C. J. (1985). The relative importance of intrinsic and extrinsic rewards as determinants of work satisfaction. *The Sociological Quarterly, 26*(3), 365–385. DOI: 10.1111/j.1533-8525.1985.tb00233.x

Muhammad, S., Khan, I., & Hameed, F. (2021). The Impact of Performance Management System on Employees' Performance. *International Journal of Business and Management Sciences, 2*(3), 38–47.

Musheke, M. M., & Phiri, J. (2021). The effects of effective communication on organizational performance based on the systems theory. *Open Journal of Business and Management, 9*(2), 659–671. DOI: 10.4236/ojbm.2021.92035

Mushtaq, S., & Akhtar, S. (2024). Sustainable HRM strategies, enhancing organizational resilience and advancing sustainability goals. *Journal of Management Development, 43*(5), 712–726. DOI: 10.1108/jmd-02-2024-0072

Muster, V., & Schrader, U. (2021). Green work life balance: A new perspective for green HRM and environmental management. *German Journal of Human Resource Management, 35*(1), 72–97. DOI: 10.1177/2397002220957396

Mutingi, M., Mapfaira, H., & Monageng, R. (2014). Developing performance management systems for the green supply chain. *Journal of Remanufacturing, 4*, 1–20. DOI: 10.1186/s13243-014-0006-z

Nadeem, S. (2025). *Selecting Green Wall Adoption Best Solutions Using Optimization*. American University in Cairo. PDF: https://fount.aucegypt.edu/cgi/viewcontent.cgi?article=3542&context=etds

Nahari, M. (2024). Vague sustainability terms and consumer confusion. *The Journal of Consumer Research*, *46*(2), 215–231.

Nakra, N., & Kashyap, V. (2024). Responsible leadership and organizational sustainability performance: Investigating the mediating role of sustainable HRM. *International Journal of Productivity and Performance Management*, *74*(2), 409–426. DOI: 10.1108/IJPPM-03-2023-0115

NASA. (2020). *Scientific consensus: Earth's climate is warming*. National Aeronautics and Space Administration.

NASA. (2021). *Global climate change: Vital signs of the planet*. https://climate.nasa.gov/

NASA. (n.d.). *Arctic sea ice: Vital signs of the planet*. NASA Climate. Retrieved March 5, 2025, from https://climate.nasa.gov/vital-signs/arctic-sea-ice/?intent=121

Naseer, S., Song, H., Adu-Gyamfi, G., Abbass, K., & Naseer, S. (2023). Impact of green supply chain management and green human resource management practices on the sustainable performance of manufacturing firms in Pakistan. *Environmental Science and Pollution Research International*, *30*(16), 48021–48035. DOI: 10.1007/s11356-023-25409-7

National Geographic. (2022, January 21). *Should you buy carbon offsets for your air travel?* National Geographic. Retrieved March 5, 2025, from https://www.nationalgeographic.com/travel/article/should-you-buy-carbon-offsets-for-your-air-travel

National Oceanic and Atmospheric Administration (NOAA). (2022). *Sea level rise and coastal flooding impacts*. https://www.noaa.gov/

Nawaz, M. A., Seshadri, U., Kumar, P., Aqdas, R., Patwary, A. K., & Riaz, M. (2021). Nexus between green finance and climate change mitigation in N-11 and BRICS countries: Empirical estimation through difference in differences (DID) approach. *Environmental Science and Pollution Research International*, *28*(6), 6504–6519. DOI: 10.1007/s11356-020-10920-y PMID: 32997248

Ngare, I., Wemali, E., Gichuki, C., Gichuki, N., Kipkemoi, I., & Karanja, J. (2018). Integrating environmental concerns in enterprises and businesses for sustainable development. *International Journal of Environmental Sciences*, •••, 3.

Nobanee, H. (2024). Digital Entrepreneurship and Greenwashing: Addressing Ethical Challenges in Sustainable Business. *Available at SSRN* 5073927.

Nordhaus, W. (2018). *The climate casino: Risk, uncertainty, and economics for a warming world.* Yale University Press.

Northouse, P. G. (2021). *Leadership: Theory and practice.* Sage Publications., DOI: 10.4135/9781071800539

NRDC. (2020). By 2070, more than 3 billion people may live outside the human climate niche. Retrieved from https://www.nrdc.org/stories/2070-more-3-billion-people-may-live-outside-human-climate-niche

Obeid, R., Smith, A., & Gonzalez, A. (2024). Consumer skepticism and trust erosion in corporate sustainability claims. *Sustainability Science, 19*(4), 42–59.

Ockenfels, A., & Pahle, M. (2025). *Das Pariser Abkommen droht zu scheitern. Was jetzt?* Zeitschrift für Wirtschaftspolitik., DOI: 10.1515/zfwp-2025-2002

OECD. (2019). *Taxing energy use 2019: Using taxes for climate action.* Organisation for Economic Co-operation and Development.

OECD. (2021). The role of government in sustainable business. Retrieved from https://www.oecd.org

Okay, N. C., Sencer, A., & Taskin, N. (2024). Quantitative indicators for environmental and social sustainability performance assessment of the supply chain. *Environment, Development and Sustainability*. Advance online publication. DOI: 10.1007/s10668-024-05210-3

Ones, D. S., & Dilchert, S. (2012). Environmental sustainability at work: A call to action. *Industrial and Organizational Psychology: Perspectives on Science and Practice, 5*(4), 444–466.

Ore, O., & Sposato, M. (2022). Opportunities and risks of artificial intelligence in recruitment and selection. *The International Journal of Organizational Analysis, 30*(6), 1771–1782. DOI: 10.1108/IJOA-11-2021-2933

Ottman, J. A. (2011). *The new rules of green marketing: Strategies, tools, and inspiration for sustainable branding.* Berrett-Koehler Publishers.

Özen, E., Singh, A., Taneja, S., Rajaram, R., & Davim, J. P. (2024). *Innovating Sustainability Through Digital Circular Economy.* IGI Global., DOI: 10.4018/979-8-3373-0578-3

Pacala, S., & Socolow, R. (2004). Stabilization wedges: Solving the climate problem for the next 50 years with current technologies. *Science, 305*(5686), 968–972.

Pagell, M., & Shevchenko, A. (2014). Why research in sustainable supply chain management should have no future. *The Journal of Supply Chain Management, 50*(1), 44–55.

Paillé, P., Valeau, P., & Renwick, D. (2020). Leveraging green human resource practices to achieve environmental sustainability. *Journal of Cleaner Production, 260*, 121137. DOI: 10.1016/j.jclepro.2020.121137

Paine, L. S. (2023). What Does "Stakeholder Capitalism" Mean to You. *Harvard Business Review*, (September-October), 108–119.

Paleari, S. (2022). The impact of the European Green Deal on EU environmental policy. *Journal of Environment & Development, 31*(2), 196–220.

Panagoda, S. S., Ranasinghe, H., Perera, V., Sandunika, I., Tilanka, G., Alwis, S., & Dilka, S. (2023). Cement manufacturing process and its environmental impact. *J. Res. Technol. Eng, 4*(3), 161–168.

Parmesan, C., & Yohe, G. (2003). A globally coherent fingerprint of climate change impacts across natural systems. *Nature, 421*(6918), 37–42. PMID: 12511946

Patagonia, Inc. (2023). *Our Responsibility Programs.* Retrieved from https://www.patagonia.com/our-responsibility-programs.html

Pavláková Dočekalová, M., & Kocmanová, A. (2016). Composite indicator for measuring corporate sustainability. *Ecological Indicators, 61*, 612–623. DOI: 10.1016/j.ecolind.2015.10.012

Pavlova, M. (2018). Fostering inclusive, sustainable economic growth and "green" skills development in learning cities through partnerships. *International Review of Education, 64*, 339–354.

Pearson, J. (2010). Turning point. Are we doing the right thing? Leadership and prioritisation for public beneft. *Journal of Corporate Citizenship, 2010*(37), 37–40. DOI: 10.9774/gleaf.4700.2010.sp.00006

Peng, Y., Lu, X., & Zhou, L. (2024). False labels: A growing problem in corporate sustainability. *International Journal of Business Ethics, 27*(3), 301–315.

Pinkse, J., & Kolk, A. (2012). Addressing the climate change—sustainable development nexus: The role of multistakeholder partnerships. *Business & Society, 51*(1), 176–210.

Pinto, L. (2020). Green supply chain practices and company performance in Portuguese manufacturing sector. *Business Strategy and the Environment, 29*(5), 1832–1849.

Piwowar-Sulej, K. (2020). Pro-environmental organizational culture: Its essence and a concept for its operationalization. *Sustainability, 12*(10), 4197. DOI: 10.3390/su12104197

Porter, M. E., & Kramer, M. R. (2011). Creating shared value. *Harvard Business Review, 89*(1/2), 62–77.

Porter, M. E., & Van der Linde, C. (1995). Toward a new conception of the environment-competitiveness relationship. *The Journal of Economic Perspectives, 9*(4), 97–118.

Prasad, R. S. (2013). Green HRM-partner in sustainable competitive growth. *Journal of Management Sciences and Technology, 1*(1), 15–18.

Priyadarshini, C., Chatterjee, N., Srivastava, N. K., & Dubey, R. K. (2023). Achieving organizational environmental citizenship behavior through green transformational leadership: A moderated mediation study. *Journal of Asia Business Studies, 17*(6), 1088–1109. DOI: 10.1108/JABS-05-2022-0185

Putra, A., Aini, S., Maulida, C., Qhadaffi, L. A. J., & Associates. (2025). *Impact of global climate shifts on the biodiversity and functionality of marine zooplankton communities.* Journal of Tropical Biology. Retrieved from https://journal.ugm.ac.id/v3/jtbb/article/view/14575

Rahman, M. J., Rana, T., & Zhu, H. (2025). *The Relationship between Carbon Accounting, Carbon Dioxide Emissions, and Climate Change.* Google Books: https://books.google.com/books?hl=en&id=TfNGEQAAQBAJ

Rahman, M., Aziz, S., & Hughes, M. (2020). The product-market performance benefits of environmental policy: Why customer awareness and firm innovativeness matter. *Business Strategy and the Environment, 29*(5), 2001–2018. DOI: 10.1002/bse.2484

Rai, V., & Singh, A. (2026). The Digital Renaissance of Hospitality: Workforce Mastery in the Smart Revolution. *Smart Operations and Enhancing Guest Experience in the Hospitality Industry*, 125-144.

Ramísio, P. J., Pinto, L. M. C., Gouveia, N., Costa, H., & Arezes, D. (2019). Sustainability strategy in higher education institutions: Lessons learned from a nine-year case study. *Journal of Cleaner Production, 222*, 300–309. DOI: 10.1016/j.jclepro.2019.02.257

Rasul, G., & Sharma, B. (2016). The nexus approach to water–energy–food security: An option for adaptation to climate change. *Climate Policy, 16*(6), 682–702.

Ratić, M., Ćulibrk, J., Tasić, N., Petrović, M., & Medić, N. (2024). Integrating sustainability and circular economy principles in leadership amidst uncertainty. In *Proceedings of the International Conference on Business, Management, and Economics Engineering (FUTURE-BME 2024)* (Paper 015). https://doi.org/DOI: 10.24867/FUTURE-BME-2024-015

Raza, M. A., Aman, M. M., Rajpar, A. H., Bashir, M. B. A., & Jumani, T. A. (2022). Towards achieving 100% renewable energy supply for sustainable climate change in Pakistan. *Sustainability, 14*(24), 16547.

Remmer, S., & Gilbert, D. U. (2025). Causal Mechanisms in Corporate Social Responsibility (CSR) Management: A Meta-Synthesis of Micro-CSR Research. *Journal of Management Studies.*

REN21. (2021). *Renewables 2021: Global Status Report.* Renewable Energy Policy Network for the 21st Century.

Renwick, D. W. S., Redman, T., & Maguire, S. (2013). Green Human Resource Management: A review and research agenda. *International Journal of Management Reviews, 15*(1), 1–14.

Renwick, D. W., Redman, T., & Maguire, S. (2013). Green human resource management: A review and research agenda. *International Journal of Management Reviews, 15*(1), 1–14.

Report, S. (2020, October 7). *Laughing gas: A growing climate problem.* Stanford News. Retrieved March 5, 2025, from https://news.stanford.edu/stories/2020/10/laughing-gas-growing-climate-problem#:~:text=In%20the%20industrial%20era%2C%20carbon,over%20a%20100%2Dyear%20period

Rockström, J., Steffen, W., & Noone, K.. (2009). A safe operating space for humanity. *Nature, 461*(7263), 472–475.

Rockström, J., Steffen, W., Noone, K., Persson, Å., Chapin, F. S., Lambin, E. F., & Foley, J. A. (2009). Planetary boundaries: Exploring the safe operating space for humanity. *Ecology and Society, 14*(2), 32. DOI: 10.5751/ES-03180-140232

Rogers, D. S., Lambert, D. M., Croxton, K. L., & García-Dastugue, S. J. (2012). Supply chain management: Processes, partnerships, performance. Supply Chain Management Institute.

Roscoe, S., Subramanian, N., Jabbour, C., & Chong, T. (2019). Green human resource management and the enablers of green organisational culture: Enhancing a firm's environmental performance for sustainable development. *Business Strategy and the Environment*, *28*(5), 737–750. DOI: 10.1002/bse.2277

Rouholamin, A., Varposhti, S., Alavian, A., & Talebi, S. (2023). Relationship Between Green Intellectual Capital, Green Human Resource Management, Sustainable Supply Chain Management, and Sustainable Performance. *Journal of Organizational Behavior Research*, *8*(2-2023), 91-106. https://doi.org/DOI: 10.51847/teEVWNYvUZ

Rowland, C., & Hall, R. (2014). Management learning, performance and reward: Theory and practice revisited. *Journal of Management Development*, *33*(4), 342–356. DOI: 10.1108/JMD-08-2012-0110

Saberi, S., Kouhizadeh, M., Sarkis, J., & Shen, L. (2019). Blockchain technology and its relationships to sustainable supply chain management. *International Journal of Production Research*, *57*(7), 2117–2135.

Sachdeva, C., & Singh, T. (2024). Green transformational leadership and pro-environmental behavior: Unravelling the underlying mechanism in the context of hotel industry. *The International Journal of Organizational Analysis*, *32*(2), 255–271.

Sachs, J. D. (2015). *The age of sustainable development*. Columbia University Press.

Sæther, S. R. (2021). Climate policy choices: An empirical study of the effects on the OECD and BRICS power sector emission intensity. *Economic Analysis and Policy*, *71*, 499–515. DOI: 10.1016/j.eap.2021.06.011

Sahlin, J., & Angelis, J. (2019). Performance management systems: Reviewing the rise of dynamics and digitalization. *Cogent Business & Management*, *6*(1), 1642293. DOI: 10.1080/23311975.2019.1642293

Sajjad, A., Eweje, G., & Raziq, M. M. (2024). Sustainability leadership: An integrative review and conceptual synthesis. *Business Strategy and the Environment*, *33*(4), 2849–2867. DOI: 10.1002/bse.3631

Saleh, I., Abu Afifa, M., & Alkhawaja, A. (2025). Environmental, social, and governance (ESG) disclosure, earnings management and cash holdings: Evidence from a European context. *Business Ethics, the Environment & Responsibility*, *34*(2), 295–308.

Sánchez-García, E., Montalvo-Falcón, J. V., Marco-Lajara, B., & Martínez-Falcó, J. (2024). Guiding organizations toward sustainable success: The strategic role of leadership in environmental corporate governance in the wine industry. *Corporate Social Responsibility and Environmental Management*, *31*(6), 6438–6456. DOI: 10.1002/csr.2925

Saqib, A., Hussain, I., & Mefteh-Wali, S. (2024). Do stock returns respond to physical and transition climate risks? Evidence from emerging BRICS economies. *Journal of Environmental Management, 372*. Advance online publication. DOI: 10.1016/j.jenvman.2024.123303 PMID: 39561456

Saratun, M. (2016). Performance management to enhance employee engagement for corporate sustainability. *Asia-Pacific Journal of Business Administration, 8*(1), 84–102. DOI: 10.1108/APJBA-07-2015-0064

Sarkodie, S. A., Ahmed, M. Y., & Owusu, P. A. (2023). Advancing COP26 climate goals: Leveraging energy innovation, governance readiness, and socio-economic factors for enhanced climate resilience and sustainability. *Journal of Cleaner Production, 431*, 139757.

Satrovic, E., Usman, O., & Alola, A. A. (2025). Drivers of eco-efficiency and the criticality of environmental policy stringency in OECD countries. *Journal of Cleaner Production*, •••, 145157.

Saunila, M., Ukko, J., & Kinnunen, J. (2024). Sustainability partnership as a moderator in the relationship between business sustainability and firm competitiveness. *Business Strategy and the Environment, 33*(2), 123–133. DOI: 10.1002/bse.3493

Savitz, A. W., & Weber, K. (2013). *The triple bottom line: How today's best-run companies are achieving economic, social, and environmental success—and how you can too*. Wiley.

Saxena, M., & Khanna, R. (2024). Green pay and reward systems: Antecedents for sustainable employee environmentism in green human resource management. *Journal of Southwest Jiaotong University, 59*(2), 270–294.

Sayım, K. Z. (2010). Pushed or pulled? Transfer of reward management policies in MNCs. *International Journal of Human Resource Management, 21*(14), 2631–2658. DOI: 10.1080/09585192.2010.523580

Schaltegger, S., & Wagner, M. (2006). Integrative management of sustainability performance, measurement and reporting. *International Journal of Accounting. Auditing and Performance Evaluation, 3*(1), 1–19. DOI: 10.1504/IJAAPE.2006.010098

Schaltegger, S., & Wagner, M. (2011). Sustainable entrepreneurship and sustainability innovation: Categories and interactions. *Business Strategy and the Environment, 20*(4), 222–237.

Schiehll, E., & Morissette, R. (2000). Motivation, measurement and rewards from a performance evaluation perspective. *Revista de Administração Contemporânea, 4*, 7-24. https://doi.org/DOI: 10.1590/S1415-65552000000300002

Schlesinger, W. H., & Bernhardt, E. S. (2020). *Biogeochemistry: An Analysis of Global Change*. Academic Press.

Scoones, I. (2007). Sustainability. *Development in Practice*, *17*(4-5), 589–596. DOI: 10.1080/09614520701469609

Scoones, I. (2016). The politics of sustainability and development. *Annual Review of Environment and Resources*, *41*(1), 293–319. DOI: 10.1146/annurev-environ-110615-090039

Scott, W. R., & Davis, G. (2015). *Organizations and organizing: Rational, natural and open systems perspectives*. Routledge., DOI: 10.4324/9781315663371

Searcy, C. (2012). Corporate sustainability performance measurement systems: A review and research agenda. *Journal of Business Ethics*, *107*, 239–253. DOI: 10.1007/s10551-011-1038-z

Sebhatu, S. P. (2009). *Sustainability Performance Measurement for sustainable organizations: beyond compliance and reporting*. Service Research Centre Karlstad University.

Sehajpal, S., & Sidhu, N. K. (2024). Neuroscience, nursing, and financial innovation: A synergistic approach to optimized patient care and economic well-being. In *Neuromarketing's Role in Sustainable Finance*. IGI Global., DOI: 10.4018/979-8-3693-9117-4.ch008

Seuring, S., & Müller, M. (2008). Core issues in sustainable supply chain management – A Delphi study. *Business Strategy and the Environment*, *17*(8), 455–466.

Shahedul Quader, M., Kamal, M. M., & Hassan, A. E. (2016). Sustainability of positive relationship between environmental performance and profitability of SMEs: A case study in the UK. *Journal of Enterprising Communities: People and Places in the Global Economy*, *10*(2), 138–163. DOI: 10.1108/JEC-05-2014-0008

Shah, M. I., Khan, Z., Moise, M. L., & Abbas, S. (2022). Tourism adaptability amid the climate change and air pollution in BRICS: A method of moments quantile regression approach. *Environmental Science and Pollution Research International*, *29*(57), 86744–86758. DOI: 10.1007/s11356-022-21725-6 PMID: 35794334

Shahzad, M., Qu, Y., Zafar, A. U., Rehman, S. U., & Islam, T. (2020). Exploring the influence of knowledge management process on corporate sustainable performance through green innovation. *Journal of Knowledge Management*, *24*(9), 2079–2106. DOI: 10.1108/JKM-12-2019-0655

Sharma, S., Gupta, C., Khanna Malhotra, R., & Upreti, H. (2024). Sustainable Banking Practices: Impact, challenges and opportunities. In P. P.K. (Ed.), *E3S Web of Conferences* (Vol. 556). EDP Sciences. https://doi.org/DOI: 10.1051/e3sconf/202455601031

Shindell, D. T., Faluvegi, G., & Koch, D. M.. (2009). Improved attribution of climate forcing to emissions. *Science*, *326*(5953), 716–718.

Shrestha, M., & Kumar, R. (2022). Performance management in digital transformation: A strategy for sustainability. In *Handbook of Research on Digital Transformation Management and Tools* (pp. 1–23). IGI Global Scientific Publishing., DOI: 10.4018/978-1-7998-9764-4.ch001

Silvestre, W., Begnini, S., & Abreu, I. (2025). The paradox of trust: How leadership, commitment, and inertia shape sustainability behavior in the workplace. *Administrative Sciences*, *15*(7), 254. Advance online publication. DOI: 10.3390/admsci15070254

Simpson, G., & Simpson, R. (2022). Managing the performance management process: The importance of employees' performance in determining the organisation's viability and efficiency. *Journal of Human Resource & Leadership*, *6*(4), 42–56. DOI: 10.53819/81018102t2100

Singh, A. (2025). India's Peaks Under Pressure: The Impact of Climate Change and. *Balancing Mountain Tourism, Cultural Heritage, and Environmental Stability*, 309.

Singh, A., & Bathla, G. (2023). Environmental, social, and governance (ESG) measures and their impact on insurance industry: A global perspective. *The impact of climate change and sustainability standards on the insurance market*, 417-427.

Singh, A., & Sihmar, P. (2024). Resource Management. *Green HRM Awareness and Training in Higher Education Institutions*, 1.

Singh, M., & Singh, A. (2025). The Role of Artificial Intelligence in Driving Sustainability in the Tourism Industry: A Marketing Perspective. In *Marketing Technology-Infused Hospitality: Upskilling Frontline Employees for Competitiveness* (pp. 115-132). IGI Global Scientific Publishing.

Singh, A., Kumar, P., & Taneja, S. (2024). Artificial Intelligence and Machine Learning-Powered Smart Finance. In Taneja, P. K. S., & Singh, A. (Eds.), *IGI Global*. IGI Global., DOI: 10.4018/979-8-3693-3264-1

Singh, A., & Kumar, V. (2025). Sustainable Waste Management in Tourism and Hospitality Sectors. In *Sustainable Waste Management in the Tourism and Hospitality Sectors* (pp. 27–44). IGI Global Scientific Publishing.

Singh, A., & Sihmar, P. (2025). Higher education institution sustainability: Role of green human resource management. In *Green HRM awareness and training in higher education institutions* (pp. 1–14). IGI Global.

Singh, S. K., Del Giudice, M., Chierici, R., & Graziano, D. (2020). Green innovation and environmental performance: The role of green transformational leadership and green human resource management. *Technological Forecasting and Social Change*, *150*, 119762. DOI: 10.1016/j.techfore.2019.119762

Smil, V. (2017). *Energy and civilization: A history*. MIT Press.

Soundararajan, V., Brown, J. A., & Wicks, A. C. (2018). Can multi-stakeholder initiatives improve global supply chains? *Business Ethics Quarterly*, *29*(3), 385–412.

Sposato, M. (2024a). Are leaders born or made? Asking the right question. *Development and Learning in Organizations*, *38*(3), 1–3. DOI: 10.1108/DLO-01-2024-0001

Sposato, M. (2024b). Leadership training and development in the age of artificial intelligence. *Development and Learning in Organizations*, *38*(4), 4–7. DOI: 10.1108/DLO-02-2024-0002

Srivastava, S., Pathak, D., Soni, S., & Dixit, A. (2024). Does green transformational leadership reinforce green creativity? The mediating roles of green organizational culture and green mindfulness. *Journal of Organizational Change Management*, *37*(3), 619–640. DOI: 10.1108/JOCM-09-2023-0364

Steffen, W., Richardson, K., & Rockström, J.. (2015). Planetary boundaries: Guiding human development on a changing planet. *Science*, *347*(6223), 1259855.

Stern, N. (2007). *The economics of climate change: The Stern Review*. Cambridge University Press.

Stiglitz, J. E. (2012). *The price of inequality: How today's divided society endangers our future*. W.W. Norton & Company.

Subrahmanyam, S. (2025). Analytics: Trends, Challenges, and Prospects of Adopting Artificial Intelligence in Human Resource Management. In Kapoor, T., Singh, A., Sahay, K., Kaur, H., & Verma, R. (Eds.), *Sustainable Management Practices for Employee Retention and Recruitment* (pp. 51–72). IGI Global Scientific Publishing., DOI: 10.4018/979-8-3693-8679-8.ch003

Subrahmanyam, S. (2025). Using AI-Driven Analysis for Training and Leadership Growth. In Mrabet, J., Boudagga, S., Sharma, R., & Bansal, R. (Eds.), *Harnessing Emotion AI for Customer Support and Employee Wellbeing* (pp. 321–352). IGI Global Scientific Publishing., DOI: 10.4018/979-8-3373-3658-9.ch013

Subrahmanyam, S., Aishwaryalaxmi, N. S., Khalife, D., Shaikh, I. A. K., Faldu, R., & Asthana, N. (2024, April). Impact of Knowledge Management and Big Data Analytics Capabilities on Firm Performance. In *2024 Ninth International Conference on Science Technology Engineering and Mathematics (ICONSTEM)* (pp. 1-5). IEEE. https://doi.org/DOI: 10.1109/ICONSTEM60960.2024.10568874

Sudmeier-Rieux, K., Fernández, M., Gaillard, J. C., Guadagno, L., & Jaboyedoff, M. (2016). Introduction: exploring linkages between disaster risk reduction, climate change adaptation, migration and sustainable development. In *Identifying Emerging Issues in Disaster Risk Reduction, Migration, Climate Change and Sustainable Development: Shaping Debates and Policies* (pp. 1–11). Springer International Publishing.

Suleman, A. R., Amponsah-Tawiah, K., & Ametorwo, A. M. (2024). The role of employee environmental commitment in the green HRM practices, turnover intentions and environmental sustainability nexus. *Benchmarking*, *31*(9), 3055–3078.

Sulistiawan, J., Herachwati, N., & Khansa, E. J. R. (2024). Barriers in adopting green human resource management under uncertainty: The case of Indonesia banking industry. *Journal of Work-Applied Management*.

Sullivan, R., & Gouldson, A. (2017). The governance of corporate responses to climate change: An international comparison. *Business Strategy and the Environment*, *26*(4), 413–425.

Suprayitno, D., Iskandar, S., Dahurandi, K., Hendarto, T., & Rumambi, F. J. (2024). Public policy in the era of climate change: Adapting strategies for sustainable futures. *Migration Letters : An International Journal of Migration Studies*, *21*(S6), 945–958.

Su, X., Xu, A., Lin, W., Chen, Y., Liu, S., & Xu, W. (2020). Environmental leadership, green innovation practices, environmental knowledge learning, and firm performance. *SAGE Open*, *10*(2), 2158244020922909. DOI: 10.1177/2158244020922909

Tachizawa, E. M., & Wong, C. Y. (2015). The performance of green supply chain management governance mechanisms: A supply network and complexity perspective. *The Journal of Supply Chain Management*, *51*(3), 18–32.

Taneja, S., Gupta, S., Singh, A., Bhatnagar, M., & Gaur, A. (2023). Green management - A new paradigm in the world of business. *Green Management - A New Paradigm in the World of Business*, 1–230. https://doi.org/DOI: 10.52305/DLIH6622

TANEJA. S., KUMAR, P., REEPU, KUKRETI, M., & OZEN, E. ¨. (2024). *Data Alchemy in the Insurance Industry*. Emerald Publishing Limited. https://doi.org/ https://doi.org/10.1108/978-1-83608-582-920241033

Tang, G., Chen, Y., Jiang, Y., Paillé, P., & Jia, J. (2018). Green human resource management practices: Scale development and validity. *Asia Pacific Journal of Human Resources*, *56*(1), 31–55.

Task Force on Climate-related Financial Disclosures. TCFD. (2017). *Final report: Recommendations of the task force on climate-related financial disclosures*. https://www.fsb-tcfd.org/recommendations

Teixeira, A. A., Jabbour, C. J. C., de Sousa Jabbour, A. B. L., Latan, H., & De Oliveira, J. H. C. (2016). Green training and green supply chain management: Evidence from Brazilian firms. *Journal of Cleaner Production*, *116*, 170–176. DOI: 10.1016/j.jclepro.2015.12.061

Tesla, Inc. (2023). *2023 Tesla Impact Report*. Retrieved from https://www.tesla.com/impact and https://www.tesla.com/ns_videos/2023-tesla-impact-report.pdf

Thangaraju, K., & Palani, P. (2024). Integrating Sustainability into Ai-Driven HRM Practices: Examining Employee Engagement and Organizational Performance in the IT Industry. *Pakistan Journal of Life and Social Sciences*, *22*(2). Advance online publication. DOI: 10.57239/PJLSS-2024-22.2.001008

The Guardian. (2025). *Coral bleaching spreads in Ningaloo Reef as marine heatwave intensifies*. Retrieved from https://www.theguardian.com/environment/2025/feb/26/coral-bleaching-spreads-in-ningaloo-reef

The Guardian. (2025). Storm Éowyn leaves 768,000 buildings without power in Ireland. Retrieved from https://www.theguardian.com/world/2025/jan/24/storm-eowyn-homes-without-power-ireland-record-gusts

The Guardian. (2025, January 24). *Storm Éowyn leaves homes without power in Ireland as record gusts recorded*. Retrieved March 5, 2025, from https://www.theguardian.com/world/2025/jan/24/storm-eowyn-homes-without-power-ireland-record-gusts

The Times. (2025). *Storm Éowyn leaves 768,000 buildings without power in Ireland*. Retrieved from https://www.thetimes.co.uk/article/storm-eowyn-damage-ireland

Thom, F. (2025). *Fundamental Changes in Food: Improving the Toolbox of Simulation Modeling*. University of Göttingen. Online Link: https://ediss.uni-goettingen.de/handle/11858/15823

Tran, N. K. H. (2023). An empirical investigation on the impact of green human resources management and green leadership on green work engagement. *Heliyon*, *9*(11).

Tshikovhi, N., Santos, A., Zou, X., Netswera, F., Yarygina, I. Z., & Divi, S. (2023). COVID-19 and Climate Change in BRICS Nations: Beyond the Paris Agreement and Agenda 2030. In *COVID-19 and Climate Change in BRICS Nations: Beyond the Paris Agreement and Agenda 2030*. Taylor and Francis., DOI: 10.4324/9781032643182

U.S. Department of Energy. (2023). *Inflation Reduction Act of 2022*. U.S. Department of Energy. Retrieved March 5, 2025, from https://www.energy.gov/lpo/inflation-reduction-act-2022

U.S. Securities and Exchange Commission. SEC. (2022). *Proposed Rule on Climate-Related Disclosures*. Retrieved from https://www.sec.gov/news/press-release/2022-46

Udeagha, M. C., & Ngepah, N. (2023). Towards climate action and UN sustainable development goals in BRICS economies: Do export diversification, fiscal decentralisation and environmental innovation matter? *International Journal of Urban Sustainable Development*, *15*(1), 172–200. DOI: 10.1080/19463138.2023.2222264

Umar, M., Ahmad, A., Sroufe, R., & Muhammad, Z. (2024). The nexus between green intellectual capital, blockchain technology, green manufacturing, and sustainable performance. *Environmental Science and Pollution Research International*, •••, 1–13. PMID: 38285260

UNEP. (2019). *Global Waste Management Outlook 2019*. United Nations Environment Programme.

UNEP. (2020). *Emissions Gap Report 2020*. United Nations Environment Programme.

UNEP. (2020). *Sustainability and reporting trends in 2020: Preparing for the future*. United Nations Environment Programme.

UNESCO. (2020). *Education for sustainable development: A roadmap*. United Nations Educational, Scientific and Cultural Organization.

UNFCCC. (2015). *The Paris Agreement*. United Nations Framework Convention on Climate Change.

Unilever. (2010). *Sustainable Living Plan*. Retrieved from https://www.unilever.com/planet-and-society/

United Nations Women (UN Women). (2023). *Gender and Climate Change: Women's Role in Achieving a Sustainable Future*. Retrieved from https://www.unwomen.org/en

United Nations. (UN Women). (2023). Gender and Climate Change: Women's Role in Achieving a Sustainable Future. Retrieved from https://www.unwomen.org/en

Van Rooijen, S. N., & Van Wees, M. T. (2006). Green electricity policies in the Netherlands: An analysis of policy decisions. *Energy Policy*, *34*(1), 60–71.

Venema, H. D., & Rehman, I. H. (2007). Decentralized renewable energy and the climate change mitigation-adaptation nexus. *Mitigation and Adaptation Strategies for Global Change*, *12*(5), 875–900.

Veza, I., Asy'ari, M. Z., Idris, M., Epin, V., Fattah, I. R., & Spraggon, M. (2023). Electric vehicle (EV) and driving towards sustainability: Comparison between EV, HEV, PHEV, and ICE vehicles to achieve net zero emissions by 2050 from EV. *Alexandria Engineering Journal*, *82*, 459–467.

Vig, S., & Datta, M. (2020). Failure of corporate governance or business ethics: The case of ICICI Bank. Business, economics and sustainable development, the emerging issues, 109-119. DOI: DOI: 10.1007/978-981-15-6232-4_8

Vila-Vázquez, G., Castro-Casal, C., Álvarez-Pérez, D., & del Río-Araújo, L. (2018). Promoting the sustainability of organizations: Contribution of transformational leadership to job engagement. *Sustainability*, *10*(11), 4109.

Warhurst, A. (2002). Sustainability indicators and sustainability performance management. *Mining, Minerals and Sustainable Development [MMSD] project report, 43*, 129.

White, K., Hardisty, D. J., & Habib, R. (2019). The elusive green consumer. *Harvard Business Review*, *97*(4), 124–133.

Wilkinson, A., Hill, M., & Gollan, P. (2001). The sustainability debate. *International Journal of Operations & Production Management*, *21*(12), 1492–1502. DOI: 10.1108/01443570110410865

World Bank. (2023). *Carbon Pricing Dashboard*. Retrieved from https://carbonpricingdashboard.worldbank.org/

World Commission on Environment and Development (WCED). (1987). *Our Common Future*. Oxford University Press.

Wright, P. M., & McMahan, G. C. (1992). Theoretical perspectives for strategic human resource management. *Journal of Management*, *18*(2), 295–320.

Wu, Q., Cherian, J., Samad, S., Comite, U., Hu, H., Gunnlaugsson, S. B., & Sial, M. S. (2021). The role of CSR and ethical leadership to shape employees' pro-environmental behavior in the era of industry 4.0. A case of the banking sector. *Sustainability*, *13*(17), 9773. DOI: 10.3390/su13179773

Yafi, E., Tehseen, S., & Haider, S. A. (2021). Impact of green training on environmental performance through mediating role of competencies and motivation. *Sustainability, 13*(10), 5624.

Ye, W., & Li, Z. (2024). The impact of socially responsible human resource management on employees' environmental citizenship behavior: The role of psychological safety and contingent reward leadership. *SAGE Open, 14*(2). Advance online publication. DOI: 10.1177/21582440241254272

Yousfani, K., & Shaikh, E. (2023). Identifying Factors Affecting Green HRM Practices And Its Impact On Organizations' Sustainability Of Multinational Corporations (Mncs). *NUST Business Review, 5*(2).

Zada, M., Zada, S., Dhar, B. K., Ping, C., & Sarkar, S. (2025). Digital leadership and sustainable development: Enhancing firm sustainability through green innovation and top management innovativeness. *Sustainable Development, 33*(5), 6704–6716. DOI: 10.1002/sd.3488

Zameer, H., Wang, Y., & Yasmeen, H. (2024). Strengthening green competitive advantage through organizational learning and green marketing capabilities in a big data environment: A moderated-mediation model. *Business Process Management Journal, 30*(6), 2047–2072.

Zhang, D., Zheng, M., Feng, G. F., & Chang, C. P. (2022). Does an environmental policy bring to green innovation in renewable energy? *Renewable Energy, 195*, 1113–1124.

Zhang, H., & Huang, H. (2024). Low-Carbon Transformational Leadership: Conceptualization, Measurement, and Its Impact on Innovation Outcomes. *Sustainability, 16*(24), 10844. Advance online publication. DOI: 10.3390/su162410844

Zhang, J., Ul-Durar, S., Akhtar, M. N., Zhang, Y., & Lu, L. (2021). How does responsible leadership affect employees' voluntary workplace green behaviours? A multilevel dual process model of voluntary workplace green behaviours. *Journal of Environmental Management, 296*, 113205. DOI: 10.1016/j.jenvman.2021.113205

Zheng, Q., Li, J., & Duan, X. (2023). The impact of environmental tax and R&D tax incentives on green innovation. *Sustainability, 15*(9), 7303.

Zhong, J., Shao, X., Xiao, H., Yang, R., & An, X. (2023). The research on the green leadership: A systematic review and theoretical framework. *Environment, Development and Sustainability, 27*(1), 377–408. DOI: 10.1007/s10668-023-03960-0

Zhou, Y. (2023). Worldwide carbon neutrality transition? Energy efficiency, renewable, carbon trading and advanced energy policies. *Energy Reviews, 2*(2), 100026.

Zhou, Y., Zhang, Y., & Montoro-Sánchez, Á. (2011). Utilitarianism or romanticism: The effect of rewards on employees' innovative behaviour. *International Journal of Manpower*, *32*(1), 81–98. DOI: 10.1108/01437721111121242

Zoogah, D. B. (2011). The dynamics of Green HRM behaviors: A cognitive social information processing approach. *German Journal of Human Resource Management*, *25*(2), 117–139. DOI: 10.1177/239700221102500204

Zou, Y., Liu, Z., Chen, Y., Wang, Y., & Feng, S. (2024). Crop rotation and diversification in China: Enhancing sustainable agriculture and resilience. *Agriculture*, *14*(9), 1465.

About the Contributors

Muhammad Ashfaq is currently working as Professor/Dean, Faculty of Management Studies, the University of Faisalabad-Pakistan. He completed his graduation in 1985 and earned MS degree in Agricultural Economics in 1988. He won the scholarship in 1990-91, from USAID for training in Agricultural Marketing and Sales from Colorado State University, USA. Later on he won scholarship from the World Bank for higher studies in Australia and went to University of New England, Armidale Australia in 1995 and successfully completed his PhD degree. There he carried out his doctoral level research on the key issue of agricultural policy in Pakistan. During his stay in Australia, he also actively participated in various socio-cultural activities. In the year 2006, he also won a prestigious Higher Education Commission funded fellowship and went to University of Laval, Canada for his post doctorate studies. Dr. Ashfaq has rendered services as consultant to various national and international organizations. He is also acting as PhD/MS supervisor for ICDD, Kassel University, Germany. He has also won the TWAS Associateship Award in 2011 and Australian Alumni Excellence Award in 2013 for Innovation and Research. He has supervised 12 PhD students and more than 150 MS/M.Phil/MSc students. He is the author/co-author of 5 books; 14 chapters and more than 100 research papers published in peer reviewed journals. He is also working as editor/section editor of different journals. He has completed 32 research projects as PI/Co-PI which were funded by national and international organisations like USAID, ACIAR, ICARDA, IFPRI, ILRI, AgMIP, JBIC, ADB, JICA, TiKA, UK Met Dept. and Govt of Pakistan.

Muhammad Tariq Iqbal Khan is currently working as a Lecturer in Economics at the Government Postgraduate College, Jaranwala. He started his research career as an Internee at the Technology transfer Institute. He joined Higher Education Department Punjab on 18-08-2012, as Lecturer in Economics. He has more than 12 years of teaching and research experience. He passed B.Sc. (Hons.) Agricultural and Resources Economics in 2011 from the University of Agriculture, Faisalabad. During his education, He actively participated in university-level competitions at the University of Agriculture, Faisalabad. He five times won the Quiz Competition at the University of Agriculture, Faisalabad during graduation. He got M.Phil. Economics degree from the Government College University, Faisalabad in 2016. In 2017, He enrolled Ph.D. Economics from Government College University, Faisalabad. During his career, he has published 55research publications in refereed International and national journals, including 18 research papers that have been published in impact factor journals. The cumulative impact factor of his research is 200, having 1866 citations and an h-index 20. His current research focus is on disaster management, marketing efficiency, environmental economics, and agricultural economics.

***

Joji Abey is acting chairperson of Department of Business Management at the College of Business Administration, Kingdom University, Bahrain. She holds the prestigious Chartered Manager designation from the Chartered Management Institute (CMI), UK, and a Fellow of the Higher Education Academy (HEA)UK. A life member of the Indian Commerce Association since 2019, she has published numerous articles in Scopus-indexed journals and presented papers at various international seminars and conferences

Aishwarya Arya is pursuing her Ph.D. in Management. She accomplished her Bachelor's and Master's from Institute of Hotel Management and Catering Technology, Lucknow in Hospitality Administration thereafter another Master's in Travel & Tourism Management. She started her academic career in the year 2019 from IHM Lucknow, previously worked with Babu Banarsi Das University, Lucknow as Assistant Professor. She presented and attended many National and International Conferences, seminars and webinars. Her area of specialization includes Room Division, Hospitality Management & Sales and Marketing. She has patents published in Under the office of the controller General of Patents, Designs and Trademarks Department of Industrial Policy and Promotions, Ministry of Commerce and Industry, Government of India.

Abhishek Khandelwal, Assistant Lecturer, Food Craft Institute Khajuraho (Ministry of Tourism, Govt. of Madhya Pradesh), is pursuing a PH.D. in Hospitality management, holding Master's in Tourism & Travel Management and bachelor's in hospitality & Hotel Administration from Dr. Ambedkar Institute of Hotel Management Catering & Technology, Chandigarh. He is having 3 years of Hotel & Hospitality experience, where he is associated with JW Marriott, New Delhi &

Noor Mahal, Karnal. His 7 years of Academic career has been very satisfying and he continuously upgraded his skills and knowledge by attending and organizing many courses, training programs, National & International seminars, workshops. His area of specialisation is Room Division, Food & Beverage Service & Management.

R.Bhuvaneswari working as an Assistant Professor of English at Dr.Mahalingam College of Engineering and Technology, Pollachi. To her credit she completed 10 NPTEL / SWAYAM and published few articles in Scopus indexed journals.

R. Velmurugan is presently working as an Associate Professor in Commerce, Karpagam Academy of Higher Education, Coimbatore. He started his career as a Project Fellow under UGC Sponsored Major Research Project at NGM College, Pollachi. He has 18 years of teaching experience has 15 years of research experience. To his credit he produced 22 Ph.D. Scholars and 5 M.Phil. Scholars. At present 1 scholar submitter his thesis, 6 Ph.D. Scholars are under his guidance. Moreover, to his credit he published 163 articles in various International Journals, of which 56 articles are published in Scopus Indexed Journals and 15 articles in Web of Science and 76 articles in edited volumes. He also participates and presented 27 papers in various International Seminar / Conference and 54 National Seminar / Conference. He acted as an external examiner for various M.Phil. and Ph.D. Viva voce examinations. He acted as a Vice Chancellor nominee for Doctoral Committee Expert at Bharathiar University, Coimbatore and Periyar University, Salem. He acted as a Chairperson at various National Level Conferences. He also acted as an Expert for Syllabus Scrutiny Committee, Question Paper Scrutiny Committee and Expert members for Board of Studies at various reputed Colleges. Considering his meritorious contribution, he received awards like Best Paper Presenter, Best Research Supervisor and Best Faculty Awards from various research foundations and councils. He too acted as a resource person for 27 Guest Lectures, 30 Workshops and 10 faculty development programmes.

Amrik Singh is working as Professor and HOD in the School of Hotel Management and Tourism at Lovely Professional University, Punjab, India. He obtained his Ph.D. degree in Hotel Management from Kurukshetra University, Haryana, India. He has published 115 Scopus-indexed publications. He has published 12 patents, and 01 patent has been granted in the inter-disciplinary domain. Dr. Amrik Singh participated and acted as a resource person in various national and international conferences, seminars, research workshops, and industry talks. He has published 20 books with IGI Global Publication, Emerald Publication, Bentham Publication, Nova Publication etc., on trending areas such as Food waste, Artificial Intelligence, Mountain Tourism, Virtual tourism, Healthcare Tourism, Guest behavior in tourism and hospitality. His area of research interest are Accommodation management, sustainability, Green practices, Human Resource Management in Hospitality, Artificial Intelligence, AR VR in hospitality, Hotel pricing, Employee Disability, Work life balance, women empowerment etc. He is currently guiding 8 Ph.D. scholars, and 5 Ph.D. scholars have been awarded Ph.D.

Martin Sposato is an Associate Professor at the College of Business, Zayed University, Dubai, UAE. He holds a PhD in Leadership. His research interests include leadership, artificial intelligence in organizations, postcolonialism as a theoretical lens, Indigenous (Chinese) theories and conceptualizations of Human Resource Management (HRM), gender in organizations, and reflexivity in research.

J.Sudarvel is working as an Assistant Professor of Management at Karpagam Academy of Higher Education. He produced five Ph.D. Scholars and four Ph.D. Scholars are under his guidance. Further, to his credit he published more than five articles in Scopus indexed journals and published more than 50 articles in international journals. He is acting as a Subject Experts in Board of studies of various autonomous colleges.

Mohammad Badruddoza Talukder is an Associate Professor & Chairman at the College of Tourism and Hospitality Management at the International University of Business Agriculture and Technology (IUBAT) in Dhaka, Bangladesh. Holding a PhD from Lovely Professional University, India, he is renowned for his innovative integration of cutting-edge technologies such as Artificial Intelligence (AI), the Internet of Things (IoT), and Blockchain into tourism and hospitality management. His research explores the transformative impact of these technologies on smart tourism, cloud kitchens, and the broader industry landscape. As a prolific author, Dr. Talukder has published numerous scholarly articles and books, with forthcoming titles that address AI innovations in hospitality, the evolution of cloud kitchens, and the enhancement of digital literacy within hospitality education. His work shapes contemporary tourism and hospitality practices, skillfully marrying technological advancements with practical industry applications. Since 2008, Dr. Talukder has imparted his knowledge through a diverse array of courses in tourism and hospitality across various universities in Bangladesh.

Index

A

B

C

E

G

M

N

O

P

R

S

www.ingramcontent.com/pod-product-compliance
Lightning Source LLC
LaVergne TN
LVHW080902300926
849162LV00003B/525
9798337324791